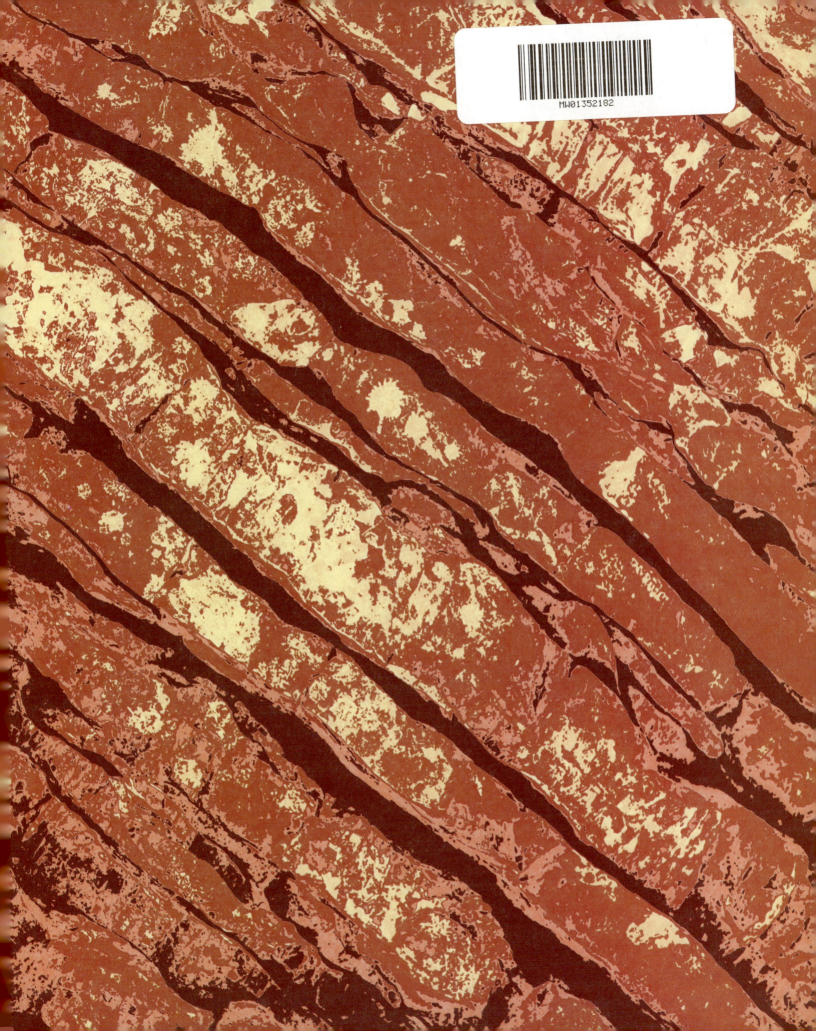

CARVED IN STONE

*Geological Evidence
of the Worldwide Flood*

CARVED IN STONE

*Geological Evidence
of the Worldwide Flood*

Timothy Clarey

Dallas, Texas
ICR.org

CARVED IN STONE
Geological Evidence of the Worldwide Flood
by Dr. Timothy Clarey

First printing: March 2020

Copyright © 2020 by the Institute for Creation Research. All rights reserved. No portion of this book may be used in any form without written permission of the publisher, with the exception of brief excerpts in articles and reviews. For more information, write to Institute for Creation Research, P. O. Box 59029, Dallas, TX 75229.

Series concept and direction: Jayme Durant, ICR Director of Communications
Senior Editor: Beth Mull
Editors: Truett Billups, Michael Stamp, Christy Hardy
Graphic Designer: Susan Windsor
Cover image: Antelope Canyon, Arizona, BigstockPhoto

All Scripture quotations are from the New King James Version.

ISBN: 978-1-946246-25-7
Library of Congress Catalog Number: 2020931327

Please visit our website for other books and resources: ICR.org

Printed in the United States of America.

DEDICATION

And Joshua said to all the people, "Behold, this stone shall be a witness to us, for it has heard all the words of the Lord which He spoke to us. It shall therefore be a witness to you, lest you deny your God." (Joshua 24:27)

To all future generations who love the Lord and His Word: Remember, the stones still bear witness to the Flood. And a special dedication to Drs. Henry M. Morris III and John Morris for their godly guidance and wisdom.

The Lord Jesus said that if men should refuse to praise Him and "should hold their peace, the stones would immediately cry out" (Luke 19:40). Yet even though the whole creation—in its beauty, complexity, and providential orderliness—gives continual praise to its Creator, men perversely have "worshipped and served the creature [or more aptly stated, the creation] more than the Creator, who is blessed for ever" (Romans 1:25).

 —Henry M. Morris, Praise from the Creation, *Days of Praise*

TABLE OF CONTENTS

Foreword 15
Acknowledgments 17

1. **Introduction** 18
   Geology: The Science of the Solid Earth 20
   It's a Battle of Worldviews 20
   Formulation of a New Flood Model 24
   A Little Background 26
   All Beginning to Make Sense 28

2. **Sedimentary Rocks** 30
   Rapid Erosion Is the Rule 32
   Classification of Clastic Sedimentary Rocks 35
   Classification of Biochemical Sedimentary Rocks 36
   Rapid Deposition of Clay and Limestone 38
   Limestone Can Form Quickly 40
   Sedimentary Structures 43

3. **Megasequences and Methods** 48
   Are Megasequences Valid? 51
   Idealized Megasequences 51
   Correlation of Megasequences Globally 53
   Methods Used and Maps Prepared 54
   Conclusion: Megasequences Are Valid 56

4. **Geologic Time** 60
   Does the Age of the Earth Matter? 62
   Principles of Relative Time 65
   Principles of Absolute Time 71
       Carbon-14 Dating 72
       RATE Project Carbon-14 Analyses 72
       Assumptions in All Radioisotope Methods 74
   Determining the Date of a Fossil 78
   Testing the Accuracy of Absolute Dating Methods 80
       Examples from Mount St. Helens and Other Historic Eruptions 80
       Examples from Wyoming and Grand Canyon 81
   Original Tissues in Fossils 81
   Conclusion: Tests Invalidate Radioisotope Dating Methods 84

5. Types of Fossils and the Fossil Record 90
- How Are Fossils Formed? 91
- Bias to the Fossil Record 92
- Types of Fossils 95
- Original Preserved Tissues in Dinosaurs and Other Fossils 98
- Brief Overview of the Fossil Record 101
- The Great Secular Conundrum: The Cambrian Explosion 104
- Pivotal Shifts in the Fossil Record 107
- Are There Inverted and Out-of-Place Fossils? 111
- Conclusion: The Flood Explains the Fossil Record 112

6. Plate Tectonics and Catastrophic Plate Tectonics 114
- Introduction to Plate Tectonics 115
  - Continental Drift 116
  - Seafloor Spreading 119
  - Modern Plate Tectonics 120
- Could the Pre-Flood Oceans Become Post-Flood Continents? ... 122
- The Low-Velocity Zone 123
- Types of Plate Boundaries 123
- Plate Tectonics Theory Explains the Geology We Observe Globally ... 127
- Secular Science Cannot Explain the Origin of Continents 128
- Modern Plate Movement Rates 131
- Catastrophic Plate Tectonics 132
  - Runaway Subduction and Superfaults 133
  - Additional Empirical Evidence for CPT 139
  - CPT Explains the Pattern of Earthquakes and Volcanoes 142
  - CPT Explains the Flooding of the Continents 143
  - CPT Explains Why the Plates Are Moving Slowly Today 145
  - CPT Explains the Conditions Necessary for the Ice Age 146
- Plate Movements During the Days of Peleg? 147
- Conclusion: The Reason for Catastrophic Plate Tectonics 149

7. **The Pre-Flood Geologic Configuration** 152
   - Pre-Flood Earth Structure 153
   - Pre-Flood Continental Configuration 157
   - Pre-Flood Sediments 160
   - Flood Alteration of Pre-Flood Sediments 164
   - Location of the Pre-Flood/Flood Boundary 165
   - Other Late Precambrian Rocks 167
   - Pre-Flood Oceans 168
   - Pre-Flood Atmosphere 168
   - Conclusion: The Pre-Flood Environment Was Very Different 169

8. **The Flood Begins: The Fountains of the Great Deep** 172
   - Mantle Hotter at the Beginning of the Flood 173
   - Peak in Volcanism and Massive Release of CO_2 in the Pre-Sauk 175
   - The Seventh Megasequence: The Pre-Sauk 178
   - Plate Configuration for the Pre-Sauk 178
     - Pre-Sauk Across North America 179
     - Pre-Sauk Across South America 186
     - Pre-Sauk Across Africa and the Middle East 188
   - How Did Subduction Begin? 190
   - Human Perspective 190
   - Conclusion: Explaining the Purpose of the Fountains 192

9. **Rising Water: Sauk Megasequence** 194
   - Initial Flooding of the Continents 194
   - What Do the Rocks Show? 197
     - North America 197
     - South America 199
     - Africa 201

 Current Data for the Sauk Megasequence 201
 Summary of the Rock Data 202
 Continental Configuration for the Sauk Megasequence 204
 Solving the Mystery of the Paleozoic Intracratonic Basins 205
 Resolving the Sheet Sand Enigma 210
 Human Perspective 211
 Conclusion: The Sauk Marks the Start of the Flooding 212

10. **More Rising Water: Tippecanoe Megasequence** 216
 Flooding Level Increases 216
 What Do the Rocks Show? 218
  North America 218
  South America 220
  Africa 222
 Current Data for the Tippecanoe Megasequence 224
 Summary of the Rock Data 224
 Continental Configuration for the Tippecanoe Megasequence 226
 Limited Flooding of the Land During the Early Flood (First 40 Days) 226
 Human Perspective 231
 Conclusion: Flooding Continued in the Tippecanoe 231

11. **Water Continues to Rise: Kaskaskia Megasequence** 234
 Flooding Again Increases 234
 What Do the Rocks Show? 234
  North America 234
  South America 238
  Africa 240
 Current Data for the Kaskaskia Megasequence 242
 Summary of the Rock Data 242
 Continental Configuration for the Kaskaskia Megasequence 244

The 40 Days of Rain	246
Why Dinosaur Fossils Are Found Only in Later Flood Rocks	246
Why Aren't There Dinosaurs in Grand Canyon?	248
The Mystery of Devonian Black Shales	250
Human Perspective	253
Conclusion: The Water Continued to Rise in the Kaskaskia	253

12. Things Go from Bad to Worse: Absaroka Megasequence ... 256

Flooding of the Land Begins	256
What Do the Rocks Show?	258
North America	258
South America	260
Africa	262
Current Data for the Absaroka Megasequence	264
Summary of the Rock Data	265
Continental Configuration for the Absaroka Megasequence	266
World-Changing Tectonic Activity	268
Not Just a Coincidence	271
Cyclothems: What Are They and How Do They Form?	275
Mixing of Land and Marine Fossils Common in the Same Rocks	276
Human Perspective	279
Conclusion: God Had a Plan Through the Chaos	279

13. Covering the Highest Hills: Zuni Megasequence ... 282

The High-Water Mark Is Reached	282
What Do the Rocks Show?	284
North America	284
South America	286
Africa	288
Current Data for the Zuni Megasequence	290
Summary of the Rock Data	291
Continental Configuration for the Zuni Megasequence	295
The K-T or K-Pg Extinction? The Chicxulub Crater	298
Dinosaurs in Marine Sediments: A Worldwide Phenomenon	301
Another Whopper Sand?	304
Human Perspective	307
Conclusion: The Zuni Is the High-Water Point of the Flood	308

14. The Receding Phase: Tejas Megasequence ... 312
- The Waters Recede ... 313
- What Do the Rocks Show? ... 316
  - North America ... 316
  - South America ... 318
  - Africa ... 320
- Current Data for the Tejas Megasequence ... 322
- Summary of the Rock Data ... 322
- Continental Configuration for the Tejas Megasequence ... 327
- Why Are the Rocky Mountains So Wide Compared to Other Ranges? ... 330
- Whopper Sand in the Gulf of Mexico ... 334
- Ogallala Sand Across the Great Plains ... 336
- Upper Cenozoic Flood/Post-Flood Boundary: Introducing the N-Q ... 339
- Grand Canyon Carved by Flood Runoff in Tejas ... 342
- Human Perspective: Where Are the Human Fossils? ... 348
- Conclusion: The Tejas Is the Receding Phase ... 349

15. The Post-Flood Ice Age ... 354
- The Tejas Megasequence Ends with a Bang ... 355
- The Ice Age Begins ... 356
- What Caused the Ice Age? ... 359
  - Subduction Zone Volcanoes ... 361
  - Hotter Oceans ... 365
- Milankovitch Theory Refuted ... 367
- Only One Ice Age ... 370
- What Was the Purpose of the Ice Age? ... 371
  - Land Bridges ... 373
  - Man's Disobedience ... 374
- Ice Age Footprints Found Along Coasts ... 374
- Conclusion: The Ice Age Was an Essential Ending to the Flood ... 375

16. Megasequences Validate the Global Geologic Column ... 378
- Using Megasequences to Study the Geologic Column ... 380
  - Method 1: Construction of Basal Lithology Maps ... 381
  - Method 2: Construction of Maps of Unique Sediments of Semi-Regional Extent ... 381
- Discussion of the Results ... 382
  - Basal Lithology Maps ... 382
  - Unique Sedimentary Units ... 387

Overthrusts Do Not Invalidate the Geologic Column 389
    Flood Solution to Overthrusts 393
Conclusion: The Geologic Column Is Global Because
  the Flood Was Global 396

17. Unlocking the Mystery of the Pre-Flood World 400
The Pre-Flood World 403
    Shallow Seas 403
    Lowland Areas 404
    Upland Areas 404
Pangaea, Not Rodinia 405
Garden of Eden 406
ICR Discovery Center Globe 407
Fossils Confirm Pre-Flood Global Greenhouse 407
Stromatolites: Evidence of Pre-Flood Hydrology 409
    Stromatolites Create an Evolutionary Conundrum 410
    Stromatolites Are Living Fossils 411
    Stromatolite Fossils Confirm the Presence of Springs
      in the Pre-Flood World 411
    Biblical Account Confirmed by Science 413
Conclusion: The Pre-Flood World Is Coming into Focus 414

18. Flood-Provided Energy Resources: Oil and Coal 418
Oil Is from the Flood 418
Rapidly Forming Oil Supports Flood Time Frame 426
Coal: Plant Material Buried in the Flood 431
Sinking the Floating Forest Hypothesis 435
    In Situ or Not In Situ? 437
    Criteria for In Situ Trees 438
    Fossil Grove Site, Glasgow, Scotland 438
    Lycopod Trees Were Not Hollow 440
    Flood Model for Fossil Grove Site 441
    Sinking the Floating Forest 441
Massive Lower Cenozoic Coal Beds Are from the Receding Flood 442
Conclusion: Oil and Coal Formed by Flood Processes 443

19.	**Evidence of a Young Earth and Recent Flood** 446
	Millions or Thousands of Years: Does It Matter?................................446
	Evidence of a Young Solar System 449
	Spiral Galaxies 449
	Blue Stars 450
	Comets 452
	Planetary Magnetic Field Decay 453
	Evidence of a Recent Flood 454
	Original Tissues in Fossils 454
	Carbon-14 in Fossils Also Indicates a Young Age 455
	Lack of Erosion (Time) Between Layers 459
	Cold Subducted Slabs Deep in the Mantle 462
	Conclusion: The Flood Was Recent 465
20.	**It All Makes Sense** 468
	Review of the Rock Data 468
	A New Global Sea Level Curve Based on Rock Data 474
	Cultural Evidence for a Global Flood 476
	Flood Traditions 477
	Chinese Language 478
	Globally Similar Building Techniques 481
	The Geological Data All Make Sense 482
	Geological Evidence Shows a Global Flood 482
	Evidence Further Supports a Recent Flood 485
	The Ark as a Type of Salvation 485
	Conclusion: It All Points to Christ 486

Image Credits 488

Index 489

FOREWORD

The Institute for Creation Research has long been known for its attention to scientific research from a biblical and creationist standpoint. I first met geologist Dr. Timothy Clarey in concert with ICR's FAST (Flood-Activated Sedimentation and Tectonics) research project. Dr. Clarey had held a productive position in oil exploration for a major oil company and then a college professorship. He was affiliated with ICR as Research Geologist on a field project in Wyoming and Alaska. I got to know him as a thorough young-earth creationist and knowledgeable Christian.

At one of the FAST project meetings, the geologists among us were introduced to the concept of the Column Project. Geological theorists have long focused primarily on local rock outcrops and well logs to determine prospects and their economic feasibility, but geologists have little reason to expand their scope globally. Since we were engaged in discerning the effects of the global Flood of Noah's Day, it only made sense to look at rock units from a global perspective. Several of those in attendance felt the need to launch such a project, but one by one they concluded that it would be an all-consuming effort, and so it was pushed to the back of the project list.

But in the providence of God, things change. Dr. Clarey joined the ICR science faculty as Research Associate in 2013. I remember meeting with him soon after and proposing the Column Project. With his past oil field experience spanning the United States and familiarity with numerous rock units, he soon adopted the project as his own. He acquired new information to add to the breadth of knowledge he already possessed. He began writing articles for geological journals and creationist publications, prepared several museum displays, participated in TV and radio interviews, and spoke in a wide variety of venues.

As his data from global sources took shape, Dr. Clarey began to notice many things no one before had the option of studying—things like the timing of plate movements, patterns of dinosaur locations, special rock locations, etc. He identified many global trends spanning the entire geologic column. He applied this newly acquired global data to the problem of the global Flood, and he began to derive answers to longstanding questions. Already, many plaguing problems have acquired solutions!

Throughout this book you will encounter much geological knowledge. It is written for nonspecialists who have an interest in and prior familiarity with these concepts. In short, it's readable. The book's first third provides the background concepts on geology and biblical knowledge necessary for the second part. Here you will see what and how Dr. Clarey has begun to "unpeel the onion" and reveal groundbreaking perceptions never before available. *Carved in Stone* will prove a challenge to those who have adopted the thinking of the mainstream. It will prove an encouragement to those who have known there must be a solution but were unable to find it. It will thrill the many who have by faith accepted Scripture's teaching but have lacked the opportunity to go deeper. May it bear good fruit of eternal value.

John Morris, Ph.D.
President Emeritus
Institute for Creation Research

ACKNOWLEDGMENTS

Most importantly, I must acknowledge and thank God for providing everything in my life, especially my loving wife, Reneé. She is worth more to me than all precious gems (Proverbs 31:10). This work could not have been completed without her continued support. I am also grateful for the tiny glimpses of wisdom that God revealed to me during this project. It is all for His glory.

As I began this work, God brought Davis J. Werner, an ICR volunteer, to assist in this project. He has become an integral part of this entire project, taking it upon himself to learn the ins and outs of the Rockworks computer software. Without him, there would be no maps or ways to visualize the data. He is a daily witness to the exciting discoveries we share. It is a joy watching him learn more about the science of geology as we jointly conduct this research. I owe him more than he knows. We are truly a team.

I have also been assisted by a local Texas high school student, Timothy Piscis, who has researched and located numerous publications and articles containing detailed geologic columns. His help has been of great assistance, and he has saved me a lot of precious time. I would also like to thank a student from Florida, Jesse Dieterle, who has helped this project greatly by plotting the locations of many of the Precambrian stromatolite and banded iron locations around the globe. Without these willing and dedicated students, this project would have taken even more time to complete.

1 Introduction

Summary: Geology is the study of Earth, its rocks, and its history. Geologists operate much like detectives, piecing together clues to arrive at the best explanation for the observable data. Unfortunately, much of the rock record we observe is not being actively repeated. The rocks often reveal things that happened in the past but are not happening today. These observations contrast sharply with the traditional uniformitarian thought taught to most geologists. This doctrine teaches that all rocks can be explained by studying the processes that currently occur—slow and gradual deposition and erosion. But what if the events that created the rocks only happened once in history? This book presents a new model of geology from a biblical perspective. It explains most of Earth's geologic features using the global Flood of Genesis and presents new data-based evidence from all over the world.

This book is about real, touchable, drillable rocks. Few books are written on Flood geology, and even fewer are based on the actual rock data. Rocks are facts for geologists. Fossils found in the rocks and sedimentary structures, like laminations and cross-bedding, are also factual data. And the global patterns we can observe from the rocks give us great insight into the origin of the rocks. This book reports the results of a data-based study of the sedimentary rocks across multiple continents. It examines the rocks that are in place today and utilizes repeatable and real data.

No matter your background, this book will challenge you to rethink your views as we examine and expose the sedimentary rock record continent by continent, layer by layer. It is not a book of fiction but a book that reveals what the rocks really show. The results are the same whether you believe that Genesis records a global flood or a local flood, and whether you believe the earth is thousands of years old or billions. However, be forewarned, the data show undeniable evidence of a recent global geologic event.

A young-earth or an old-earth geologist could compile similar or even identical data and produce a similar result. The rocks do not lie. Most of the rock data is the result of measured outcrops, oil wells, and seismic data collected between oil wells. About 80 to 85% of the data set for this book is based on the rocks in place. The remaining 15 to 20% involved a bit of interpretation and geologic insight, particularly in some locations where the wells fail to penetrate all of the sedimentary section. For example, across much of Egypt the oil wells rarely penetrated past the "pay zones." I had to interpolate the deeper sedimentary rock section from surrounding data sets.

I spent over eight years with a major oil company learning to prospect for oil and develop oil and gas fields. During these years, I familiarized myself with the geology of many locations across the United States. I feel God used these years in industry to introduce me to the diverse geology across many geologic provinces and to train me in the use of industry data. Unfortunately, many academic geologists do not have oil industry training, nor do they understand how to utilize oil well and seismic data in their research. I couldn't have done this research without my past training in the oil industry.

When we began this study several years ago, we didn't know where the data would lead us. Dr. Henry Morris III, CEO of the Institute for Creation Research (ICR), hired me in 2013 and just said, "Go." It was his decision to work on what we refer to as the Column Project. It has become the greatest and most rewarding research project with which I have been associated in my 35 years as a professional geologist. And it confirms the truth of God's Word. There really was a global flood.

Geology: The Science of the Solid Earth

Geology is the study of the solid earth, its rocks, and its history. Geology differs from the other natural sciences because it is mostly forensic or historical. Most other natural sciences—like chemistry, biology, and physics—use a higher percentage of repeatable experimentation. Geologists have to observe what is here today and try to figure out how it may have gotten there in the past. We use the rocks and fossils as the pages in Earth's history book. As we go deeper in the pages, we go further back in time.

Geologists operate much like detectives, piecing together the clues to arrive at the best explanation for the observable data. Unfortunately, much of the rock record we observe is not being actively repeated. The rocks often reveal things that happened in the past that are not happening today. These observations contrast sharply with the traditional uniformitarian thought taught to most geologists. The philosophy of uniformitarianism stresses that "the present is the key to the past." Geologists who hold to this view think they can explain all of the rocks by studying processes that occur in the world today, like studying current rivers or volcanic eruptions to understand past river systems and volcanoes. However, what if the events that created the rocks only happened once in history? Uniformitarianism fails if we cannot find a modern event to explain what we observe. Where in the world today do we find 10,000 feet of pure salt being deposited like we observe below the Red Sea? Where are 100- to 200-foot-thick coal seams being deposited like we see in Wyoming? We need to recognize that there were past events that may only have happened once in history, like the global Flood.

It's a Battle of Worldviews

The battle for science today is a battle of worldviews, and geology is at the forefront. Young-earth geologists accept God's Word as truth, and they accept the book of Genesis as a true historical account of the creation and the Flood. After all, it was written by the One eyewitness to both events.

Young-earth geologists believe there was a global flood about 4,500 years ago. In complete contrast, whether they accept some of the truth of the Bible or not, old-earth geologists have fabricated their own alternative story of Earth's origin and the origin of its rocks. Sadly, this view has become mainstream science and is taught exclusively in all public institutions and even most religious institutions.

Francis Bacon (1561–1626) is often credited with devising the scientific method in which experimentation and collected data are used to make conclusions. He accepted the Flood narrative and the 6,000-year-old age of the earth. Unfortunately, few scientists today are using the scientific method as outlined by Bacon. They practice what has been called *verification science* where they merely attempt to verify what they deem to be already known.[1] This type of science leaves little room for falsification and true testing of hypotheses and theories.

Keep in mind that prior to the early 19th century most scientists were young-earth creationists. They believed the earth was about 6,000 years old and there was a global flood. They believed the Bible was historically accurate. Deep time was not readily accepted until midway through the 19th century about the time Darwin proposed his

Francis Bacon

theory of evolution. Evolutionists found they needed both concepts for their alternative version of origins. They needed deep time (millions of years) to give their proposed evolutionary theory a chance at being accepted. So, the geologists looked at the thick layers of sedimentary rocks and tried to imagine the amount of time it took for them to form using rates they could observe today. This is the central theme of *uniformitarianism*. Uniformitarian scientists insist that nearly all the rocks can be explained by the same processes observed today, with exceptions for episodes of more rapid activity and a few local catastrophes. This version of uniformitarianism is called *actualism*.

For these reasons, many 19th-century geologists quickly became convinced that the sedimentary rock record must have formed over millions of years at slow uniformitarian rates. They forgot there was a global flood, just as God predicted they would. The apostle Peter wrote:

> For this they willfully forget: that by the word of God the heavens were of old, and the earth standing out of water and in the water, by which the world that then existed perished, being flooded with water.
>
> (2 Peter 3:5-6)

Soon after this, all sorts of doubt about the historical accuracy of the Bible began to infiltrate the sciences. People wondered if God really meant what He said in Genesis. Did God really say there was a global flood? These questions are eerily reminiscent of the questions the serpent asked Eve in Genesis 3. And these are the same doubt-filled questions being asked today by scientists and theologians. I am frequently asked, "Where is the evidence for the Flood?" And I just answer, "Look down at the ground below your feet." Most places in the world are covered by thousands of feet of sedimentary rock filled with billions of fossils. The evidence of the global Flood is at your feet. We just need to be receptive of the truth of the rock record.

And yet old-earth geologists look at the same rocks and insist there was no such thing as a global flood. They claim the earth was never completely flooded at any point in its history. But, as this book will reveal, they have never looked at the rock record across multiple continents simultaneously. They merely accept the secular story as taught to them by their geology professors, who were taught the same story by their professors, and on and on. They don't necessarily question what they are told. They just accept the words and tales of what is called science.

These rock units in Grand Canyon can be interpreted in more than one way based on what presuppositions the investigator held before their interpretation began

Sadly, most geologists today believe in an old earth because of this indoctrination. They believe radioisotope dates "prove" an old age. In the process, they have convinced themselves that the majority of scientists cannot be wrong. However, many are never taught about the assumptions that go into the determination of every rock age estimation. I will cover more on this in my chapter on geologic time. I hope this book opens the minds of old-earth geologists to the truth.

I have not yet studied the entire globe. We plan to finish the remaining continents. In the meantime, I will tell two sides of the account. The first is a sense of personal discovery as the research progressed, a play-by-play, if you will. The second is an examination of the actual rock data continent by continent, revealing the results of the first half of our global study. Although this book is written from a young-earth perspective,

I hope both young- and old-earth geologists recognize the abundance of factual rock data that is presented. These data are the same no matter what your worldview is. This is truly data-driven science.

Formulation of a New Flood Model

In the last 25 years, the creation geology community has not progressed appreciably beyond the Flood model of catastrophic plate tectonics (CPT).[2] A comprehensive Flood model that explains the distribution of global sedimentation patterns, the fossil record, and the timing of uplifts has remained problematic. This book presents a new and novel Flood model that is based on analysis of rock columns across multiple continents, developing a framework to which future studies can be linked. It builds on the CPT conceptual model but then takes it a step further by explaining the rock record in greater detail. A primary source of information to begin this study was the American Association of Petroleum Geologists-sponsored COSUNA (Correlation

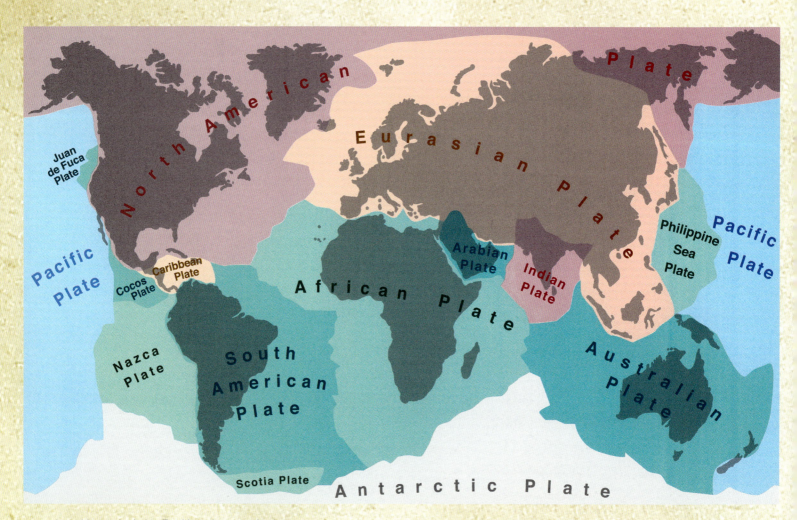

A world map depicting Earth's major tectonic plates

of Stratigraphic Units of North America) data set for the United States.[3,4] These data consist of compiled stratigraphic columns, providing both lithology and thicknesses, at hundreds of locations across the U.S. Most of these columns were compiled by state and government geologists and supplemented by input from the oil and gas industry. Other data were compiled directly from oil well logs and from government reports, especially in western Canada.[5]

The incentive to conduct this study goes back to the encouragement of Dr. John Morris, who saw the need to examine the stratigraphic columns in order to better understand how the Flood progressed from start to finish.[6] He also saw the necessity to examine the columns using the concepts of sequence stratigraphy, namely by mapping the extent of the six megasequences globally.[7] Without a data-driven model, we cannot expect to improve our scientific understanding. The present model utilizes nearly 2,000 stratigraphic columns across North America, South America, and Africa, including the Middle East and parts of western Europe. All data were entered into a computer database (Rockworks 17) that allowed sediment volumes, thickness maps, and stratigraphic relationships to be examined in detail. The rocks identified at each site are factual data. It is only the interpretation of time within and between the rock layers that creates a difference between the views of uniformitarian and creation scientists.

The results are a challenge to all scientists. The creation community needs to take a fresh look at all available geologic data and create a global Flood model that isn't based on merely tweaking secular ideas but is instead a new conceptual model based on tangible geological data. This book is an attempt to develop a Flood model from that perspective. Secular ideas are employed but not necessarily

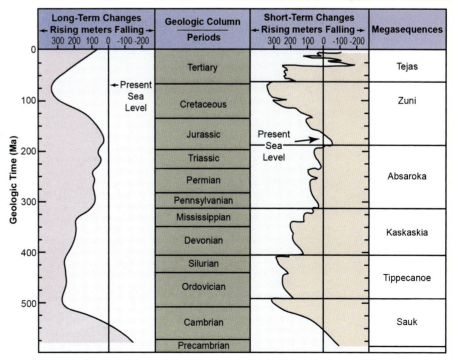

Secular chart showing presumed geologic time, global sea level, and the six megasequences (after Vail and Mitchum[8])

Author standing on the coast of Kodiak Island, Alaska

adhered to. The science is the same. It only differs in the interpretation of the data. Results of this study are presented in a biblically based, global Flood model that assumes the Flood was real and occurred about 4,500 years ago. These results, because they are based on the rocks that are actually in place, are empirical and repeatable. Someone else could spend years of their life gathering and compiling the data and arrive at nearly the same results. These data are about 80 to 85% repeatable and real. These are the rocks that exist across the continents.

A final goal is to provide the Christian community and creation scientists with a better understanding of the progression of the Flood across the globe, tied directly to the biblical narrative of God's judgment in Genesis 6–8. In a sense, this data set will help build a chapter-by-chapter model of how the Flood changed the surface of the earth. In our human minds we tend to make the Flood small since we have never witnessed a catastrophe of this extent. Only the Bible and the rocks left behind can reveal the awesome scale of the Flood event.

A Little Background

After I joined ICR, some people with a secular worldview labeled me a nonscientist, primarily because it was now known that I do not subscribe to the worldview of evolutionary science. Instead, I hold to the Bible as truth first and foremost. Science merely confirms the Bible. I still practice the same science I did before. When I was with the oil company, I was successful at finding oil even though I did not believe in the great ages. The practice and science of finding oil are still the same. I even found a new pay zone in an older field that everyone else had overlooked.

> "My science is still the same science. The rocks are still the same rocks. My observations of them are exactly the same."

I also had a very successful academic career. I published numerous peer-reviewed papers and abstracts from my research. My Ph.D. dissertation was well-received by my committee, and I was able to author or co-author four peer-reviewed papers from its results. My GPA was as high or higher than my peers at every level, Ph.D. included. I was considered a good student and a good scientist. I write this not to brag but to let you know I was blessed by God with great academic credentials. I was even the top student in my field course at Indiana University, besting many students from premier institutions across the country. But alas, as soon as it was known I was a creationist, I was scratched off the scientist list.

What changed? My practice and approach to science are exactly the same. I still gather and collect observations like all other geologists do. It is just now known that I do not hold to the old-earth stories and evolutionary views of secular geology. My science is still the same science. The rocks are still the same rocks. My observations of them are exactly the same.

Author standing on the coast of Kodiak Island, Alaska

All Beginning to Make Sense

Before we can dive into the stratigraphic columns, I need to explain some basics of geology. In particular, we need to understand a certain amount of information about sedimentary rocks to make the data more meaningful (chapter 2). I also need to define what a megasequence is and say a little about the reasons we chose this methodology to study the sedimentary story of the Flood (chapter 3). I will then review geologic time and the methods geologists use to try and determine the age of the rocks (chapter 4). If you recall, secular geologists need vast amounts of time to give their evolutionary worldview a chance, however remote. We will briefly examine the secular attempts to make the rocks of the earth very old.

I will also discuss the fossil record and the various types of fossils found in sedimentary rocks (chapter 5). I then explain why I believe that catastrophic plate tectonics is the best choice for a Flood model instead of some of the other models presented in the creation literature (chapter 6).

Chapter 7 will describe my thoughts on the pre-Flood geologic situation, including the internal makeup of the earth. This chapter is more speculative than the others since we know so little about the "world that then existed" (2 Peter 3:6).

> *God's providence for the post-Flood world abounds by providing energy resources through fossil fuels that were created by the global Flood event.*

Following that I will describe in detail the account of the Flood, examining it megasequence by megasequence and doing my best to tie it to the biblical narrative (chapters 8–14). We will look at the progression from the very first day when the fountains of the great deep burst open to the last day when God tells us the earth was dry. The record of the Flood is revealed in the progressive patterns of the sedimentary rocks on all continents. We just need to see what the rocks really show. We will then discuss the causes and the reasons for the post-Flood Ice Age (chapter 15).

Chapter 16 will describe the global nature of the geologic column and show that the pattern to the rocks and fossils derive from the global nature of the Flood. Similar observable patterns within the megasequences also allowed reconstruction of the very first, data-based, pre-Flood world map, showing its general highs and lows and shallow seas (chapter 17). Chapter 18 will address the geologic resources God provided through the chaos of the global Flood. God's providence for the post-Flood world abounds by providing energy resources through fossil fuels that were created by the global Flood event.

Chapter 19 summarizes the evidence showing the Flood took place about 4,500 years ago and the earth really is only about 6,000 years old, just as the Bible describes in Genesis. The genealogies are there for a reason.

As this research journey unfolded, I began to see how the overall geology of the world (the rocks, fossils, and tectonic plates) all started to make sense. Patterns emerged again and again that could only be explained by a recent, global flood. And they all fit perfectly within a biblical worldview (chapter 20). The rocks don't lie!

References
1. Feduccia, A. 2012. *Riddle of the Feathered Dragons: Hidden Birds of China*. New Haven, CT: Yale University Press.
2. Austin, S. A. et al. 1994. Catastrophic Plate Tectonics: A Global Flood Model of Earth History. In *Proceedings of the Third International Conference on Creationism*. R. E. Walsh, ed. Pittsburgh, PA: Creation Science Fellowship, 609-621.
3. Childs, O. E. 1985. Correlation of stratigraphic units of North America-COSUNA. *American Association of Petroleum Geologists Bulletin*. 69 (2): 173-180.
4. Lindberg, F. A., ed. 1986. *Correlation of Stratigraphic Units of North America (COSUNA): Correlation Chart Series*. Tulsa, OK: American Association of Petroleum Geologists.
5. Mossop, G. D. and I. Shetsen. 1994. *Geological Atlas of the Western Canada Sedimentary Basin*. Calgary, AB, Canada: Canadian Society of Petroleum Geologists and Alberta Research Council.
6. Morris, J. D. 2012. *The Global Flood: Unlocking Earth's Geologic History*. Dallas, TX: Institute for Creation Research.
7. Sloss, L. L. 1963. Sequences in the Cratonic Interior of North America. *Geological Society of America Bulletin*. 74 (2): 93-114.
8. Vail, P. R. and R. M. Mitchum Jr. 1979. Global cycles of relative changes of sea level from seismic stratigraphy. *American Association of Petroleum Geologists Memoir*. 29: 469-472.

2 Sedimentary Rocks

> **Summary:** Earth's rocks are largely composed of three types: sedimentary, igneous, and metamorphic. Sedimentary rocks are mostly laid down by water, igneous form from cooled magma or lava, and metamorphic are formed through heat and pressure.
>
> Most of the data in this book came from stratigraphic columns of sedimentary rock. These columns are like vertical lines running down into the ground that tell what kind of rock lies at what depth. Oil companies drill stratigraphic columns—wells—all over the world and record their data.
>
> Many secular scientists believe that sedimentary rocks require many years to form or that they could not have been deposited under flood-like conditions. But empirical experiments demonstrate otherwise. Flood geology is constantly gaining more evidence.

When oil companies drill into the ground, they record data that are used to produce stratigraphic columns. The rocks themselves do not tell us how quickly they formed, but they have left behind a lot of clues. I believe most of the sedimentary rocks can be shown to easily fit within the one-year Flood event.

Most of the stratigraphic columns used in this study were composed of sedimentary rocks. However, we also kept track of volcanic deposits (igneous rock) in our lithologic data since there are often significant amounts of ash and lava at many locations.

We also kept track of the youngest Precambrian sediments in our data compilation. These rock units represent deposits that secular geologists commonly refer to as Neoproterozoic and Mesoproterozoic. We lumped all of these sediments into a separate megasequence and entered them into our database when present. It is not immediately clear whether all of the Precambrian sediments (sometimes altered through

Cooled lava flow in Etna Park, Sicily

metamorphism during the Flood event) are pre-Flood or whether they're part of the earliest Flood sediments and the "fountains of the great deep" eruptions. We will look into this a bit later.

Many sedimentary rocks have indicators called *sedimentary structures* that tell us they were deposited by water. They contain features like cross-beds (discussed below) that even preserve the direction of current flow. Sedimentary rocks are the products of weathering and/or breakdown of other rocks or earth materials. There are several types of sedimentary rocks, but we will lump them into two major categories: 1) clastic and 2) biochemical. Clastic rocks are derived from clastic sediments. Some

Kinds of Rocks

Sedimentary: Formed largely by waterflow

Igneous (such as basalt): Formed by volcanic eruptions or cooled magma

Metamorphic: Formed through heat and pressure

books call these detrital sediments and therefore detrital rocks, but we'll stick to the older term "clastic."

Clastic sediments result from physical weathering of pre-existing earth materials. As rocks are exposed to surface conditions, they freeze, thaw, get heated by the sun, expand from unloading, and get worked over by biological activity such as tree roots and burrowing creatures. These processes generally break the rocks into smaller fragments, or *clasts*. With increasing distance of transport from a source area (erosion), the clasts become better sorted by size and more rounded. To a geologist, the degree of sorting and the roundness of the resulting sediments are important clues to the clasts' past history.

Biochemical sediments result from chemical weathering of pre-existing earth materials. As rocks are exposed to the elements, they are wetted by water. This process also occurs beneath the earth's surface because groundwater is everywhere. At the surface, rain and snowmelt cause vast amounts of fresh water to come in contact with exposed rocks. Fresh water has a greater capacity to dissolve because it is further from its saturation limit. This results in dissolution of selected minerals within the rocks. Water is the universal solvent, and it will dissolve nearly anything given enough time—even oil. Some minerals like halite and gypsum dissolve very readily, while others like quartz dissolve slowly. Generally, the more humid the climate, the more chemical weathering one can expect.

Rapid Erosion Is the Rule

Both physical and chemical weathering act concurrently on exposed rocks and earth materials. These processes are not separate, and one normally allows the other to proceed faster and more efficiently. For example, increased physical breakup can be enhanced by freezing of rainwater entering pores (tiny holes)

Sandstone, shale, limestone, and coal

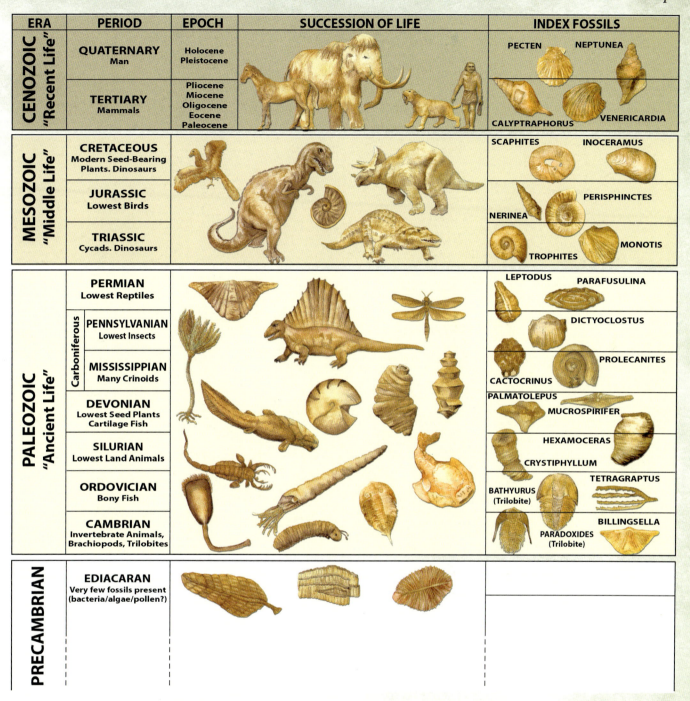

Secular geologic timescale showing representative fossils observed in each layer

created by chemical dissolution. The processes of weathering and erosion are very efficient at making and transporting sediment. Dr. John Baumgardner demonstrated that cavitation processes from tsunami waves generated by the Flood could have eroded and deposited tremendous volumes of sediment, up to 40 feet per day.[1]

It has been well known for decades that the modern rates of erosion are so fast that the continents should have been reduced to sea level long ago.[2] A recent study confirmed that the average erosion rates of outcrops are about 40 feet per million years.[3] This leaves the time needed to completely erode most continents as less than 50 million years.

Secular geologists have resorted to imaginary rescuing devices, such as episodic uplift due to tectonic forces, in order to explain the existence of today's continents. However, much of Canada and the eastern U.S. have not experienced any significant geologic uplift since the creation of the Appalachian Mountains over 250 million years ago, according to the secular timescale. Considering that much of the midwestern and southeastern states of the U.S. is less than 1,000 feet above sea level, it's a wonder there is any land above sea level at all in these regions. As secular scientists clearly demonstrated, 1,000 feet of elevation would erode away in just 25 million years at the rates calculated.[3] So, if the continents are extremely old, as secular science claims, then why are they still above sea level? Erosion rates testify to the youth of our planet and the truth of God's Word.

Even the Hawaiian Islands should not be in existence if they are millions of years old, as claimed by evolutionary scientists. Scientists studying photographs and maps

Hawaiian islands

made since 1900 found that most beaches on Kaua'i, O'ahu, and Maui experience erosion averaging 0.4 feet/year, or about five inches per year.[4] This erosion process would completely destroy the islands in only a few hundred thousand years. Doing the math, we get 76 miles of erosion in only one million years (at 0.4 feet/year), which would completely eliminate the islands—except possibly the Big Island where volcanism is still occurring. If the islands are really millions of years old, they should have eroded beneath the sea long ago. However, if they are only around 4,500 years old, as young-earth creationists believe, then the islands have experienced about a third of a mile of erosion. And that is precisely what we observe. The Hawaiian Islands really are young.

Classification of Clastic Sedimentary Rocks

Geologists have to name sedimentary rocks consistently so they can identify the same types of rocks globally. They need to know that a sandstone in South America is the same as a sandstone in North America or Europe. To do this, geologists defined a system of sedimentary rock classification.

The system classification for clastic rocks is based on careful examination of two primary textural features: 1) grain size and 2) grain shape. Clastic material is first examined for the dominant grain size of the rock specimen. Most geologists use something like the Wentworth scale, established in 1922 (Table 2.1).

The second step in naming a clastic rock involves the examination of grain shape. This step consists of looking at the relative roundness of the grains in the rock specimen. Instead of looking at the entire rock, geologists evaluate the individual clasts composing the rock. Better-rounded grains imply greater distance of travel. A conglomerate and a breccia both have large gravel-size clasts. But a conglomerate has well-rounded clasts, while a breccia possesses angular and jagged clasts. This difference tells a geologist something about the distance the gravel clasts traveled prior to solidifying into stone, or becoming *lithified*. The rounding generally increases with distance with more opportunity for bumping and grinding, resulting in less-angular clasts.

Grain Name	Grain Size
Gravel	greater than 2 mm
Sand	1/16 to 2 mm
Silt	1/256 to 1/16 mm
Clay	less than 1/256 mm

Table 2.1. Wentworth scale

As clastic sediment is transported by water and wind, its particles also become uniformly smaller. The larger clasts are left behind as the energy necessary to move the material diminishes with distance from an uplifted mountain source. The better-sorted materials by size (i.e., more uniform sizes) like beach sand imply longer distances of travel from a source area. Uniform-size clasts have generally traveled farther than clastic material of many sizes or rocks rich in gravel.

Sedimentary: Formed largely by waterflow
Clastic (physical weathering)
Gravel, Sand, Silt, Clay (such as shale)
Biochemical (chemical weathering)
Organic (rich in fossils or fossil material)
Chemical (chemical precipitation from seawater)
Halite (rock salt), gypsum, chert, limestone mud/carbonates

Table 2.2. Classification of sedimentary rocks

Fossiliferous or organic limestone

Classification of Biochemical Sedimentary Rocks

Biochemical sedimentary rocks are normally subdivided into two subgroups: 1) organic and 2) chemical.

Organic sedimentary rocks are rich in fossils or fossil material, whether plant or animal. They can contain megafossils, microfossils like chalk, or plant debris like coal. Even fossil-rich limestones fall into this category.

Author walking along limestone deposit in Petoskey, Michigan

Chemical sedimentary rocks, by contrast, usually develop as chemical precipitates from seawater. They require a chemical change, such as a pH shift or temperature decrease, to come out of solution and settle on the seafloor in layers. Interbedded layers of rock salt and rock gypsum, sometimes thousands of feet thick, are not uncommon in places like Michigan, North Dakota, beneath the Gulf of Mexico, and even under the Red Sea. Just how these extensive thicknesses of chemical sediments form is still unclear.

Uniformitarianism fails when it comes to explaining the massive beds of halite (rock salt, a type of chemical sedimentary rock) and gypsum globally. There is no large-scale example of evaporation to the point of halite or gypsum accumulation occurring anywhere on these scales in the modern world. Seawater evaporation alone apparently is not the answer because it would require tremendous volumes of seawater to cycle through the system to generate such thicknesses. Seawater must be evaporated to about one-tenth of its original volume to precipitate halite. Strict uniformitarianism obviously does not hold true.

Rock salt (above)
Gypsum (below)

Creation scientists think that deposits of halite and gypsum, and maybe even chert or flint, may be products of hydrothermal waters produced as part of the outpouring of the fountains of the great deep (Genesis 7:11). Recent findings show that great quantities of water are still trapped in the upper mantle of the earth.[5] During the Flood, some of these trapped waters may have been catastrophically extruded, precipitating great thicknesses of salt and gypsum, and likely chert, across portions of the continents, settling in the basins. Some of the salt layers even show evidence of water flow, exhibiting cross-beds in the mines in Michigan, supporting the hydrothermal model.[6]

Chert is also included here as a chemical sedimentary rock. We know it is a microcrystalline version of the mineral quartz. However, the origin of chert is a bit more mysterious. Some books give it an organic origin while others list it as both organic and chemical. Siliceous (quartz) oozes can form from biological activity such as radiolarian shell deposition on the deep ocean floor. There is additional evidence that chert can also form as an amorphous inorganic precipitate in restricted water bodies. Abundant layers of chert and rocks rich in the mineral dolomite are also more common in Precambrian and Paleozoic rocks than they are in more recent sedimentary layers. Again, strict uniformitarianism does not hold true.

Rapid Deposition of Clay and Limestone

Secular science has long taught that many sedimentary rocks were deposited slowly over vast ages. They have used the slow deposition of sediments like clay and limestone mud (micrite) as arguments for an old earth, claiming that all clays form by slowly settling out of stagnant water. People have been indoctrinated with the notion that enormous periods of time are necessary to explain these thick rock layers. And yes, we do see clay settling out of stagnant water today.

Chattanooga Shale stratification

But the rocks we observe did not form that way. Clay, the most common sediment on Earth, doesn't slowly settle out of still water to form rocks. Clay-rich rocks like shale and mudstones often exhibit fine laminations or thin-bedded layers that only form by moving water, not stagnant water. How do we know? Recent empirical evidence demonstrates that laminated clays must be deposited in energetic settings by moving water.[7] These results match the predictions of creation geologists, who interpret clay, the resulting mudstones and shales, and nearly all sedimentary rocks as rapid deposits that occurred during the year-long Flood.[8]

A second finding has uniformitarian geologists scratching their heads. Another groundbreaking paper shows that lime mud (micrite) is also deposited by moving water and not in a slow-settling process as previously thought.[9] Although some lime-rich mud rocks called *carbonates* have been interpreted to form in high-energy settings, such as fossil-rich zones and aggregate particles, carbonate mud has always been thought of as forming in "quiescent ocean settings."[10]

Carbonates comprise 20 to 25% of the total sedimentary strata and can be quite thick. For example, the Redwall Limestone in Grand Canyon is 400 to 800 feet thick, but some carbonates can exceed 3,000 feet.[10] Carbonate rocks also contain about one-third of the world's oil deposits and are important sources of construction materials.[10]

The concept of slow-forming limestone strata has been taught as fact for so long that it's ingrained in the minds of countless students going back for generations. Uniformitarians have used the presence of these rocks to criticize the biblical Flood account, pointing out that thick deposits of "quiet water" carbonates must have taken millions of years to form.

> "Clay, the most common sediment on Earth, doesn't slowly settle out of still water to form rocks. Clay-rich rocks like shale and mudstones often exhibit fine laminations or thin-bedded layers that only form by moving water, not stagnant water."

But now, all that has changed. Another long-held uniformitarian belief has been exposed as false. Flume experiments have verified that carbonate mud is not deposited slowly but rapidly by wave and current action. Laboratory experiments demonstrate that water flowing between 10 and 20 inches/second creates ripples and laminated carbonate mud layers identical to those observed in carbonate rocks.[9]

According to the authors of the study, "These experiments showed unequivocally that carbonate muds can also accumulate in energetic settings." They added, "Observations from modern carbonate environments and from the rock record suggest that deposition of carbonate muds by currents could have been common throughout geologic history."[9]

Unfortunately, secular scientists forget that the global Genesis Flood was responsible for most of the rock history of all the continents on Earth (2 Peter 3:5-6). And conditions during the Flood were anything but quiescent! Catastrophic water currents from many directions clearly washed across the continents in tsunami-like fashion, as described in Genesis 7:19: "The waters prevailed exceedingly on the earth, and all the high hills under the whole heaven were covered."

Carbonate muds were a large component of the sediments deposited by the rapidly flowing floodwaters. Again, secular scientists had to back away from their strict adherence to uniformitarianism and become more accepting of catastrophism to explain their findings. "The observations we report suggest that published interpretations of ancient muds and derived paleoceanographic conditions may need to be reevaluated."[9]

This study reaffirms the biblical Flood as a historical event. We can be assured that the thousands of feet of sedimentary rocks found across the continents stand as powerful empirical evidence against generations of uniformitarian dogma. Shales, mudstones, and carbonate rocks were deposited rapidly, not slowly, attesting to the power unleashed on the world during the Genesis Flood.

Limestone Can Form Quickly

Deposition of carbonate mud is one thing, but how long does it take for lime mud to turn to stone? Again, most uniformitarian scientists claim this is a slow process that should take many years, even thousands. But what does empirical science reveal? Does limestone take that long to form?

Many environmentalists today are concerned with anthropogenic production of carbon dioxide (CO_2) produced by burning fossil fuels. Research projects are being funded across the globe to find ways to sequester the "extra" CO_2 produced by these processes in an effort to "save our planet."

One such project was discussed in an article by Valeria Perasso of the BBC News Service.[11] She detailed the efforts of a company called CarbFix, a consortium consisting of Reykjavik Energy (Iceland), the French Centre for Scientific Research, the University of Iceland, and Columbia University (U.S.). Since 2014, CarbFix has been working with the Hellisheidi Power station about 25 miles outside of Reykjavik to conduct carbon capture experiments.

> The process starts with the capture of waste CO_2 from the steam [of the hydrothermal plant], which is then dissolved into large volumes of water. "We use a giant soda-machine," says [plant manager Dr. Edda Sif]

Basalt columns of Reynisfjara, Iceland

Aradottir....*"Essentially, what happens here is similar to the process in your kitchen, when you are making yourself some sparkling water: we add fizz to the water."*[11]

The carbonated water is then transferred to a deep-well injection site about a mile away and pumped into the local rocks at depths of about 3,200 feet.

The local rock in Iceland is a volcanic rock called *basalt*, one of the most common rocks on Earth. The minerals in basalt contain lots of elements like calcium and magnesium that readily bind with CO_2 to form carbonate minerals like calcite. And calcite is the main component of carbonate-rich rocks like limestone.

Basalt that cools near the surface often contains a high percentage of rounded holes caused by gas bubbles in the cooling magma. The result is a fairly porous rock that resembles Swiss cheese. As the injected CO_2-rich water percolates through the holes in the basalt, it dissolves some of the calcium and magnesium from the basalt and precipitates carbonate minerals.

A little over a year after its project began, CarbFix drilled down and cored the surrounding rock near the injection well. What they found surprised even them. The

Tapeats Sandstone cross-bedding, Grand Canyon

basalt rock was speckled with white minerals. Nearly all of the holes were filled with carbonate minerals. The amazing thing is how fast this process took. Sigurdur Gislason of the University of Iceland explains:

> "Before the injection started in CarbFix, the consensus within the [secular] scientific community was that it would take decades to thousands of years for the injected CO_2 to mineralise....Then we found out that it was already mineralised after 400 days."[11]

Ripples at Dinosaur Ridge, Colorado

Uniformitarian beliefs fail again. These results show that the primary minerals composing limestone can form quickly in the time frame of the global Flood described in Genesis. Thousands or millions of years are not necessary to make sedimentary rocks. Just like the recent research demonstrating that limestone rocks were deposited rapidly by energetic currents,[9] the present study reminds us that secular notions of deep time are not based on observation but imagination.

Sedimentary Structures

A sedimentary structure is an identifiable feature emplaced during deposition and/or created early in the lithification process. Examples of such features are ripple marks, cross-bed sands, raindrop prints, flute and scour marks, graded bedding, and so-called "mudcracks." The ubiquitous presence of sedimentary structures in rocks all over the world is a testament to the catastrophic burial conditions during the Flood. Most of these structures are very delicate and need to be inundated quickly and buried rapidly by new water-deposited sediments to preserve them and to prevent complete destruction by bioturbation. Ripples, flute marks, and raindrop prints are not going to last long if exposed for even a few hours or days. These structures are hard evidence of a catastrophic flood event.

Ripple marks and cross-bedded sands are fairly straightforward indicators of water flow. Ripples can be symmetrical or asymmetrical, but either way they imply the presence of moving water during deposition of the layer. *Symmetrical* ripples indicate oscillating water where currents move back and forth in approximately equal velocities. *Asymmetrical* ripples are much more common and indicate a current dominated by one direction of flow. In addition, the direction of paleocurrent flow can be estimated since we know that modern ripples migrate from the flatter side to the steeper side. Asymmetrical ripples give geologists an extra benefit: They can be used to interpret currents and flow directions.

Cross-bedded sands also indicate current flow and are distinguished by strata inclined at an angle to the bed boundaries. Geologists can use cross-beds to determine flow directions, just like ripple marks. Ripples basically create cross-bedded layers by migrating with current flow.

Raindrop prints are much less common compared to ripple marks. They occur when raindrops hit wet sand or mud and the sediment dries out before additional sediments bury the prints. Timing here is a bit tricky, hence the rarity of raindrop prints.

Flute and scour marks usually occur at the base of a sand layer as rocks and water hit irregularities in the base of a stream. This will often leave scoured areas and scoop-like "flutes" within the base of the layer and can sometimes indicate current direction.

Graded bedding is formed as fast-moving sediment slows down, such as during a landslide down a submarine canyon. When the sediment reaches the flat, deep ocean floor, the current slows and the largest and heaviest pieces settle out first, followed by progressively finer and finer particles. The resulting bed is graded or sorted by size, with the largest clasts at the base of the unit.

Mudcracks are surprisingly common, but most are probably not really mudcracks. True

Modern mudcracks and raindrop prints

mudcracks occur when wet sediments are allowed to dry out, at least partially, and cracks develop just like in dried-up mud puddles today. Most geologic mudcracks are possibly due to dewatering and expulsion processes as sediments rapidly piled on top of each other during the Flood. In this case, the mudcracks are created by vertical movement of groundwater as it escapes the compacted sediment. Many claimed mudcracks at the base of the Coconino Sandstone in Arizona are interpreted as forming by dewatering and expulsion processes.[12] Therefore, the presence of mudcracks doesn't necessarily prove extensive periods of surface exposure as most evolutionary scientists claim. Uniformitarian scientists were not there when these structures formed and are merely projecting their worldview on their interpretation.

Sedimentary structures are important clues that help with depositional interpretations. Nearly all of these structures indicate a water (marine) source for the vast majority of sediments. For example, secular science claims the Permian Coconino Sandstone across the southwestern U.S. represents an ancient aeolian or dry desert deposit.

Cross-bedded rocks, Coconino Sandstone, Colorado River

Cross-bedding in the Munising Formation, Michigan

This unit was studied extensively across Arizona by Dr. John Whitmore, Raymond Strom, Stephen Cheung, and a few others. They found numerous dolomite layers and even dolomite ooids within the Coconino Sandstone.[13] Ooids are sand-size round particles of carbonate rock that form from wave activity. Dolomite beds and carbonate ooids, in particular, form in marine waters with normal to elevated salinity.[14]

The researchers also identified mica grains (mostly muscovite) disseminated throughout the sandstone unit.[15] Mica grains are not very durable in a dry desert since they are very soft, with a hardness of about 2 to 2.5, and break down quickly during windblown transport. However, empirical research has shown that mica minerals can last much longer under water transport. Therefore, the presence of these minerals strongly suggests a marine origin for the Coconino.

Finally, the cross-bedded layers of the Coconino—for which it is so famous—have an average dip from horizontal of about 20°. This matches the angle of cross-beds formed by modern water waves.[16] In contrast, modern aeolian dunes have cross-beds that are closer to the angle of repose for dry sand, or about 25° to 30°. Compaction alone cannot explain away this difference in dip angle. It appears that the Coconino Sandstone is indeed a marine deposit, formed by water transport during the global Flood.

Nearly all sedimentary structures provide a second use to the geologist. Because they are directional features, they can be used to tell which way was originally up during the deposition of the sedimentary layers. This is especially useful to the structural geologist working with sedimentary rocks that may have been overturned (turned upside down) during later deformation. With a little practice, it becomes quite easy to tell which direction was originally up and which was down.

Contrary to uniformitarian thought, sedimentary rocks—and other geologic features—don't require millions of years to form. Empirical experiments provide solid evidence that they can, and do, form very quickly. Much evidence exists that the majority of Earth's rocks were rapidly deposited by water. This is consistent with a global flood.

Now that I have covered the basics of sedimentary rocks and sedimentary structures, we can move on to the megasequences. Geologists recognized that many sedimentary rocks were deposited in discrete packages. This opened the door to a whole new way to view sediments, and it revolutionized the science of sedimentology in the process.

References
1. Baumgardner, J. R. 2016. Numerical Modeling of the Large-Scale Erosion, Sediment Transport, and Deposition Processes of the Genesis Flood. *Answers Research Journal.* 9: 1-24.
2. Blatt, H., G. Middleton, and R. Murray. 1980. *Origin of Sedimentary Rocks*, 2nd ed. Englewood Cliffs, NJ: Prentice-Hall, Inc.
3. Portenga, E. W. and R. R. Bierman. 2011. Understanding Earth's eroding surface with [10]Be. *GSA Today.* 21 (8): 4-10.
4. Fletcher, C. H. et al. 2012. *National Assessment of Shoreline Change: Historical Shoreline Change in the Hawaiian Islands.* Open-File Report 2011–1051. Reston, VA: U.S. Geological Survey, 55.
5. Pearson, D. G. et al. 2014. Hydrous mantle transition zone indicated by ringwoodite included within diamond. *Nature.* 507 (7491): 221-224.
6. Dellwig, L. F. and R. Evans. 1969. Depositional Processes in Salina Salt of Michigan, Ohio, and New York. *American Association of Petroleum Geologists Bulletin.* 53 (4): 949-956.
7. Schieber, J. et al. 2007. Accretion of mudstone beds from migrating floccule ripples. *Science.* 318 (5857): 1760-1763.
8. Snelling, A. A. 2009. *Earth's Catastrophic Past: Geology, Creation & the Flood,* vol. 2. Dallas, TX: Institute for Creation Research, 493-499.
9. Schieber, J. et al. 2013. Experimental Deposition of Carbonate Mud from Moving Suspensions: Importance of Flocculation and Implications for Modern and Ancient Carbonate Deposition. *Journal of Sedimentary Research.* 83 (11): 1025-1031.
10. Boggs Jr., S. 2006. *Principles of Sedimentology and Stratigraphy,* 4th ed. Upper Saddle River, NJ: Pearson/Prentice Hall, 159-167.
11. Perasso, V. 2018. Turning carbon dioxide into rock – forever. *BBC News.* Posted on bbc.com 18 May 2018.
12. Whitmore, J. H. 2004. An alternative to the mud crack origin for sand-filled cracks at the base of the Coconino Sandstone, Grand Canyon, Arizona. *Geological Society of America Abstracts with Programs.* 36 (5): 55.
13. Cheung, S. et al. 2009. Occurrence of dolomite beds, clasts, ooids and unidentified microfossils in the Coconino Sandstone, Northern Arizona. *Geological Society of America Abstracts with Programs.* 41 (7): 119.
14. Danise, S. and S. M. Holland. 2018. A sequence stratigraphic framework for the Middle to late Jurassic of the Sundance Seaway, Wyoming: implications for correlation, basin evolution, and climate change. *The Journal of Geology.* 126 (4): 371-405.
15. Whitmore, J. H. and R. Strom. 2010. Petrographic analysis of the Coconino Sandstone, Northern and Central Arizona. *Geological Society of America Abstracts with Programs.* 41 (7): 122.
16. Personal communication with J. H. Whitmore, 2012.

3 Megasequences and Methods

> **Summary:** Each sedimentary rock layer across the globe is assigned to a group called a *megasequence*. Many megasequences are composed of a sandstone layer on the bottom, followed by a layer of shale, and then finally limestone on top. Each megasequence is thought to represent one depositional interval. Megasequences are not constructed based on the fossil record but rather on common erosional boundaries across the continents. Because of this, they have become the preferred method of studying sedimentary deposits.

Dr. Laurence Sloss defined megasequences as discrete packages of sedimentary rock that are bounded top and bottom by interregional erosional surfaces and are traceable on a continental scale.[1] In other words, these sequences are massive groupings of sedimentary strata that were deposited in a definable interval and often follow a distinctive pattern. Because the terminology associated with sedimentary sequences has ballooned in the past decades, some have chosen to use the term *megasequence* for strata bounded by the most prominent regional unconformities.[2] Haq and his colleagues used that term to designate their First Order sequences, or their largest scale sequences, equivalent to the aforementioned Sloss sequences.[3] Other secular and creation scientists have followed, using megasequence to describe rock-stratigraphic units traceable over vast areas bounded by unconformities (or their correlative conformities). Hereafter, this term will be used to designate the six Sloss-defined megasequences.

Many megasequences are marked by a coarse sandstone layer at the bottom (deposited first), followed by shale, and then limestone at the top (deposited last). The corresponding size of the sedimentary particles is also thought to decrease upward in each package of rock. The basal sandstone layers are thought to represent the shallowest sea level, the shale a little deeper water environment, and the limestone the deepest water environment in each sequence. By tracking these major erosional surfaces and changes in rock types, geologists were able to define each megasequence.

According to secular geologists, subsequent megasequences are

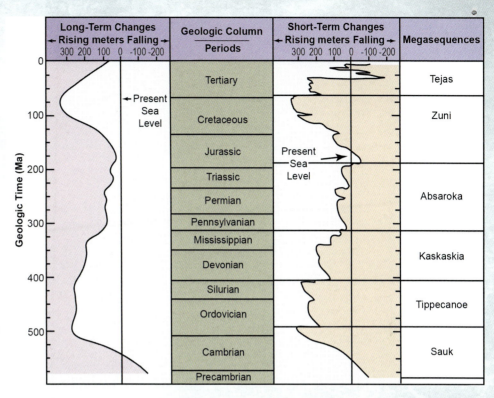

Secular chart showing presumed geologic time, global sea level, and the six megasequences (after Vail and Mitchum[4])

supposed to demonstrate this same pattern of sandstone to shale to limestone as sea levels repetitively rose and fell, again and again, over millions of years, flooding the continent up to six separate times. The upper erosional boundaries are believed to have been created as each new megasequence eroded the top of the earlier sequence as it advanced landward. Secular scientists then use these megasequences to infer past environments, and of course as an argument in support of deep time.

Because the megasequences are based on physical relationships demonstrated by observable erosion, they are independent of regional differences in nomenclature and even paleontological dating methods.[1] Therefore, many consider each megasequence boundary as a nearly common time surface (chronostratigraphic horizon). By not relying on the paleontological record, which is so intertwined in the development of the classical geologic timescale, megasequences have become the preferred method for studying the sedimentary deposits of the great Flood.[5]

Megasequences include multiple geologic systems and in many instances can be recognized by their bounding erosional surfaces and sudden changes in rock type,

Carved in Stone

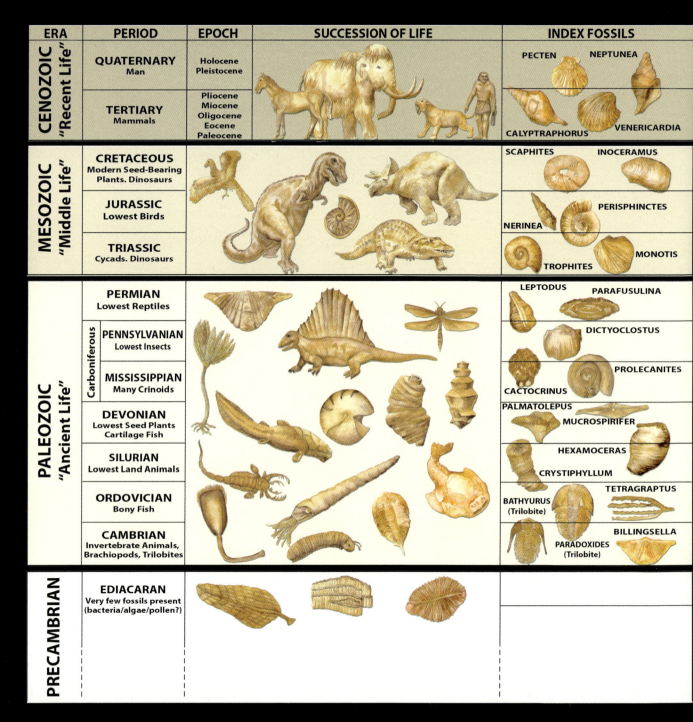

Secular geologic timescale showing representative fossils observed in each layer

independent of fossil content. Many creationists believe megasequences record the sedimentology of the Flood, while fossils record what flora and fauna were buried within each megasequence. They differ from the standard geologic timescale because they are not based solely on changes of fossil content like the eras, periods, and epochs of the traditional geologic timescale.[1]

Are Megasequences Valid?

The use of sequences or megasequences to study Flood geology has been criticized by some creation scientists.[6] These creationists claim, "The heart of the issue of using Sloss-based megasequences is their dependence on the geological timescale."[7] In other words, they claim the use of megasequences is dependent on fossil correlation and therefore is reliant on evolutionary theory.

Other creation scientists, like Dr. Marcus Ross, have countered that the global geologic column and the megasequences are based on multiple comparisons and coincidence of both paleontological and physical geologic data and have been validated for many decades.[8] Ross further emphasized, "The ability to correlate rocks on the basis of fossils contained is not dependent on evolutionary reasoning. Rather it is based on sound recognition of similar *patterns of fossils* found in disparate locations."[9] He argued that the type of rocks, and distinctive chemical signals in some of the rocks, allow consistent correlations. It is not just the fossils that are compared from place to place but many of the rocks also.[8]

To resolve this issue, we tested the megasequences against the global geologic column.[10] Our results show that the basal megasequence lithologic units (i.e., blanket sandstones) are correlative across vast regions and even continent to continent. Also, unique lithologic units, like salt and chert-rich layers, were tracked from column to column and found to remain at consistent levels within the megasequences. The patterns in the fossil record further confirmed these findings. The correlation of these stacked basal megasequence units, and other unique lithologies (i.e., salt and chert layers), confirm the validity of the geologic column and the use of megasequences on a global scale.[10] The use of megasequences does not imply a belief in evolutionary theory. Instead, megasequences are based heavily on empirical correlations of strata deposited in the Flood.

Idealized Megasequences

The most extensive sandstone layer observed in the stratigraphic column data in North America is found at the base of the Sauk Megasequence. This sandstone layer, commonly known as the Tapeats Sandstone (and equivalent), is generally agreed upon by creation geologists as the first continent-scale deposit of the advancing floodwaters. It's no surprise this sandstone is so prevalent across the entire North American conti-

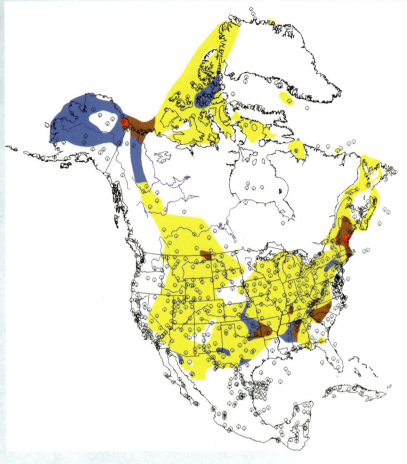

Tapeats Sandstone and equivalent units across North America. Sandstone is represented by yellow

nent. However, some subsequent megasequences start with limestone at the bottom and sandstone at the top, a complete reversal of the secular story. Others start with shale or even salt at the base and no sandstone in the sequence at all.

In a Flood model, variations in the sequence-bounding rock types make more sense. Flood geologists don't expect a complete draining off of the continent and a drop to previous sea levels between each megasequence. The Bible says, "The waters prevailed and greatly increased on the earth, and the ark moved about on the surface of the waters. And the waters prevailed exceedingly on the earth, and all the high hills under the whole heaven were covered" (Genesis 7:18-19). As we will see, the rocks confirm this exactly. The megasequences show that the floodwaters seem to have risen somewhat steadily across the continent. The waters may have backed off between megasequences

> **"Although there are erosional boundaries between the megasequences, there is no evidence of millions of years of missing time. The rocks were merely stacked one on top of another, layer by layer, sequence by sequence, as the Flood rose higher and higher."**

(assuaged a little) but may not have completely drained off the land between each cycle. Differences in rock types at different locations and at different sequence boundaries merely reflect the local conditions during the one continuous Flood event.

Although there are erosional boundaries between the megasequences, there is no evidence of millions of years of missing time. The rocks were merely stacked one on top of another, layer by layer, sequence by sequence, as the Flood rose higher and higher. All claims of vast amounts of time are uniformitarian attempts to explain the effects of the one real Flood as a series of smaller floods. Here again, uniformitarian scientists are imposing their worldview onto the rock data. They were not there during the deposition of the various layers. They merely assume the layers took great lengths of time to be deposited. The rock columns found across the globe are best explained in the context of the one-year Flood as recorded in the Bible.

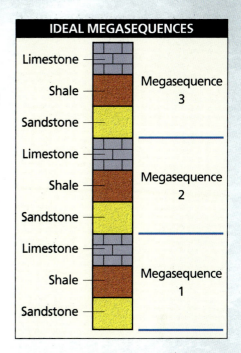

Idealized megasequence rock layers. Secular geologists believe that as sea level rises, the first rocks deposited are sandstones, followed by shale and limestone as the water becomes deeper. This cycle is repeated for each subsequent sequence. But actual rock-column data tell a different tale. The ideal cycle is observed best in the lowermost megasequence, the Sauk, which was deposited as early floodwaters spread across the continents.

Correlation of Megasequences Globally

Surprisingly, continent-scale studies of the six megasequences have not been published since the more descriptive paper of Sloss.[1] That same year, Dr. Harry Wheeler examined in some detail the regional extent of the post-Sauk and pre-Absaroka Megasequences, looking primarily at the Tippecanoe and Kaskaskia.[11] A decade later, Shell Oil published a collection of continental-scale maps for each geologic time period in 1975 but did not use the megasequence boundaries.[12]

Dr. Peter Burgess and his colleagues utilized a three-dimensional finite-element model to try and simulate the formation of the six megasequences across the interior of North America.[13] They were able to generate thickness maps for several sequences based on computer analysis of dynamic topography, or elevation differences caused by actively moving mantle material. Unfortunately, they did not compare their results to actual three-dimensional sedimentary thickness data but instead compared them to a two-dimensional cross-section, claiming partial agreement.

Although Sloss initially defined his megasequences across only the interior of

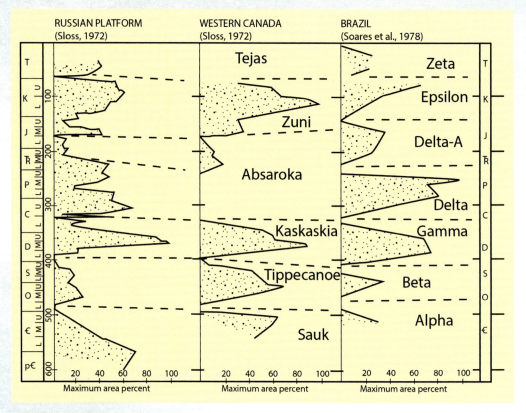

Correlation of megasequences across three continents[14]

North America, oil industry geologists quickly extended these sequence boundaries to the offshore regions surrounding North America and to adjacent continents.[2,15] Oil industry geologists have tracked the megasequence boundaries from the craton to the ocean shelves on the basis of distinctive seismic reflection patterns (many due to abrupt truncations) as well as lithologic changes in oil well bores (using downhole well logs, biostratigraphy data, and cores).[2,16] These same Sloss-megasequence boundaries were correlated to at least three other continents based on seismic data and oil well drilling results.[2,14-16] In fact, nearly identical megasequence boundaries were identified and aligned to global tectonic events in North America, the Russian Platform, Brazil, and Africa.[14]

Methods Used and Maps Prepared

Stratigraphic columns were compiled from published outcrop data, oil well boreholes, cores, cross-sections, and/or seismic data tied to boreholes. Lithologic and stratigraphic interval data (megasequence boundaries) were input into a database, allowing the creation of a three-dimensional lithologic model for each of the three continents in this study. These models also allow the correlation of rock types within individual megasequences and along their bounding surfaces.

Outcrop of lava from the Midcontinent Rift near Lake Superior

Our database consisted of selected COSUNA (Correlation of Stratigraphic Units of North America)[17,18] stratigraphic columns across the United States, stratigraphic data from the Geological Atlas of Western Canada Sedimentary Basin,[19] numerous well logs, and hundreds of other available online sources. Using these data, we constructed 710 stratigraphic columns across North America, 429 across Africa, and 405 across Central and South America from the pre-Pleistocene, meter-by-meter, down to local basement. We input detailed lithologic data, megasequence boundaries, and latitude and longitude coordinates into RockWorks 17, a commercial software program for geologic data, available from RockWare, Inc., Golden, Colorado, U.S. The chart on page 56 is an example stratigraphic column from the Michigan Basin, showing the 16 types of lithology that were used for classification and the sequences. Depths shown in all diagrams are in meters.

Each column recorded the complete record of sedimentary rocks at that location from surface to crystalline basement along with the corresponding Sloss-megasequence boundaries. Any erosional gaps in the COSUNA columns were collapsed so that only the rocks present at each location were used in the study.

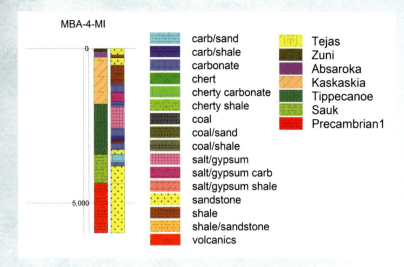

Example stratigraphic column from the Michigan Basin, showing the 16 rock types and the megasequences that were used in this study

We included volcanic deposits in our lithologic data since there are often significant amounts of ash and lava at many locations. Instead of leaving these layers out, we decided to include them in our compilations. Although they are not attributed to changes in sea level per se, they are important to the local geology and the timing of volcanic activity. RockWorks 17 also allows easy exclusion of the volcanic deposits and lava flows when doing purely sedimentological analysis.

Of particular interest were the basal rock types in each megasequence, deposited as the ocean water transgressed across the continents. The basal rock types were most likely the best preserved of any interval within each megasequence as all subsequent erosion from regressive phases eroded from the top of the megasequence down. That is not to say that all the basal rocks in each megasequence were preserved. In some locations, the regressive phase may have removed all of the preceding megasequence rock. Accordingly, maps of the basal rock type in each megasequence and stratigraphic cross-sections were constructed that allowed continent-scale correlations of the basal stratigraphy for each megasequence.

We also compiled maps of distinctive rock types, like bedded chert layers and salt-rich and gypsum-rich layers, keeping track of each by megasequence. These unique lithologic units also allowed us to test our megasequence boundary picks on a regional scale. For example, we assumed megasequence correlations were validated if the salt-rich or chert-rich layers remained in the same relative location within the megasequences, from column to column, and did not crosscut the megasequence layering up or down in the stratigraphic section. We also examined published maps of extensive and lithologically distinct rock units, like the Morrison Formation and Pierre Shale in the western U.S. These semiregional (multistate units in the U.S.) formations were also tracked within the confines of the megasequence boundaries to test the validity of the correlations.

Conclusion: Megasequences Are Valid

Now that megasequences have been defined and explained, I will move on to the basics of geologic time and how secular geology went from young-earth to old-earth. We'll see that a lot of the cause was not based directly on the rocks themselves but was triggered by a shift in worldview away from Genesis as historical truth.

References
1. Sloss, L. L. 1963. Sequences in the Cratonic Interior of North America. *Geological Society of America Bulletin.* 74 (2): 93-114.
2. Hubbard, R. J. 1988. Age and significance of sequence boundaries on Jurassic and Early Cretaceous rifted continental margins. *American Association of Petroleum Geologists Bulletin.* 72 (1): 49-72.
3. Haq, B. U., J. Hardenbol, and P. R. Vail. 1988. Mesozoic and Cenozoic Chronostratigraphy and Cycles of Sea-Level Change. In *Sea-Level Changes: An Integrated Approach: SEPM Special Publication 42.* C. K. Wilgus et al, eds. Tulsa, OK: SEPM, 71-108.
4. Vail, P. R. and R. M. Mitchum Jr. 1979. Global cycles of relative changes of sea level from seismic stratigraphy. *American Association of Petroleum Geologists Memoir.* 29: 469-472.
5. Morris, J. D. 2012. *The Global Flood: Unlocking Earth's Geologic History.* Dallas, TX: Institute for Creation Research.
6. Froede Jr., C. R., A. J. Akridge, and J. K. Reed. 2015. Can 'megasequences' help define biblical geologic history? *Journal of Creation.* 29 (2): 16-25.
7. Ibid, 21.
8. Ross, M. R. 2014. Improving our understanding of creation and its history. *Journal of Creation.* 28 (2): 62-63.
9. Ross, M. 2013. The Flood/post-Flood boundary. Letter to the Editor. *Journal of Creation.* 27 (2): 43-44. Emphasis in original.
10. Clarey, T. L. and D. J. Werner. 2018. Global stratigraphy and the fossil record validate a Flood origin for the geologic column. In *Proceedings of the Eighth International Conference on Creationism.* J. H. Whitmore, ed. Pittsburgh, PA: Creation Science Fellowship, 327-350.
11. Wheeler, H. E. 1963. Post-Sauk and pre-Absaroka Paleozoic stratigraphic patterns in North America. *Bulletin of the American Association of Petroleum Geologists.* 47 (8): 1497-1526.
12. Cook, T. D. and A. W. Bally, eds. 1975. *Stratigraphic Atlas of North and Central America.* Princeton, NJ: Princeton University Press.
13. Burgess, P. M. et al. 1997. Formation of sequences in the cratonic interior of North America by interaction between mantle, eustatic, and stratigraphic processes. *Geological Society of America Bulletin.* 109 (12): 1515-1535.
14. Soares, P. C., P. M. B. Landim, and V. J. Fulfaro. 1978. Tectonic cycles and sedimentary sequences in the Brazilian intracratonic basins. *Geological Society of America Bulletin.* 89 (2): 181-191.
15. Sloss, L. L. 1972. Synchrony of Phanerozoic sedimentary-tectonic events of the North American craton and the Russian platform. *International Geological Congress, 24th,* Montreal, Canada. 6: 24-32
16. Van Wagoner, J. C. et al. 1990. *Siliciclastic Sequence Stratigraphy in Well Logs, Cores, and Outcrops: Concepts for High-Resolution Correlation of Time and Facies.* AAPG Methods of Exploration Series, No. 7. Tulsa, OK: American Association of Petroleum Geologists.
17. Childs, O. E. 1985. Correlation of stratigraphic units of North America-COSUNA. *American Association of Petroleum Geologists Bulletin.* 69 (2): 173-180.
18. Salvador, A. 1985. Chronostratigraphic and geochronometric scales in COSUNA stratigraphic correlation charts of the United States. *American Association of Petroleum Geologists Bulletin.* 69: (2): 181-189.
19. Mossop, G. D. and I. Shetsen. 1994. *Geological Atlas of the Western Canada Sedimentary Basin.* Calgary, AB, Canada: Canadian Society of Petroleum Geologists and Alberta Research Council.

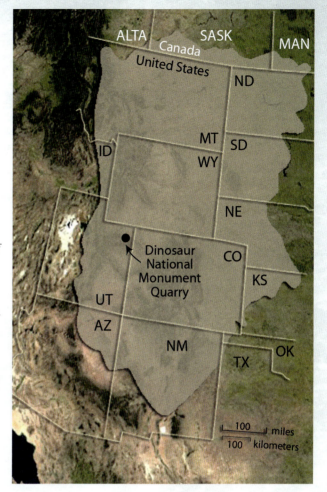

Map of the extent of the Morrison Formation

Sandstone landscape in Petra, Jordan

4 Geologic Time

> **Summary:** Europeans once assumed the Bible's chronology was historical—even scientists. However, in the 18th and 19th centuries, the theory of uniformitarianism began to rise. It taught that the majority of geologic features were formed through gradual processes. The conclusion was that they took many ages to form. Therefore, Earth had to be extremely old, contrary to what the Bible teaches.
>
> Geologic time can be measured in two ways: on a relative scale or an absolute scale. Relative time just means whether certain layers were formed before or after others. It says nothing about age. However, absolute time makes statements about the ages of the layers. It relies upon dating methods such as radiometric dating. These dating methods contradict each other in serious ways and use too many assumptions to be reliable. Additionally, the presence of short-lived materials in supposedly ancient rocks challenges their age assignments of millions of years.

In the 17th century, Archbishop James Ussher compiled historical and biblical data to make a famous calculation: Earth was created on October 22, 4004 BC. Before you disregard or ridicule his estimate, let's continue the history lesson. Ussher arrived at this date for the earth by adding up the detailed Jewish genealogies listed in the Bible, starting with Adam. For the next 200 years, this date was not openly challenged by any scientist or theologian, and was mandated by many of the world's churches.

However, in the late 18th century and into the 19th century, geologists began to develop a philosophy of uniformitarianism, which viewed the earth's processes as unchanging and progressing at rates similar to those observed in modern times. Thereafter, layers of sedimentary rock were interpreted as taking vast amounts of time to be deposited. Inches of sediment were equated to hundreds or even thousands of years.

Simultaneously, these same scientists began to question the literal history recorded in Genesis.

Within the same century, Charles Darwin published his now-famous treatise on evolution (1859). Darwin needed vast amounts of time for the natural selection process to be viable. His thinking was that small gradual changes (microevolution) would require a lot of time in order to make big changes (macroevolution). The philosophy of uniformitarianism fit perfectly into the time frame Darwin envisioned for evolution. Both required great amounts of time to explain the immense thicknesses of sedimentary rocks and to account for the microevolution of the fossils therein. Therefore, deep time and evolution exist in a kind of symbiotic relationship.

James Ussher

Most geologists of the late 19th century quickly saw the need for the earth to be much older than 6,000 years. They argued the rock layers were too thick to be merely thousands of years old. At the same time, many churches were losing their political clout and their influence over science, giving the uniformitarians and Darwinists free rein. The earth quickly aged to about 20 to 40 million years old by the late 19th century. With the discovery of radioactivity in 1896 by Henri Becquerel, the earth was allowed to age even more, and estimates of billions of years became popular by the early 20th century. The invention of the mass spectrometer in the 1930s and the development of radioisotope dating techniques further pushed the estimate to over four billion years by the mid-20th century.

At the time of this writing, secular scientists believe the earth is 4.56 billion years old. Surprisingly, this is not based on Earth's rocks but on samples from the moon and meteorites that have hit the earth. Secular science generally assumes that the earth, moon, and most meteorites formed

Charles Darwin

Henri Becquerel

about the same time. Therefore, even though the oldest minerals on the surface of the earth are dated by secularists at 4 to 4.4 billion years old, the earth is considered to be several hundred million years older.

Does the Age of the Earth Matter?

What's the bottom line? Does the age of the earth really matter? In any practical sense of geology, it doesn't make any difference if the earth is 6,000 years old or billions of years old. It truly doesn't matter to the coal geologist whether or not the coal seam is 600, 6,000, or 60 million years old, just that it exists today. It doesn't matter to the paleontologist whether or not the dinosaur bone is 6,000, 6 million, or 60 million years old, just that he/she can place it properly compared to other finds (relative dating). Geologists searching for precious minerals and oil and gas can find these resources whatever age the earth is. They may insist they can only find oil because the oil and rocks are millions of years old, but they forget that just because these resources exist doesn't mean

Coal seam showing the typical flat top and flat bottom

> "Secular scientists cannot tell you, with certainty, that the earth isn't 6,000 years old! They cannot independently verify any absolute ages older than about several thousand years ago without direct comparison to a historical artifact such as a coin."

they are that age. Oil doesn't require that much time to form. I will further discuss the impossibility of old oil in chapter 18.

Secular scientists cannot tell you, with certainty, that the earth isn't 6,000 years old! They cannot independently verify any absolute ages older than about several thousand years ago without direct comparison to a historical artifact such as a coin. And even tree ring interpretations can get sketchy since more than one tree ring can form in a given year. All dates older than this have to rely on a radioisotope method for any hope of verification. And we will see below that all radioisotope methods are based on the same series of unverifiable assumptions. You cannot use one unverifiable technique to verify another unverifiable technique. That is not falsifiable science.

John Eddy, an astronomer, was quoted as saying:

> I suspect that the Sun *is* 4.5 billion years old. However, given some new and unexpected results to the contrary, and some time for some frantic recalculation and theoretical readjustment, I suspect that we could live with Bishop Ussher's value for the age of Earth and Sun. I don't think we have much in the way of observational evidence in astronomy to conflict with that. Solar physics now looks to paleontology for data on solar chronology.[1]

Why all the controversy about the age of the earth, you ask? It comes down to worldview. If you ignore the clear evidence of God throughout creation (Romans 1:18-20), then you must develop an alternative and seemingly viable story. Secular science has had to make the earth really old and assume that somehow, somewhere, over vast amounts of time, life began from nothing and evolved into all the biological diversity we see today. Time becomes their "black box" that allows anything to happen, no matter how remote the possibility.

If you read the book of Genesis as true history, as intended, and count up the genealogies, you'll find the earth really is only about 6,000 years old. In contrast, if Genesis is merely viewed as a myth or as a collection of fictional tales, then a man named Adam never existed. The creation just becomes one of the many stories not meant to be taken literally. And with no Adam or his sin and no curse, there is no need for a Savior. The secular worldview destroys the very basis of why we need

Halley's comet.

salvation through Christ. Jesus is the Second Adam who offered Himself as a sacrifice to take away the sins of the world—mine, yours, and Adam's. It is no surprise that many secular scientists abhor Genesis and try to discount it. It makes them accountable to a Supreme Being and accountable for their own personal sinfulness. They also pass up the opportunity for salvation by faith due to the grace of God (Ephesians 2:6-10). Christ has provided the path of redemption. All we have to do is believe.

One of the first steps down this slippery slope to secular humanism is the belief that science has proven that the earth is billions of years old. That sets the hook. Sadly, many well-meaning old-earth geologists cannot get past what they have been taught in college. They cannot see past the dogma of old ages and the effect this has on their worldview. They believe secular science cannot be wrong since the system seems to work. They can still find oil, as I mentioned above. But that doesn't mean they are correct in the old ages.

In fact, science confirms the young age of the earth. As we discuss in chapter 19, the empirical evidence found as soft tissue preservation in fossils, the numerous comets in our solar system, the rapid decay of the earth's magnetic field, and many more findings show the earth cannot be any older than thousands of years old, like the book of Genesis says. God told us the truth. We just have to accept it and throw off the binding shackles of deep time. There was a real Adam created by God just 6,000 years ago.

Principles of Relative Time

Geologic time can be measured in two very different ways. It can either be viewed on a relative scale or an absolute scale. Relative time places geologic events in an order from youngest to oldest, or vice versa. This type of time measurement is concerned only with determining the proper order of past events. Absolute time, by contrast, attempts to assign a numerical value to each event in terms of years before present. This method generally involves the use of radioactive isotopes and also gives the relative order of past events.

Relative time merely concerns itself with the proper ordering of

Nicolas Steno

past events. This was the original method of determining geologic time. It is still the most useful, practical, and cheapest method, requiring nothing more than brainpower.

Geologists use several principles to help them determine the order of events. Some of these go back to the 17th-century writings of the Danish scientist Nicolas Steno, also known as Niels Steensen. Many of his ideas were forgotten but were later resurrected and added to by Scotsmen James Hutton and Charles Lyell in the later 18th century and early 19th century, respectively. We will review some of these principles.

Superposition: This principle states that sedimentary rock layers are deposited in order, from oldest at the bottom to youngest on the top. It's not very profound, but it was one of the early principles of the fledgling science of geology.

Original Horizontality: This states that sedimentary rocks are normally deposited nearly horizontally, at least at the time of original deposition. Therefore, if sedimentary rocks are tilted or folded, one can assume that the deformation occurred after the sedimentary rocks were already in place. Again, this principle is not very profound, but it is important to recognize original horizontality when dating deformational events.

Crosscutting Relations: This principle states that igneous intrusions and/or faults cannot crosscut rocks that do not already exist. This is one of the more utilized methods of dating and sequencing geologic events. Nearly every mountain range has numerous faults and intrusions, allowing geologists to precisely determine when faulting or intrusions entered the picture and when the activity ceased.

Unconformities: These surfaces are usually marked by an erosional event of some kind, removing rock of some type. They normally form a roughly horizontal surface but can have considerable relief. Some claimed unconformities may be caused by a temporary hiatus in deposition and may or may not represent any real erosion. Vast, regional, and even intercontinental unconformities are what bounds the tops and bottoms of each megasequence. There are three basic types of unconformities: angular unconformities, nonconformities, and disconformities.

Angular unconformities are identified by the nonparallel nature of the overlying sedimentary layers compared to the sedimentary layers underneath the unconformity surface. This variety implies a period of deformation or uplift prior to the erosional event, followed by deposition of new sediments on top of the unconformity. However, the time between deformation, erosion, and renewed deposition is not specified and

The Great Unconformity in Grand Canyon. Tapeats Sandstone on Precambrian crystalline rock.

could represent just hours or days in a global flood scenario. Because of the angular difference of the rock strata above and below the unconformity surface, these unconformities are quite easily recognized.

Nonconformity surfaces exist whenever sedimentary rocks are deposited directly on either igneous or metamorphic terrains. By definition, there had to be a period of erosion to expose the igneous or metamorphic rocks at the surface prior to the deposition of sediments. However, a large igneous intrusion can sometimes be confused for a nonconformity. Geologists must look for evidence of heat alteration (contact metamorphism) in the sedimentary rocks directly above or along the upper edge of the igneous rock body. This usually eliminates any potential confusion and allows the geologist to tell if the sedimentary layers were deposited on top of the igneous or metamorphic rocks or if the igneous rock intruded into the strata at a later time.

Disconformities are the most difficult type of unconformity to identify, and some are probably fabrications steeped in evolutionary theory with very little physical

Siccar Point, Scotland, illustrating an angular unconformity

The Great Unconformity in Grand Canyon illustrating a nonconformity

evidence for their existence. Disconformities are identified solely on the basis of fossil evidence or lack thereof. A disconformity is placed between two sedimentary units if the corresponding index fossils indicate another type of index fossil is missing or is absent. For example, if the lower sedimentary layer contains a fossil indicative of the Upper Triassic, such as the dinosaur *Coelophysis* (because it has only been found in Upper Triassic rocks to date), and the next layer above contains an Upper Cretaceous dinosaur, such as *Triceratops* (which has only been found in Upper Cretaceous rocks to date), secular scientists would infer that the entire Jurassic section is missing and place a disconformity between the two rock layers.

Often, disconformities show no real evidence of erosion at all, making one wonder if there really is an unconformity or just a lack of fossils within the supposed missing time period. It is likely that many disconformities were identified using the older Darwinian, gradual style of evolution, and not the punctuated equilibrium style, where major changes in the fossils are the norm. Unfortunately, the theory of evolution requires secular scientists to invoke a disconformity to explain the missing fossil lineages. Another possibility is that there really was no erosional event at all—and that secular theories require revision. A global flood would tend to deposit fossils rapidly on top of one another, with no transitions. And some layers found in one place might be missing in another. The Flood didn't just deposit continually as the waters assuaged (Genesis 8:3), meaning they went back and forth. These oscillating waves may have eroded some of the freshly deposited sediments before depositing more on top.

Principle of Faunal Succession: This principle invokes the use of index fossils and groups of fossils to correlate rocks from place to place and establish their relative time

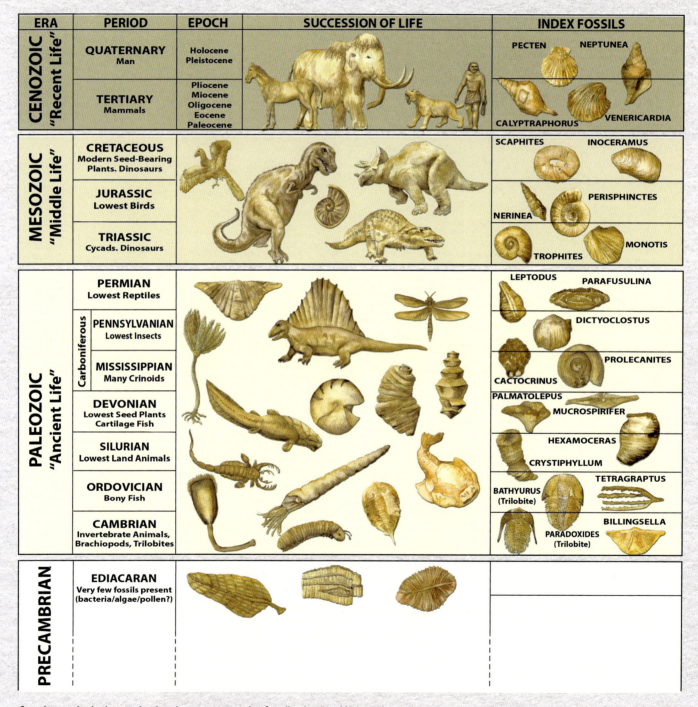

Secular geologic timescale showing representative fossils observed in each layer

frame. It established an order to the fossils and allowed secular geologists to create the global geologic column. This principle was developed by William Smith during his construction of the first geologic map of Britain in 1815. He is known as the Father of Biostratigraphy because he established the process of dating rocks by comparing fossils he found to index fossils he had found earlier in a particular depositional order.

Unfortunately, this concept was taken too far by secular science and was used to establish the geologic timescale by simply adding the millions of years to the relative order of the geologic column. For this reason, this is the most misunderstood of all the principles of relative time.

The bottom line is that this principle only suggests there is a recognizable order to the fossils. It does not imply evolution or vast amounts of time. In fact, a globally progressive flood would be expected to inundate and bury similar fossils from the same environments at about the same time and in the same order in the rock record.

Principles of Absolute Time

Since the middle of the 20th century, secular scientists have claimed to be able to make determinations of absolute time using radioactive isotopes of certain elements. Commonly used radioactive elements are isotopes of uranium, thorium, rubidium, strontium, argon, and potassium. Most of the time, only igneous rocks are analyzed. Sedimentary rocks may contain the proper radioactive elements, but they are usually mixtures of rock types from various locations and therefore give mixed ages if analyzed. There are several methods for determining absolute ages of selected sedimentary rocks,[2] but most are susceptible to much greater error and some can only be used on fairly recent sediments. Metamorphic events often "reset" the radioactive clocks, making age determinations trickier than with igneous rocks.

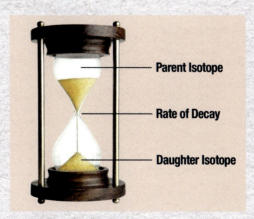

Absolute dates are based on the decay rate of unstable (radioactive) isotopes. Isotopes are elements that have the same proton number but different atomic masses (different number of neutrons). Decay rates are often directly measurable and are assumed to be constant with time (uniformitarianism again). We will see later that this does not hold true for all radioactive elements under all conditions. Decay rates are converted to a half-life, which is the time it takes for one-half of the radioactive element to decay to another element. The original radioactive element, or radioisotope, is called the parent isotope and the decay product is the daughter isotope.

Carbon-14 Dating

Let's illustrate radioactive decay with carbon-14. Although this method is rarely used in geology, it makes a good example because most people have heard of it. Carbon can come as three isotopes: carbon-12, the most common variety; carbon-13, the next-most common variety; and carbon-14, the rarest and only radioactive variety. All three carbon isotopes have six protons, which defines them as carbon. Carbon-12 has six neutrons, carbon-13 has seven neutrons, and carbon-14 has eight neutrons, making the latter unstable. Carbon-12 and carbon-13 remain the same forever. Carbon-14, by contrast, spontaneously decays into nitrogen-14 over time. Carbon-14 has a half-life of about 5,730 years, meaning that one-half of any quantity of carbon-14 will convert to nitrogen-14 in that time.[3] By determining how much carbon-14 is left in an organic sample, compared to the modern amount of carbon-14 in living organisms, we can estimate the how long ago the organism died and hence its age. The longer ago an organism died, the smaller the amount of measurable carbon-14. If we have one-quarter of what we started with, we can assume two half-lives have passed and therefore determine that the sample is 11,460 years old (5,730 + 5,730).

The most modern instrumentation can only measure small amounts of carbon-14 equivalent to about 100,000 years before present. And this is why secular geologists rarely use this method for dating since they assume most fossils and rocks are millions or even billions of years old. And any carbon-14 from fossils in those rocks should have all decayed away. However, scientists are finding plenty of measurable carbon-14 in dinosaur bones. They are also finding original proteins, blood vessels, and cells.[4]

RATE Project Carbon-14 Analyses

In the 2000s, the Radioisotopes and the Age of the Earth (RATE) research project, supported by ICR, sent numerous coal and diamond samples to be tested for carbon-14. Every one of the samples (10 coals, 12 diamonds) came back with measurable carbon-14, demonstrating these rocks and minerals are less than 100,000 years old.[5] The coal samples were from every era of the Phanerozoic, Paleozoic, Mesozoic, and

Cenozoic, dating back in presumed secular time to over 300 million years. Most diamonds are generally assumed to be Precambrian, or much older than the coal samples, and possibly a billion years old or more. Instead, the measurable carbon-14 in these samples places them closer to the biblical timeline. Secular claims of contamination are harder to justify in the diamonds since these minerals are the hardest natural substance and have a melting point of 4,000°C.[5]

Carbon-14 dating is different from all other radioisotope methods because it does not require determination of the amount of both the parent and daughter isotopes. It merely measures the amount of carbon-14 that is left in a sample compared to the modern amount measured in living organisms. All living organisms on land gain a fairly equal ratio of carbon-14 to carbon-12 through breathing, eating, and/or photosynthesis. This method also assumes that a constant amount of carbon-14 is created in the upper atmosphere and therefore assumes cosmic radiation has always been constant. However, secular science knows that that is not entirely true. That is why they tried to calibrate carbon-14 dates with tree ring data, but tree ring data are rather ambiguous because more than a single ring can form in a calendar year.[6]

Assumptions in All Radioisotope Methods

In principle, the general methodology works for all radioactive elements. It is based on a measurable scientific process and uses scientific reasoning. It can be very precise. But precision and accuracy are two different things. And testing two or three different radioisotopes can result in different dates for the same rocks, often with differences of over hundreds of millions of years. We will discuss more on this with some examples later. The assumptions in all of the methods do not warrant accuracy by any means. Secular science just gets the old ages it wants and claims they are factual. But nothing in science is factual if you cannot observe it, repeat it, or test it and produce the same results over and over again. You cannot go back in time to verify any of the radioisotope dating results.

In reality, the determination of absolute ages is very subjective and laden with unanswerable and unverifiable assumptions. Let's run through four of these assumptions. First, secular scientists assume they can measure the decay rates of radioactive elements accurately. This may seem straightforward, but many radioactive isotopes used in geology have half-lives in the 100-million-year to billion-year range. These elements decay so slowly that measuring any decay is miniscule.

Second, they assume that the decay rates never change and have remained constant throughout all of geologic time. This is probably all they can assume, following strict uniformitarianism, but how do they really know the decay rate won't change or hasn't

C-14 Results for Coal Seams

Coal Location and Geologic Era	Coal Seam	Conventional Geologic Age (Millions of Years)	C-14/C-12 (pMC ± 1σ)
Cenozoic			
Texas	Bottom	34-55	0.30 ± 0.03
North Dakota	Beulah	34-55	0.20 ± 0.02
Montana	Pust	34-55	0.27 ± 0.02
Mesozoic			
Utah	Lower Sunnyside	65-145	0.35 ± 0.03
Utah	Blind Canyon	65-145	0.10 ± 0.03
Arizona	Green	65-145	0.18 ± 0.02
Paleozoic			
Kentucky	Kentucky #9	300-311	0.46 ± 0.03
Pennsylvania	Lykens Valley #2	300-311	0.13 ± 0.02
Pennsylvania	Pittsburgh	300-311	0.19 ± 0.02
Illinois	Illinois #6	300-311	0.29 ± 0.03

Average percent modern carbon for the ten coal samples is 0.247 ± 0.025

C-14 Results for Diamonds

Country of Origin	Diamond Location	C-14/C-12 (pMC ± 1σ)
Botswana, south-central Afria	Orapa mine	0.06 ± 0.03
Botswana	Orapa mine	0.03 ± 0.03
Botswana	Lethakane mine	0.04 ± 0.03
Botswana	Lethakane mine	0.07 ± 0.02
South Africa	Kimberley mine	0.02 ± 0.03
Guinea, West Africa	Kankan placer	0.03 ± 0.03
Namibia, southwest Africa (six diamond samples)	Placer deposits	0.31 ± 0.02
Namibia	Placer	0.17 ± 0.02
Namibia	Placer	0.09 ± 0.02
Namibia	Placer	0.13 ± 0.03
Namibia	Placer	0.04 ± 0.02
Namibia	Placer	0.07 ± 0.02

Average percent modern carbon for the 12 diamonds is 0.09 ± 0.025

changed? The world has been around a long time, and we've only been measuring decay rates for about 100 years. Scientists at some of the world's leading institutions (Brookhaven National Lab, the German PTB, Baylor College of Medicine, and Purdue University) have recently reported seasonal variations in the decay rate of several elements, including manganese-54, silicon-32, radium-226, and chlorine-36.[7] They have noticed fluctuations of about 0.3% in decay rates with the seasons, possibly caused by changes in the distance to the sun throughout the year. What causes this variation is still a mystery, but it really calls into question the assumption that decay rates never change. Projecting back very small changes in the present creates great discrepancies in ages for ancient rocks.

A third assumption deals with the history of the rock or mineral sample that contains the radioisotope. Any metamorphic event, no matter how minor, can alter the apparent date of the sample. Many Precambrian-age rocks show evidence of several metamorphic events (likely caused by intense heat and pressure during the Flood). The problem is with other metamorphic events we don't see evidence for in the rocks today. These can cause the determined age to be in great error without knowing it. Some elements, like the uranium-lead system, can partially compensate for such events, but many methods cannot be corrected.[2] Isochrons may not be the solution either since they seem to be similar to mixing lines where end-member values have to be estimated.

The last assumption also deals with the history of the rock and the assumption of a closed system. To determine the age using most radioisotopes, the correct amounts of the parent and daughter isotopes in the rocks have to be determined accurately. Groundwater is everywhere, and everything leaks. Water percolates through all natural materials over time—some materials faster than others. All rocks and minerals below the water table are in constant contact with moving water. Because water is the universal solvent, it will selectively dissolve nearly every natural substance, given enough time. The key word here is *selectively*. Water can obviously dissolve some minerals like halite (salt) easier than a mineral like quartz. There are differences in solubility in all substances. Uranium is very soluble in water. Many dinosaur bones from Colorado are enriched in radioactive uranium because the porous interior allowed concentration of the element via groundwater flow. Radioactive uranium isotopes ultimately decay into various isotopes of lead. If more uranium is selectively dissolved from the rocks compared to the lead, it will look like a lot of uranium has converted to the lead isotope. This will result in ages that are much older than they should be, even millions or billions of years older. Again, the problem is in determining the exact amount of the original parent and daughter isotopes at the start and also the amount of parent and daughter isotopes that have remained in place. Absolute ages are based on knowing the start and finish time on the radioisotope clock.

Secular geologists may try to work around these assumptions by using several different isotope methods, like uranium-lead and potassium-argon. If they get similar dates from the two or more methods, they feel more confident in their age determination. However, to my knowledge, nobody has really addressed the problem of selective

Geiger counter on dinosaur bone in Colorado, Dinosaur Ridge

dissolution by groundwater. How could someone? The earth is not a closed system. There is always movement in and out of the systems. How could someone possibly claim they know the entire history of the rock and its chemical interactions with groundwater? Groundwater is constantly flowing through the subsurface, dissolving radioactive and nonradioactive elements alike. It often leaves behind other minerals to replace the ones it dissolves, leaving no real way to trace the process.

This leads us back to our discussion at the beginning of this chapter. Do we really know the earth is 4.56 billion years old as taught by secular science, or is it 6,000 years old as many creationists believe? The Bible tells us that the earth is young. There is no true, verifiable science to counter this conclusion. In fact, the measurable carbon-14 in dinosaur fossils and the original proteins, like blood vessels and collagen, in dinosaur bones match best with the age of 6,000 years. Real science confirms the Bible.

Determining the Date of a Fossil

Fossils, because they are normally found in sedimentary rocks, cannot be dated directly using radioactive isotopes. Many people have gotten the wrong impression that paleontologists can magically stick a dinosaur bone into a machine and out comes the proper age. However, creation scientists have begun to directly date dinosaur bones using carbon-14 and have found measurable carbon-14 in samples claimed to be many tens of millions of years old.[4]

How do secular scientists date fossils using absolute time if they cannot date the fossils themselves or the sedimentary rocks they are found in? Let's assume secular scientists can make reasonable estimates of absolute time, even though we know this is not true. Secular geologists are then able to use these dates to determine an absolute age, or more appropriately, an age range for fossils by using bracketing and/or correlation.

Bracketing a fossil requires something like a volcanic ash bed or a lava flow above and below the layer containing the fossil. The absolute age of the ash or lava is determined with radioisotope dating and is assumed to be correct because they are believed to be from a single igneous event like an eruption of a nearby volcano. The fossil is then assigned an age between the two calculated igneous dates, thereby bracketing the age of the fossil. More often than not, however, there will not be an ash layer or lava flow near the fossils. Then, the geologist must correlate the rocks and fossils in the area of interest with rocks and fossils elsewhere that have ash layers and/or lava flows nearby. In this way, secular scientists can compare (or correlate) bracketed fossils with unbracketed fossils and assign absolute dates to the unbracketed fossils.

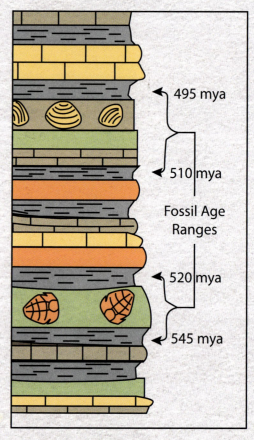

Bracketing an age for fossils using ash deposits

Once a fossil is dated this way, secularists assume the same dates for that same

1980 eruption of Mount St. Helens

fossil found elsewhere. This is another assumption that seems to work in a regional context and in principle at least, but not necessarily globally. The basis for this assumption is the principle of faunal succession, which was established before the concept of absolute time. Nonetheless, as we saw above, all radioisotope dates are suspect due to the long list of assumptions that cannot be verified.

Testing the Accuracy of Absolute Dating Methods

A true scientific test of the validity of radioisotope dating techniques should yield the correct results and also repeatable results. Unfortunately, for these methods neither is the case. Other than the historically verifiable carbon-14 dates that are fairly accurate for the past few thousand years, none of the other radioisotope methods produce accurate or highly repeatable results. These tests invalidate the current methods that geologists use to determine absolute dates.

Examples from Mount St. Helens and Other Historic Eruptions

In 1996, Dr. Steve Austin sampled the dacite lava dome from Mount St. Helens that cooled in 1986.[8] Observationally, the rocks were known to be only 10 years old. He sent off multiple samples to secular laboratories for analysis. The results of the potassium-argon (K-Ar) dating method produced dates ranging from 34,000 years old to 2.8 million years old.[8] None of these results matched close to the true ages of these rocks.

Some old-earth geologists have tried to explain these results away by claiming Dr. Austin failed to gather the correct rocks at Mount St. Helens. However, it's hard to miss a new lava dome and confuse it for older rocks. And even the claim that some of the minerals in the dacite may have cooled many years earlier, before being extruded, is falsified since Dr. Austin had whole rock analysis performed also, not just individual mineral studies. The whole rock analysis showed an age of 35,000 years.[8] Of course this age is preposterous compared to the true 10-year-old age of the dacite.

Lava and Actual Age	K-Ar Dates in Million of Years[9]
Hualalai Basalt (Hawaii, AD 1800-1801)	1.6 ± 0.16
Hualalai Basalt sample #2	1.41 ± 0.08
Mt. Etna Basalt (Sicily, 122 BC)	0.25 ± 0.08
Mt. Etna Basalt (Sicily, AD 1792)	0.35 ± 0.14
Mt. Lassen Plagioclase (California, AD 1915)	0.11 ± 0.3
Sunset Crater Basalt (Arizona, AD 1064-1065)	0.27 ± 0.09
Sunset Crater Basalt sample #2	0.25 ± 0.15

Secular Test Results of the K-Ar Method

In case you think these results were tainted by creationists (even though the results were performed by secular research laboratories), Austin reported on some earlier results by a secular scientist who also tested historic volcanic rocks using K-Ar.[9] These results were also way off track.

These results demonstrate that the K-Ar dating method should not be used when dating rocks of unknown ages. In fact, this method yields inaccurate results in every test. And yet, this technique is still commonly used to date igneous rocks today. The question is, why?

Examples from Wyoming and Grand Canyon

In case you think only the K-Ar method is tainted, consider another test. The ICR RATE project sampled many locations in and around Grand Canyon and the American West.[5] Results using multiple radioisotope methods from established secular laboratories were compared for the same rocks in several locations.[10]

The first samples were from the basement core of the Beartooth Mountains in Wyoming. The generally accepted secular age for this amphibolite rock is 2.79 billion years old.[10] However, the newly derived potassium-argon (K-Ar) dates ranged from 1.52 to 2.62 billion. The rubidium-strontium (Rb-Sr) method yielded ages from 2.515 to 2.79 billion. Samarium-neodymium (Sm-Nd) analyses of the minerals gave a date of 2.886 billion, and the lead-lead (Pb-Pb) method gave an age of 2.689 billion.[10] This shows the variety of ages we can get from different radiometric methods—as much as a billion years' difference!

Another set of samples from the Cardenas Basalt in Grand Canyon gave similar discordant results.[10] The conventional age of the Cardenas is 1.103 billion years old. However, the newly calculated results using the K-Ar method gave ages ranging from 0.516 to 1.013 billion. The Rb-Sr method yielded ages from 0.892 to 1.111 billion. The Sm-Nd method gave an age of 1.588 billion, and Pb-Pb analysis yielded an age of 1.385 billion. Radiometric methods don't agree with each other.

Original Tissues in Fossils

What do the fossils in the rocks tell us about the age of the earth? Do they give us any clues if they are millions of years old or merely thousands of years old?

As of the publication of this book, there have been about 100 peer-reviewed papers that have reported finding some type of original protein or tissues in fossils claimed to be millions of years old. And I assume there will be scores more in the next decades. Some college textbooks are calling them *chemical fossils*, openly acknowledging their existence but not mentioning their obvious age implications. Before we go too far, let's examine a bit of the history of these discoveries.

Since Dr. Mary Schweitzer's discovery of preserved original dinosaur proteins and blood cells in 2005, many secular scientists have remained skeptical. How could dinosaur fossils retain original organic material after millions of years? Following publication of her original paper,[11] some of Schweitzer's critics claimed that what she actually found was contamination from lab analysis or contamination during field collection. Others claimed these proteins were really from modern bacterial activity and merely simulated original soft tissues.[12]

In response, Dr. Schweitzer and her colleagues performed more studies on the soft tissues in question.[13,14] They even extracted collagen from a tyrannosaur leg bone and found a protein match of about 58% with bird collagen and 51% with frog and newt collagen—evidence these samples couldn't be from contamination. In further support that these are real dinosaur organic tissues, another study examined 89 amino acids extracted from a *T. rex* specimen, finding perfect matches with some modern animal proteins. Schweitzer's team demonstrated what was supposed to be impossible: These dinosaur soft tissues are real!

This debate of original tissues versus bacterial microbes spilled into the study of microscopic pigment indicators called *melanosomes*. Previously, scientists used powerful electron microscopes to examine

> *"Schweitzer's team demonstrated what was supposed to be impossible: These dinosaur soft tissues are real!"*

the fibers from the tail of the theropod *Sinosauropteryx*, finding preserved melanosomes, which contain the pigment melanin.[15] Different shapes of melanosomes produce different colors in today's animals. By comparing modern melanosomes to the ones found preserved in the fossils, scientists were able to speculate on the colors of this and other extinct animals. The melanosomes in the filaments/fibers of this *Sinosauropteryx* specimen indicate chestnut to reddish-brown bands along the tail and possibly the back.[15]

Another study found preserved melanosomes in feathers from the bird *Archaeopteryx*, showing a black color somewhat like a modern crow or raven.[16] An additional study of the feathered bird *Microraptor* also showed melanosomes indicating a black or dark blue color, but these melanosomes were more tightly packed, implying some degree of iridescence.[17]

Research on fossil ichthyosaurs, mosasaurs, and sea turtles supposedly 55 to 190 million years old has revealed actual preserved skin tissue and melanosomes that indicate a brown-black skin color.[18]

However, once again critics claim that these discoveries may not be melanosomes after all but just microbes that *look* like melanosomes, since both are similar in size and shape.[19] One way to more clearly distinguish between melanosomes and microbes is to find keratin associated with the melanosomes. Keratin is a protein that surrounds modern feather melanosomes. The discovery of melanosomes and keratin together would resolve the dilemma. Since microbes are not found embedded in keratin, both the melanosomes and associated keratin would logically have to come from the fossilized (but not mineralized) bird.

To answer these critics, lead author Yanhong Pan and co-authors, including Mary Schweitzer, examined fossilized bird feathers from Early Cretaceous system rocks in China.[20] They examined the feathers under an electron microscope, finding bundles of fibers that looked like keratin, but the team couldn't be sure from observation alone.

So they conducted a series of chemical tests on the fiber bundles and the surrounding matrix. They found "strong evidence for the retention of original and phylogenetically significant [a sample substantial enough to compare with living animals] protein components in *Eoconfuciusornis*."[20]

Pan's team concluded, "Our work represents the oldest ultrastructural and immunological recognition of avian beta-keratin from an Early Cretaceous (~130-Ma) bird."[20] In other words, the melanosomes and keratin found in this ancient bird are clearly real.

Their discovery again demonstrates the presence of preserved original proteins in fossils claimed to be many millions of years old. The authors offer no testable explanation for this preservation "miracle." Instead, they hypothesize that calcium, possibly mediated by bacterial activity, might have helped preserve the organic molecules. And yet, they found no calcium in the rock matrix surrounding the fossil, leaving any such source of calcium a mystery.[20] Their hypothesis is speculation without substance.

Secular scientists have found no viable explanation for how these proteins were preserved. They maintain the fossils are millions of years old, in spite of the contrary data.

I summarized this preservation issue in my book *Dinosaurs: Marvels of God's Design*:

> The finding of actual soft tissue in fossils that are supposedly millions of years old has evolutionists scratching their heads to explain. They are scrambling, trying to come up with "miracles" of preservation. Many secular scientists cannot fathom that dinosaurs are only thousands of years old because they "walk in the futility of their mind, having their understanding darkened, being alienated from the life of God, because of their ignorance that is in them, because of the blindness of their heart" (Eph. 4:17-18).[21]

The case for thousands and not millions of years is growing stronger with each new find.

Conclusion: Tests Invalidate Radioisotope Dating Methods

It is pretty obvious from the results I gave above that one radioisotope method cannot be used to verify and/or validate another method because few of the dates actually

> **"Falsification is a major part of the scientific method."**

agree. Rocks from the Beartooth Mountains amphibolite gave ages that varied from 1.52 to 2.886 billion years old. The Cardenas Basalt yielded ages ranging from 0.516 billion (younger than sedimentary the rocks above it, which is impossible) to 1.588 billion. In both instances, the generally accepted secular ages fell somewhere in the middle of these ranges. If absolute dates are considered factual by secular science and accurate, then why are these dates all so different? We're talking about differences of a billion years!

Second, and in every case, we see that the calculated radioisotope ages are drastically different from the known historical ages. Falsification is a major part of the scientific method.[22] And yet, test after test demonstrates the inaccuracy of the method. Instead of re-evaluating their assumptions, secular scientists continue to teach that their dates are facts that cannot be questioned.

I tend to trust empirical science over assumption-based science. To me, the repeated discoveries of original tissues in fossils demonstrate the true fallacy inherent in all absolute dating methods. Assumptions do not make for good science. Real, observable proteins like collagen, osteocyte cells, and blood vessels found in dinosaurs are tangible, touchable, and empirical. These organic materials cannot last even one million years under the best ideal conditions.

As a scientist, I prefer to trust things I can touch and see with my own eyes over results based on uniformitarian assumptions that cannot be verified. Sure, precise numbers come out of the mass spectrometer indicating millions of years, but how do we verify the accuracy of the numbers going in in the first place, let alone the numbers coming out? It's the old adage "garbage in, garbage out."

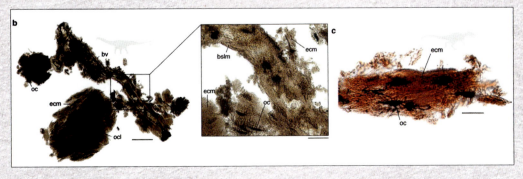

Partly polymerized soft tissues from demineralized *Diplodocus* and *Allosaurus* bone. Images show remnants of osteocytes (oc), osteocyte lacunae (ocl), a blood vessel (bv), extracellular matrix (ecm), and basal lamina (bslm).

Old-earth geologists need to explain why I should ignore empirical data found in the rocks and instead trust in their unverifiable methods. We cannot go back in time and see the original parent/daughter ratios in our samples, nor are we witnesses to what happened to that rock/mineral sample over time to see how that ratio may have changed. Empirical data always trump methods based on assumptions. The fossils don't lie.

Science does not insist these old-earth values are correct. It is scientists' blind belief in spite of the facts. What engineer would continue to build the same bridges if they kept collapsing? Shouldn't that be a clue something is wrong? It is only their secular worldview that forces them to accept these dates as accurate. Recall, secular scientists need deep time to give their theory of evolution any chance. It is that belief, and not observable science, that prevents them from being able to falsify the methodology. Deep time is not verifiable or reliable. And yet the great ages calculated by these secular scientists are found in every science textbook across the globe. They have the microphone, as I say. Who can tell them they are wrong? Violating the scientific method does not deter them. *Science* doesn't tell us the earth is billions of years old—scientists do.

> "Everything we observe demonstrates that the earth is no older than about 6,000 years."

God's Word tells us a different history. Genesis is confirmed by true science. Everything we observe demonstrates that the earth is no older than about 6,000 years. Methods of determining absolute time are no exception. If you want more information on critiques of many of the specific methods of radioisotope dating, see *Rethinking Radiometric Dating* by ICR nuclear physicist Dr. Vernon Cupps.[23]

This chapter covered the ways geologists measure time, both relative and absolute. Relative time is without a doubt the most useful method in any practical sense. Determining the relative order of events is all that is needed to study the stratigraphy of the continents and the individual megasequences. The principle of superposition allows us to track the depositional order of the megasequences at each column location. The identification of regional and intercontinental unconformities allows us to identify and correlate the megasequence boundaries. Methods of absolute time are unnecessary in this type of research. It is all about relative time and the order of deposition and burial in the rising waters of the Flood.

We already possess one of the best and most accurate ways of telling time, but unfortunately it is too readily dismissed. Scientists ignore and scoff at the accuracy of the Bible, in particular the first 11 chapters of Genesis. They prefer their own stories of deep time and evolution instead. Archaeological discoveries again and again show the Bible was right. The lost cities described are found. Dinosaurs even ate grass, as described in Job 40—the behemoth is clearly a sauropod.

Many believe in the accuracy of the biblical narratives of Abraham, and maybe even Joseph, but consider most of the earlier book of Genesis inaccurate. Ask yourself, when does true history begin in the Bible? Why do so many consider a global flood to be merely a tall tale? The record of the Flood and the genealogies listed in the book of Genesis are historical records that should be considered factual. Why do so many prefer so-called "science" over a written historical document? When has the accuracy of radioactive dating methods ever been proven true, at least for dates beyond 4,000 years ago? Instead, we see the inaccuracies of dating methods when applied to rocks of known age. Do not disregard the accuracy of the book of Genesis. It was written to tell us the history of the earth by the one witness who was there, God Himself. And as this present book shows, there is still great geological evidence of that worldwide Flood. We just need an open mind to see the truth.

The next chapter turns to a brief overview of fossils and the types of fossils in the individual megasequences. We will see that there is a pattern to the fossil record just like the principle of faunal succession indicates. Each megasequence has its own unique types of fossils. Later, we will discuss the reasons for this observation, but not until after we examine the megasequences in better detail.

References
1. As quoted in Kazmann, R. G. 1978. It's about time: 4.5 billion years. *Geotimes*. 23: 18-20.
2. Faure, G. 1986. *Principles of Isotope Geology*, 2nd ed. New York: John Wiley & Sons. Emphasis in original.
3. Godwin, H. 1962. Half-life of radiocarbon. *Nature*. 195: 984.
4. Thomas, B. and V. Nelson. 2015. Radiocarbon in dinosaur and other fossils. *Creation Research Society Quarterly*. 51 (4): 299-311.

5. Vardiman, L., A. A. Snelling, and E. F. Chaffin, eds. 2005. *Radioisotopes and the Age of the Earth: Results of a Young-Earth Creationist Research Initiative*. San Diego, CA: Institute for Creation Research and Chino Valley, AZ: Creation Research Society.
6. Glock, W.S., R.A. Studhalter, and S.R. Agerter. 1960. Classification and Multiplicity of Growth Patterns in the Branches of Trees. *Smithsonian Miscellaneous Collections*. 140 (1): 1-292.
7. Castelvecchi, D. 2008. Half-life (more or less). *Science News*. 174 (11): 20.
8. Austin, S. A. 1996. Excess Argon within mineral concentrates from the new dacite lava dome at Mount St. Helens. *Creation Ex Nihilo Technical Journal* (*Journal of Creation*). 10 (3): 335-343.
9. Dalrymple, G. B. 1969. $^{40}Ar/^{36}Ar$ analyses of historic lava flows. *Earth and Planetary Science Letters*. 6: 47-55.
10. Morris, J. D. 2007. *The Young Earth: The Real History of the Earth—Past, Present, and Future*. Green Forest, AR: Master Books.
11. Schweitzer, M. H. et al. 2005. Soft Tissue Vessels and Cellular Preservation in *Tyrannosaurus rex*. *Science*. 307 (5717): 1952-1955.
12. Kaye, T. G., G. Gaugler, and Z. Sawlowicz. 2008. Dinosaurian soft tissues interpreted as bacterial biofilm. *PLOS ONE*. 3 (7): E2808.
13. Schweitzer, M. H. et al. 2007. Analyses of Soft Tissue from *Tyrannosaurus rex* Suggest the Presence of Protein. *Science*. 316 (5822): 277-280.

Hunan Province, China. Sandstone pillars formed from mechanical weathering.

14. Asara, J. M. et al. 2007. Protein Sequences from Mastodon and *Tyrannosaurus rex* Revealed by Mass Spectrometry. *Science*. 316 (5822): 280-285.
15. Zhang, F. et al. 2010. Fossilized melanosomes and the color of Cretaceous dinosaurs and birds. *Nature*. 463 (7284): 1075-1078.
16. Carney, R. M. et al. 2011. New evidence on the colour and nature of the isolated *Archaeopteryx* feather. *Nature Communications*. 3: 637.
17. Li, Q. et al. 2012. Reconstruction of *Microraptor* and the Evolution of Iridescent Plumage. *Science*. 335 (6073): 1215-1219.
18. Lindgren, J. et al. 2014. Skin pigmentation provides evidence of convergent melanism in extinct marine reptiles. *Nature*. 506 (7489): 484-488.
19. Moyer, A. E. et al. 2014. Melanosomes or Microbes: Testing an Alternative Hypothesis for the Origin of Microbodies in Fossil Feathers. *Scientific Reports*. 4: 4233.
20. Pan, Y. et al. 2016. Molecular evidence of keratin and melanosomes in feathers of the Early Cretaceous bird *Eoconfuciusornis*. *Proceedings of the National Academy of Sciences*. 113 (49): E7900-E7907.
21. Clarey, T. 2015. *Dinosaurs: Marvels of God's Design*. Green Forest, AR: Master Books, 49.
22. Feduccia, A. 2012. *Riddle of the Feathered Dragons: Hidden Birds of China*. New Haven, CT: Yale University Press.
23. Cupps, V. R. 2019. *Rethinking Radiometric Dating: Evidence for a Young Earth from a Nuclear Physicist*. Dallas, TX: Institute for Creation Research.

5 Types of Fossils and the Fossil Record

Summary: Making a fossil requires special conditions. To avoid being eaten by scavengers, a creature's remains need to be buried rapidly and deeply. Catastrophic activity is a prerequisite. Billions of fossils are found all over the world. Big creatures such as sauropods and mosasaurs indicate huge catastrophic activity.

Many observations contradict evolutionary expectations. In addition to containing short-lived original tissues, many fossils appear suddenly—fully formed—in the fossil record with no obvious ancestors. They are followed by stasis. In other words, they don't change throughout many supposed millions of years. The fossil record doesn't show slow, gradual evolution. The most obvious example is the Cambrian Explosion, which represents the sudden appearance of most of the major groups of creatures, complete with extremely complex features. Creation scientists interpret the Cambrian as the initial onset of the global Flood of Genesis.

Fossils are defined as the remains and/or traces of ancient life. This generally means they have to be old enough to be considered prehistoric, Ice Age, or older. The study of fossils is called *paleontology*. Don't confuse paleontology with archaeology. Archaeology is the study of human artifacts, like pottery and Egyptian mummies. A paleontologist is a geologist who specializes in the study of fossils.

Fossils do not form under normal conditions. Yet, billions of fossils are scattered all around the world, from Antarctica to the Ukraine. This is just what you'd expect from a global, catastrophic flood. John R. (Jack) Horner and James Gorman in their book *Digging Dinosaurs* discussed a single hill in Montana that contains the remains of approximately 10,000 individual duck-billed dinosaurs, made up of over 30 million fossil fragments of a single species.[1] They named this dinosaur *Maiasaura*, or "mothering dinosaur," because they believe it cared for its hatchlings for an extended period of time. *Maiasaura* is a plant-eating hadrosaur dinosaur that is found in rocks of the Upper Cretaceous system.

Dr. Art Chadwick of Southern Adventist University has reported between 10,000 to 25,000 individual dinosaurs in a single Upper Cretaceous bone bed in the Powder River Basin of eastern Wyoming. He and his students excavate thousands of bones per summer season from this site. He believes the deposit was a debris flow that poured the dinosaurs into a shallow ocean environment.[2] Today, the bones are found mixed with a few marine fossils in a 0.5-meter-thick mudstone layer in the Lance Formation.

Jumbled together in a bone bed near Agate Springs, Nebraska, are remains from an amazing assortment of mammals that date from the final stages of the great Flood of Noah's day. No such deposit is forming today. The present is not the key to the past, as uniformitarian thinking insists.

It takes special conditions to preserve and make a fossil. Things don't just die and magically become a fossil with time. Organisms have to be buried both fast and deep. Scavengers normally eat the choice parts of any dead animal, and then aerobic bacteria complete the decomposition process. "Dust to dust" is an appropriate phrase in most cases.

How Are Fossils Formed?

How are things fossilized? The most important condition is a rapid burial. Catastrophic activity is a prerequisite. Where and when does this occur? Large floods are most commonly envisioned when it comes to land animals and plants. However, landslides, sometimes set off by earthquakes or heavy rains, can also bury creatures sufficiently. Just sinking to the bottom of a lake or sea is not enough because scavengers and aerobic bacteria are there too. How many of the remains of *Titanic*'s victims survived? Not a single bone was found in the wreckage when it was discovered in 1985. So, we not only need a rapid burial, but we need a rather deep burial to cut off the oxygen supply or at least limit it. This step is necessary to slow and prevent complete disintegration by bacteria. Secularists think most dinosaurs were catastrophi-

All dinosaurs that were not on the Ark died in the Flood. Those animals were buried rapidly by massive sand and mud layers transported by the rising water. As the floodwaters receded, uplift and water erosion exposed the bones at or near the surface.

cally buried during rainy seasons and/or river flood events. Creation scientists conclude that dinosaurs, and nearly all fossils, were deposited in the one year-long great Flood, just thousands of years ago.

> "How are things fossilized? The most important condition is a rapid burial. Catastrophic activity is a prerequisite."

Fossils are found in many types of sedimentary rocks. They are commonly found in shales and limestones. They are less common in sandstones, even though most dinosaur fossils are found in higher-energy sandstones. Shales and limestones allow better preservation of fossil details because the rocks are usually finer-grained. Creatures buried in sand are more readily dissolved by groundwater. Sand is very porous and more groundwater is able to flow through it, compared to clay and lime mud, lessening the chance of preservation.

Bias to the Fossil Record

Many secular geologists think that the fossil record (the collective information gained from all discovered fossils) is incomplete. Part of the reason for this is the selective nature of fossilization. Hard parts are obviously easier to preserve than soft parts. Therefore, the soft tissue and inner organs of most vertebrates are rarely found, although that seems to be changing in recent years. We are now finding more soft tissue than ever expected in some fossils. This also carries over to marine invertebrates, where only the outer shells are typically preserved as fossils. The fossil record is obviously biased in favor of hard shells and bones. In fact, many mammals are known and identified only from their teeth because enamel is so strong and durable compared to bone.

The fossil record is also biased toward shallow marine organisms living in areas of high sediment influx during the Flood. And terrestrial organisms that lived near water had a better chance of preservation because they were more likely to be transported into deeper water and buried rapidly. Many black shales appear to have been deposited so rapidly that they even preserved high amounts of marine algae and also many soft-bodied organisms (such as the Burgess Shale in British Columbia, Canada). Many of these black shales became source rocks for petroleum after the Flood.

Fossils tell us a lot about the levels of the Flood. The types of fossils we find are great clues to the materials being carried by the floodwaters at any given time. Most fossils show up fully formed in the rocks and remain the same for some time, then often seem to disap-

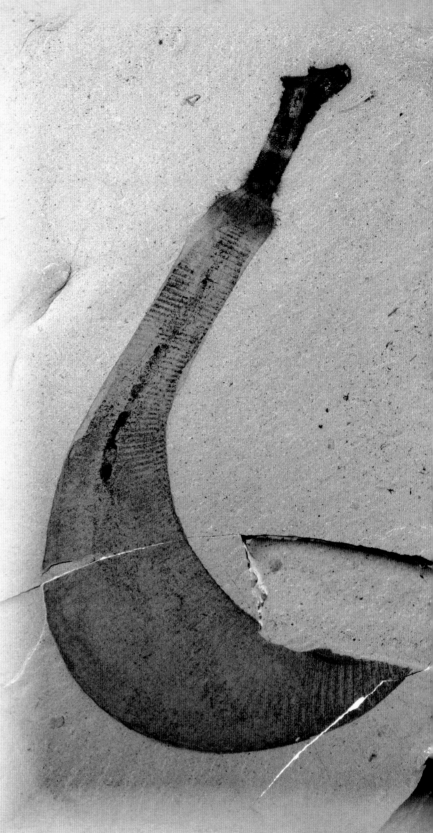

Holotype of *Ottoia tricuspida* from the Burgess Shale

The name *Sinosauropteryx prima* means "first Chinese lizard-wing." This specimen, on display at ICR headquarters in Dallas, Texas, was found in the famous Jiufotang Formation in central Asia, supposedly deposited about 100 million years ago, from which numerous outstanding fossils have been recovered. Tiny grains dominate the deposit, allowing even ephemeral features to be preserved. Note the fish fossil nearby. This type of dinosaur did not live in the water, but both were encased in watery sediments. The fish was caught writhing in suffocation, and the dinosaur was gasping for breath.

pear in the rock record, or go extinct. That is the typical fossil pattern: show up suddenly fully made, stay the same, then disappear. This pattern is observed throughout the Paleozoic, Mesozoic, and Cenozoic. As a Flood geologist, this makes perfect sense to me. Floodwaters would tend to transport different organisms as the tsunami-like waves washed across the continents from slightly different directions and water levels got higher. You would expect to find the sudden appearances of different organisms stacked one on top of the other, which is exactly what we observe.

Another thing I noticed quite commonly as I compiled the stratigraphic columns across the continents is the mixing of environments. Land and marine animals are often found together in a single rock layer, and sometimes freshwater and saltwater animals are found within a single rock layer, or both. For example, six species of sharks are found in the exact same rock layer (the Hell Creek Formation) with many *T. rex* fossils in Montana. We will discuss more on this in later chapters that deal with animals buried in specific megasequences like the Absaroka and the Zuni.

Types of Fossils

There are many different types of fossils. Most can be lumped into the following general categories: 1) petrifaction (which includes permineralization and replacement), 2) recrystallization, 3) carbonization, 4) molds and casts, 5) trace fossils, and 6) preservation of original tissues.

Petrifaction means to change the original organism to stone. This can be accomplished in several ways but usually involves the complete removal of all organic material associated with the dead organism. Groundwater typically dissolves away the original parts while simultaneously depositing inorganic minerals, preserving many of the details of the dead organism. These chemical deposits, such as quartz and pyrite, leave behind exact

Petrified wood

replicas of the fossil for geologists to study. Petrified wood is an example of this type of fossilization. Most dinosaur bones are also at least partially petrified.

Permineralization is a variety of petrifaction in which chemical sediment is deposited within pore spaces and/or soft tissues of the original shell or bone. The resulting fossil is sort of like a "negative" of the original, yet some of the fine details, such tree rings, are often preserved. A second variety, replacement, is a process in which the original organic material is removed and simultaneously replaced by molecular exchange with inorganic compounds. This results in an altered "positive" of the original organism minus the organic compounds.

Recrystallization occurs commonly in shelled invertebrates such as mollusks and corals. Many marine organisms construct their original shells from aragonite, one of the two crystal forms of calcium carbonate (calcite is the other form). Aragonite is relatively unstable and will quickly convert to the more stable form of calcium carbonate, calcite, following burial. The resulting fossil normally loses some of its original detail in the conversion process.

Carbonized plant fossil (left)
Recrystallized coral fossil (right)

Carbonization is a very typical type of fossilization for plants and selected marine organisms like graptolites. Commonly, plant leaves and other parts of a plant are buried, and during lithification the organic material "cooks" or distills and leaves behind a thin, black carbon residue.

Molds and casts are the impressions, or shapes, of ancient life. Often the actual organism dissolves away, leaving only the imprint of the organism in the rock. Molds are created in this way. There can be internal molds, which give the internal imprint of an organism, and external molds, which give the imprint of the outside of the organism. Dinosaur skin prints are a type of external mold. Seashells will commonly create an internal mold, giving details about the animal that the external shape doesn't reveal. Casts are a secondary process in which a mold must first be created by dissolution of the original shell or bone. The cast is made from later infill of the hollowed-out mold. The result is similar to an internal mold and can be hard to distinguish. It is often easiest to just recognize them as either molds or casts.

Cast of a gastropod fossil

Fossil coprolite

Trace fossils include tracks, burrows, trails, and anything else that indicates the activity of an organism without being a direct part of the organism. Other trace fossils include coprolites, or fossilized feces, and gastroliths, or gizzard stones. Footprints and even dung will not be preserved without a rapid burial. The Flood again provides us with the best explanation for these types of fossils.

Preservation is the retention of original organic parts. This type of fossil was thought to be extremely rare. Examples include the frozen wooly mammoths found in Siberia and Alaska. Insects in amber (fossilized tree sap) are often preserved also.

Even the enamel in mammal and dinosaur teeth may be preserved. Extreme examples of preserved fossils include Devonian-age plant material found in northern Ontario, Canada, where the original plant cuticle appears to have remained intact.[3]

Another surprising find was announced in 2017 by a team of scientists led by Dr. Erik Gulbranson. They claimed to have discovered a forest of fossil trees in Antarctica. Gulbranson, a paleoecologist at the University of Wisconsin-Milwaukee, claims this is the "oldest polar forest on record from the southern polar region."[4] The trees were found in Antarctica's Transantarctic Mountains and include a mix of evergreens, deciduous trees, and gingkoes. Gulbranson and his team claim the fossil trees they found are about 280 million years old, which places them in Permian system strata. The discovering scientists found the trees were so rapidly buried in volcanic ash that they contained fossilized plant cells that are virtually mummified, preserved "down to the cellular level." *LiveScience* reported:

> The plants are so well-preserved in rock that some of the amino acid building blocks that made up the trees' proteins can still be extracted, said Gulbranson, who specializes in geochemistry techniques.[4]

The scientists did not adequately explain how the amino acids and plant cells were preserved for several hundred million years. Even in their worldview, Antarctica has not remained frozen for 280 million years.

Original Preserved Tissues in Dinosaurs and Other Fossils

A more significant and groundbreaking discovery involving preserved organic tissues within dinosaur bones was published in *Science* in 2005[5] and was reported by Jack Horner and James Gorman in their book *How to Build a Dinosaur*.[6] These publications discussed how Dr. Mary Schweitzer found preserved soft tissue in the bone of a *T. rex* while conducting a histologic investigation (microscopic study of bone structure). She has not been able to extract complete DNA sequences by any means but has partially extracted some protein and collagen. These matched about 58% with chicken collagen, thus verifying these proteins are real and not contamination.[7] She even found red blood cell-like structures still in the blood vessels and osteocyte cells with apparently intact nuclei.

Since this announcement in 2005, scores of peer-reviewed scientific papers have documented preserved proteins and soft tissues in fossils that are found in rocks as deep as the Cambrian system, likely within the earliest Flood sediments. If these fossils are really 500 million years old as secular scientists claim, then they have a lot of explaining to do, as do old-earth creationists. Proteins, collagens, and other soft tissues cannot last for millions of years.

T. rex Sue

At a lecture I attended in Dallas on March 29, 2018, Dr. Schweitzer revealed that the famous *T. rex* known as Sue at the Field Museum in Chicago also had original blood vessels that were soft and hollow and "looked like all the world like they have red blood cells." She further commented that she has examined other rock formations as far back as Triassic strata, including many different depositional environments, and there is "more soft tissue and molecules than I ever expected." And she concluded her lecture by admitting she "had no idea how fossils form at the molecular level." All she knew was the fossils had to be "buried fast and deep, in 10 years or less."

Soft tissue continues to be an embarrassment to the uniformitarian community and the old-earth geologists. Secular science has yet to offer a viable explanation for the preservation of these molecules and original tissues and proteins. Schweitzer claims to have preserved a few ostrich blood vessels in an iron-rich solution, but the test was only two years and was merely qualitative. She made a visual determination of the preservation of the blood vessels after two years and did not conduct any chemical tests.[8]

Now it appears the secular community is trying to explain away all these discoveries with a different tactic. They maintain that the preserved tissues are still many millions of years old, and it is just the unknown preservation process that eludes them.[6] Decades of studies of protein decay rates by physical chemists apparently amount to very little when they conflict with the secular worldview and its need for deep time. Science has demonstrated that these proteins and tissues cannot last even one million

> "Science has demonstrated that these proteins and tissues cannot last even one million years under the best conditions. And yet, the secular community continues to insist these preserved tissues are 68 to over 500 million years old."

years under the best conditions. And yet, the secular community continues to insist these preserved tissues are 68 to over 500 million years old. Who's ignoring the empirical data now?

Brief Overview of the Fossil Record

The fossil record is one of sudden appearance, stasis, and then often disappearance, or extinction. This is the same pattern we observe in every geologic subdivision of the geologic column, including the systems and erathems. Secular geologists like to call the systems *periods* and the erathems *eras* since they believe these rock layers represent actual periods and eras of time in the past. Creation geologists think these may represent merely days or weeks during the year-long Flood. The fossil record is viewed as the order of burial.

Fossils are so important to the geologic column that each subdivision of the column was divided on the basis of abrupt fossil changes in the rock layers. As you go up or down the geologic column, different fossils appear and disappear. Secular geologists and many creation geologists think the layers that contain the same organisms were buried at similar times in the past. The scientists obviously differ on how long ago they believe these layers were deposited and on the conditions of burial.

The biggest change in fossils, where fossils suddenly appear in the rock record in great and diverse numbers, is designated by the Phanerozoic Eonothem, or "visible life eon." This point in the rock record also coincides with a new erathem and a new system called the Paleozoic Erathem and the Cambrian system, respectfully. Below this point, the rocks are lumped into the collective and generic Precambrian, which has also been further divided into three *eonothems*, or *eons*. Since we are dealing with the fossil record here, we can ignore these subdivisions and note that these rocks do indeed contain some fossils, but most are microfossils and/or algal-type fossils like stromatolites. Most of the Precambrian fossils are likely pre-Flood. For all practical purposes, the fossil record starts in the Phanerozoic Eonothem, Paleozoic Erathem, and Cambrian system. This coincides with the onset of the flooding of great portions of the continents and will be discussed in more detail in chapter 9.

Changes in the fossils in a vertical sense that are not big but still significant represent the boundaries of erathems or eras. The Paleozoic Erathem contains primarily marine fossils, but toward the top, in the Pennsylvanian system, we see more land animals and plant fossils suddenly appearing in great numbers in the rocks. The Mesozoic Erathem contains mostly reptile fossils and dinosaurs. And the Cenozoic Erathem contains a multitude of mammal fossils of various types. However, all three of these erathems contain billions of marine fossils mixed in with the terrestrial fossils. The mixing of land and marine environments is extremely common in the rock record and will be discussed later. This mixing is powerful evidence of a global flood.

> "The mixing of land and marine environments is extremely common in the rock record.... This mixing is powerful evidence of a global flood."

Smaller changes in the fossils were designated as systems or periods. These are what subdivide the erathems. Each represents a change in the fossils in a vertical sense. Many of the boundaries of these systems and erathems coincide with what the secular community considers extinction events. These so-called extinctions are where the fossils change abruptly and some organisms disappear upward within the rock record.

There are five, and now possibly six, major extinction events within the Phanerozoic Eonothem.[9] Secular scientists try to identify the causes for these rapid changes in the fossils and for the disappearance of major groups of fossils. Meteor impacts and rapid climate changes caused by volcanism or other factors have been suggested. However, most of these so-called extinction events are still a mystery to the secular scientists.

Creation scientists do not consider these as true extinction events at all. Instead, these horizons are interpreted as major shifts in the burial pattern of fossils during the Flood. So-called extinctions are merely the level at which certain fossils were no longer being actively buried, so they disappear upward in the geologic column. It may be that at these levels the environments that contained these animals and/or plants were already inundated, preventing any further burial in younger rocks. We will look at this in more detail in later chapters.

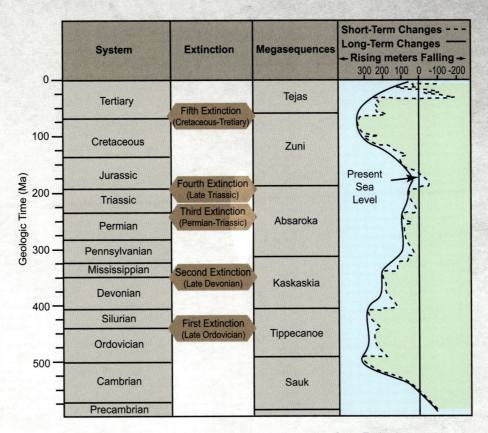

Secular geologic timescale showing the purported Five Great Extinctions

One pattern I noticed was that many of the major extinction events do seem to coincide with the six megasequences.[10] In fact, most of the major extinction horizons often fall within the middle or toward the top of the megasequence boundaries.

> However, because the megasequences are defined on the basis of major shifts in rock types, it should be no surprise that some of these changes correspond to rapid shifts in the fossil content also. The fossils deposited are dependent on the tectonic forces at work, currents, waves and the height of relative sea level.[11]

These extinctions may represent the high-water point of a megasequence or may represent the end of a megasequence cycle. Only the Absaroka Megasequence contains two of the so-called major extinctions. Reasons for this are not immediately clear. It may be because the Absaroka is the megasequence in which great numbers of land animals and land plants suddenly appear. The Absaroka seems to represent a pivotal moment in the Flood. This will be examined in more detail in chapter 12.

However, that is not to say the Flood didn't cause extinctions. Many of the presumably unique pre-Flood environments were likely destroyed by the Flood's tectonic activity. This caused a lot of marine animals to go extinct during or shortly after the Flood. For example, animals like trilobites and many of the Paleozoic brachiopods and corals appear to be extinct today. The exact reason for this is unclear. God did not tell us He saved any marine animals on the Ark. However, He did make sure that the oceans would be teeming with life after the Flood. There are still plenty of fish, corals, and mammals in our oceans.

The Great Secular Conundrum: The Cambrian Explosion

As mentioned above, the fossil record shows a sudden start in the Cambrian system called the Cambrian Explosion. There are some fossils below this level in the rock record, but most of these earlier fossils are soft-bodied or algal in nature. Secular scientists have written many books and articles to try and explain the sudden appearance of almost every major animal group—all at once—during the deposition of the Cambrian system. To date, no explanation has been satisfactory from an evolutionary viewpoint. Secular scientists remain baffled by the sudden appearance of such diverse forms of life without any ancestors in the rocks below. It is almost like they appear out of nowhere.

Lobster fossil

Trilobite fossil

Trilobites and arthropods are two groups of animals that suddenly appear in sedimentary rocks as part of the Cambrian Explosion.[12] They appear fully formed and functional, and the older rock layers below them contain no hint of any ancestors to these animals.[13,14]

Trilobites are members of the phylum Arthropoda, which includes spiders, insects, and crustaceans. Today, members of this group make up at least 85% of the species on Earth and live in every environment. Insects alone account for over 870,000 of these species.[12] God designed all arthropods with an exoskeleton (i.e., an outer skeleton) that is segmented into appendages. In Greek, *arthron* means joint and *podos* means foot. This exoskeleton does not grow as the animal grows but rather has to be shed—molted—as the animal matures.

Although arthropods dominate the biomass of the planet today, their fossil record is much more limited, with only about 30,000 fossilized arthropod species identified.[12] Because most arthropods have an exoskeleton of organic chitinous cuticle, they decompose easily and don't preserve well as fossils. A few arthropod groups, like the trilobites whose shells were calcified, were preserved in the Flood. There are over 2,000 genera (the plural of genus, the category above species) of trilobites in the fossil record and thousands of named species.[12]

Creation scientists explain the sudden appearance of these Cambrian system fossils as evidence of the first major advance of the floodwaters across the continents at the onset of the Flood. This advance of the sea is part of the Sauk Megasequence (chapter 9). Imagine huge tsunami-like waves overtaking millions of crea-

Brachiopod fossil

Trilobite fossil

tures. It seems likely that shallow marine organisms were the first types of animals buried by these raging floodwaters.[12]

A 2018 report published in *Proceedings of the National Academy of Sciences* details a comprehensive new study of the global Ediacaran sediments—those found just below the Cambrian.[15] The goals of the research included searching for arthropod fossils in Ediacaran sediments and explaining why there are few, if any, arthropods below the Precambrian/Cambrian boundary.

It has been argued that the absence of trilobites and other shelled fossils from the Ediacaran is merely due to the lack of preservational opportunities in the earlier rock beds. However, the scientists found there were similar opportunities for preservation in both the Cambrian and Precambrian rocks. Yet, euarthropod fossils and trace fossils are found only in the Cambrian sediments, and euarthropod trace fossils are "strikingly absent" in the Ediacaran system.[15]

This study clearly demonstrates that the arthropods, like the trilobites, have no earlier ancestors in the rock record. They show up suddenly as part of the Cambrian Explosion just like all other major animal groups. As Allison Daley and her colleagues have shown, the preservation opportunities in the Cambrian and the Ediacaran were exactly the same, but no arthropods appear in the rock record until the Cambrian.[15] This study deepens the Cambrian Explosion enigma for secular scientists.

Trilobite fossils produce some of the strongest evidence of catastrophic burial. A 2013 study found that many of these creatures were inundated rapidly while they were still alive![16] Numerous specimens are found in a rolled-up position—like giant roly-polies—to protect themselves from danger. However, a few varieties lacked the design to "lock their shells" and hold the ball-shaped position; these arthropods had to rely on their internal muscles to hold themselves in a rounded shape.

Recently, evolutionists were surprised when they discovered many of these trilobites with non-locking shells in a rolled, protective position. Javier Ortega-Hernandez and his co-authors reported:

> After death, the muscles responsible for flexing the trunk would have relaxed, causing the carcass to return to the outstretched position. Thus, the best possibility of preserving rolled olenellids [this type of trilobite] would require rapid burial of live individuals.[16]

"After death, the muscles responsible for flexing the trunk would have relaxed, causing the carcass to return to the outstretched position. Thus, the best possibility of preserving rolled olenellids would require rapid burial of live individuals."

The global Flood remains the best explanation for the fossil record, the Cambrian Explosion, and the rapid burial of trilobites. Animal fossils merely appear in rocks in their order of burial. Evolutionary scientists can continue to search for ancestors in the rock record, but it appears they will always come up empty-handed.

Pivotal Shifts in the Fossil Record

After the Cambrian, most of the Paleozoic Erathem fossil record remains dominated by shallow marine animals buried in floodwaters. During the Devonian system we find the first few amphibian fossils, and then a few reptiles show up

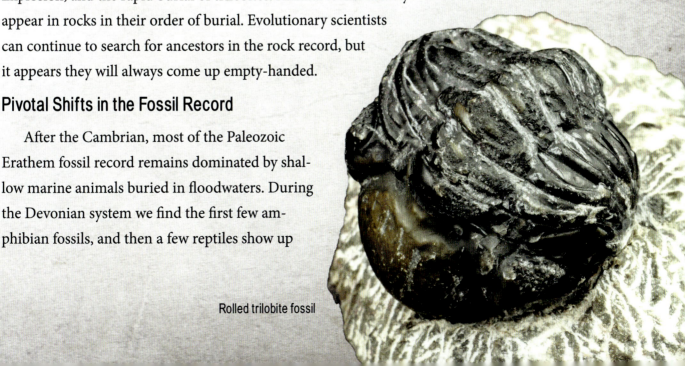

Rolled trilobite fossil

in the strata above that in the Mississippian. It isn't until the Upper Ordovician and the Silurian systems that a few land plants show up as fossils. These were primarily seedless vascular plants. Higher in the rock record, seeded plants and gymnosperms show up as fossils in the Upper Devonian and Mississippian strata but were only a maximum of 30 feet tall. Right on top of those, in Pennsylvanian strata, we find prolific coal beds made of lycopods, trees, and ferns. Some of these tree fossils were upward of 100 feet tall. At the top of the Paleozoic, we find prolific numbers of reptiles in Permian strata, including the mammal-like reptiles and an abundance of gymnosperm plant fossils. The end of the Paleozoic marks the end of the fossil record for the trilobites and many invertebrate marine animals, including many varieties of corals, bryozoans, and brachiopods.

Exposed Mesozoic rocks near Cody, Wyoming

Mesozoic fossils are much different from the Paleozoic, although there is some overlap. Most of the land flora and fauna seem to fit wetland-type environments. The lowermost system of the Mesozoic is known as the Triassic. Rocks of this system contain the first fossil mammals and dinosaurs. Large swimming reptiles and flying reptiles first appear in rocks at this level. As we move up to the Jurassic system, we find many new dinosaurs, including most of the large sauropod dinosaurs and the first bird, *Archaeopteryx*. Marine ammonites and other cephalopods are commonly found in Jurassic rocks. New varieties of corals and cycad land plants also appear in great numbers. At the top of the Mesozoic, during the deposition of the Cretaceous, we find many additional types of dinosaurs, like large herds of hadrosaurs (duck-billed dinosaurs) and ceratopsians (horned dinosaurs). We also find most of the raptor-like dinosaurs and large predators like *T. rex*. Flowering plants, including grasses, first appear as fossils in the Cretaceous strata. At the end of the Cretaceous, the large reptiles disappear in the rock record, including the dinosaurs, swimming reptiles, and flying reptiles. Many marine animals disappear at this level too, like the ammonites.

The rocks of the Cenozoic Erathem exhibit another major shift in the fossil record. Marine fossils found in these layers include many invertebrate groups like the foraminifera, bryozoans, corals, and mollusks. Angiosperms (flowering plants) dominated the fossil plant record of the Cenozoic. In fact, the sudden appearance of so many angiosperm plants in the Cenozoic, without ancestors, was called an "abominable mystery" by Charles Darwin. He recognized the plant fossil record of the Cenozoic was a serious problem for his fledgling theory of evolution because there were no transitional forms.[17]

Insect fossil

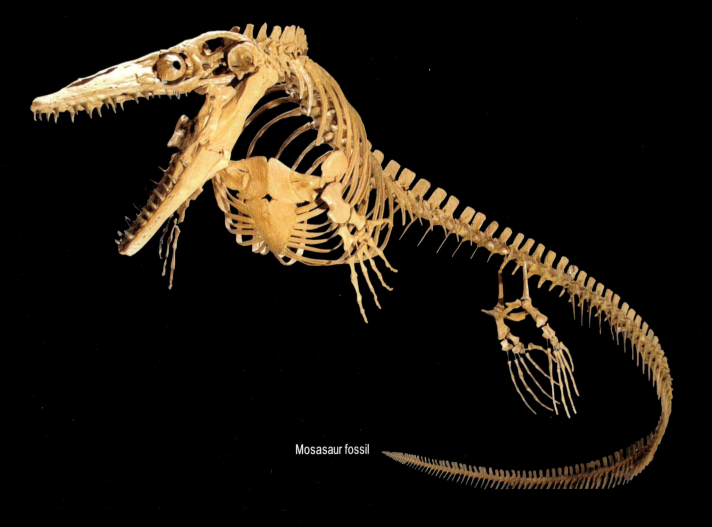

Mosasaur fossil

The majority of the land animals in the Cenozoic are mammals, such as horses, pigs, and camels. Also, large marine mammals show up in great numbers in Cenozoic strata, like the whales. The largest and most extensive coal seams appear in Cenozoic rocks. Some of these coal seams are 60 miles by 60 miles and up to 200 feet thick. These coals are not made of lycopods like the Pennsylvanian coals but are instead made of angiosperms and conifers like the *Metasequoia*. Toward the top of the Cenozoic, in rocks of the Quaternary system, we find the mammoths, mammals, and the Ice Age flora and fauna.

Throughout the fossil record, we always find marine organisms. Dr. Nathaniel Jeanson plotted the occurrences of many of the marine and land animal groups found in the Paleobiology Database. He found that virtually 100% of the animals within the Lower Paleozoic Erathem were exclusively of marine origin. This pattern continued upward until the Upper Carboniferous (Pennsylvanian), when land animals began to make their first significant number of fossil occurrences. The rock layers above the Pennsylvanian continued to show more and more land animal occurrences but were always mixed with significant numbers of marine animals. This persistent mixing of land and marine fossils in the same rocks is a recognizable pattern in the fossil record that must be explained in any model.

Berlin specimen of the bird *Archaeopteryx*

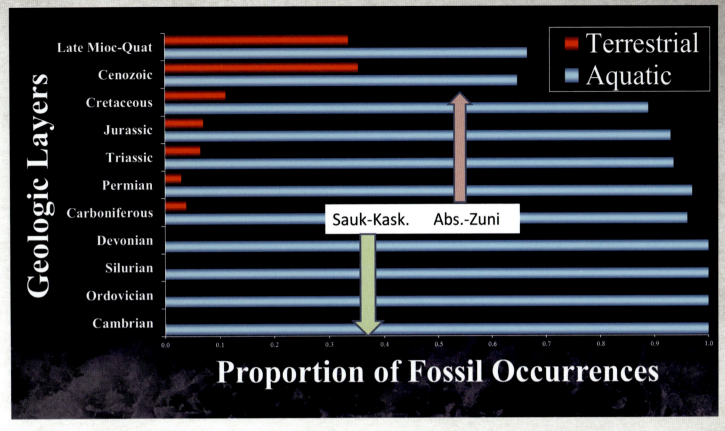

Proportion of terrestrial vs. aquatic animals across the stratigraphic column. Data include all major animal phyla in the Paleobiology Database. Note that the Sauk–Kaskaskia Megasequences (Cambrian–Mississippian) are nearly exclusively aquatic (marine). Also note that the proportion of land animals increases upward starting in the Absaroka Megasequence (Pennsylvanian) and are always mixed with marine animals. Courtesy of Nathaniel Jeanson.

Are There Inverted and Out-of-Place Fossils?

Creationist authors have claimed that many locations contain out-of-place fossils or inverted fossils.[18-23] Many of these criticisms are directed at rocks that geologists term *overthrust faults*. The mechanical difficulty of moving large, coherent sheets of rock great distances down fairly flat slopes has never been fully explained in the secular geologic literature.[24] Lithified sedimentary rock will not fold and behave plastically at surface conditions,[25] yet we see the clear geometric results in overthrust belts around the globe. Creationists in the past have been right to criticize secular explanations for overthrusts.

When Drs. John Whitcomb and Henry Morris published their classic book *The Genesis Flood* in 1961,[18] no one understood overthrust faults. The "rules" of overthrusting were not established until the 1970s by the oil industry geologists after hundreds of wells were drilled and thousands of miles of seismic data were collected.[26,27] Today creationists must accept the results of the oil industry data proving the existence of nearly every overthrust fault, including the Heart Mountain Fault and the Lewis Overthrust discussed in Whitcomb and Morris' book. The rock data demand it. Creationist authors who claim all overthrusts are not real but are merely examples of out-of-place fossils or inverted fossils are incorrect.[28] Instead, creationists should embrace these features because only the conditions of the Flood can offer an explanation for their existence. Overthrusts are not examples of out-of-place fossils but are examples of real catastrophic activity that took place during the Flood. Chapter 16 contains a more thorough discussion of overthrusts.

Conclusion: The Flood Explains the Fossil Record

The global pattern of fossils cannot be denied. Why certain animals and plants are only found in certain rock layers is still largely unresolved. Creation scientists have often speculated and proposed various ideas to try and explain the patterns we observe in the fossil record. Among these ideas are hydrodynamic selectivity and sorting by size, fossil composition, and settling velocity.[18] Other factors relate to mobility,[18] and possible factors like ecological zonation have also been considered.[29] One of the goals of the present study was to examine rock data across multiple continents and see

Canadian Rockies

which of these factors best explains the fossil record. If we follow the data, they should lead us to the best available solution. In subsequent chapters you will see what I have concluded about the order of the fossils and the reasons we get such a defined pattern. There is a definite global signature to the fossils and definite causes. Answers are being revealed.

References
1. Horner, J. R. and J. Gorman. 1988. *Digging Dinosaurs*. New York: HarperCollins Publishers.
2. Personal communication with Art Chadwick.
3. Lemon, R. R. 1993. *Vanished Worlds: An Introduction to Historical Geology*. Dubuque, IA: Wm. C. Brown Publishers.
4. Pappas, S. 280-Million-Year-Old Forest Discovered in…Antarctica. *LiveScience*. Posted on livescience.com November 15, 2017, accessed November 17, 2017.
5. Schweitzer, M. H. et al. 2005. Soft-tissue vessels and cellular preservation in *Tyrannosaurus rex. Science*. 307: 1952-1955.
6. Horner, J. R. and J. Gorman. 2009. *How to Build a Dinosaur: Extinction Doesn't Have to Be Forever*. New York: Dutton.
7. Schweitzer, M. H. et al. 2007. Analyses of soft tissue from *Tyrannosaurus rex* suggest the presence of protein. *Science*. 316: 277-280.
8. Schweitzer, M. H. et al. 2014. A role for iron and oxygen chemistry in preserving soft tissues, cells and molecules from deep time. *Proceedings of the Royal Society B*. 281 (1775): 20132741.
9. Pimiento, C. et al. 2017. The Pliocene marine megafauna extinction and its impact on functional diversity. *Nature Ecology & Evolution*. 1: 1100-1106.
10. Clarey, T. 2015. *Dinosaurs: Marvels of God's Design*. Green Forest, AR: Master Books.
11. Ibid, 158.
12. Clarey. T. 2014. Trilobites: Sudden Appearance and Rapid Burial. *Acts & Facts*. 43 (2): 13.
13. Thomas, B. Is the Cambrian Explosion Problem Solved? *Creation Science Update*. Posted on ICR.org December 12, 2011, accessed May 23, 2018.
14. Thomas, B. Cambrian Fossil Intensifies Evolutionary Conundrum. *Creation Science Update*. Posted on ICR.org September 26, 2014, accessed May 23, 2018.
15. Daley, A. C. et al. 2018. Early fossil record of Euarthropoda and the Cambrian Explosion. *Proceedings of the National Academy of Sciences*. 115 (21): 5323-5331.
16. Ortega-Hernandez, J., J. Esteve, and N. J. Butterfield. 2013. Humble origins for a successful strategy: complete enrolment in early Cambrian olenellid trilobites. *Biology Letters*. 9 (5): 20130679.
17. Tomkins, J. P. and T. Clarey. 2018. Darwin's Abominable Mystery and the Genesis Flood. *Acts & Facts*. 47 (6): 16.
18. Whitcomb, J. C. and H. M. Morris. 1961. *The Genesis Flood: The Biblical Record and Its Scientific Implications*. Nutley, NJ: The Presbyterian and Reformed Publishing Company.
19. Lammerts, W. E. 1966. Overthrust faults of Glacier National Park. *Creation Research Society Quarterly*. 3 (1): 61-62.
20. Lammerts, W. E. 1972. The Glarus overthrust. *Creation Research Society Quarterly*. 8 (4): 251-255.
21. Burdick, C. L. 1969. The Empire Mountains–a thrust fault? *Creation Research Society Quarterly*. 6 (1): 49-54.
22. Burdick, C. L. 1974. Additional notes concerning the Lewis thrust-fault. *Creation Research Society Quarterly*. 11 (1): 56-60.
23. Burdick, C. L. 1977. Heart Mountain revisited. *Creation Research Society Quarterly*. 13 (4): 207-210.
24. Briegel, U. 2001. Rock mechanics and the paradox of overthrusting tectonics. In *Paradoxes in Geology*. U. Briegel and W. Xiao, eds. Amsterdam, Netherlands: Elsevier, 231-244.
25. Snelling, A. A. 2009. *Earth's Catastrophic Past: Geology, Creation & the Flood*. Dallas, TX: Institute for Creation Research.
26. Royse Jr., F., M. A. Warner, and D. L. Reese. 1975. Thrust belt structural geometry and related stratigraphic problems, Wyoming-Idaho-Northern Utah. *Rocky Mountain Association of Geologists 1975 Symposium*, 41-54.
27. Boyer, S. E. and D. Elliott. 1982. Thrust systems. *American Association of Petroleum Geologists Bulletin*. 66: 1196-1230.
28. Matthews, J. D. 2011. The stratigraphic geological column—a dead end. *Journal of Creation*. 25 (1): 98-103.
29. Clark, H. W. 1968. *Fossils, Flood, and Fire*. Escondido, CA: Outdoor Pictures.

6 Plate Tectonics and Catastrophic Plate Tectonics

> **Summary:** Have you ever noticed how South America and Africa fit together like a puzzle? Some scientists speculated that continents moved around, while others scoffed and said that was impossible. Alfred Wegener was the most important early voice in what became known as *plate tectonics*. Satellites eventually confirmed that Earth's plates are indeed moving.
>
> What exactly is plate tectonics? Imagine that all land and seafloor on Earth are split into huge puzzle-like pieces that are constantly moving—but very slowly today. Magma from deep inside the earth seeps up between the cracks and cools, creating new seafloor at ocean ridges. Seafloor is pulled back into the earth at ocean trenches.
>
> The origin of the continental crusts baffles secular scientists. However, biblical thinking attributes them to God's work during the creation week. Before the great Flood, many of the continents were connected in one big supercontinent. During the Flood, the continents broke apart and rapidly moved into their current locations. Plate tectonics fits perfectly with a biblical model of geology.

Earth is the only planet in our solar system that has tectonic plates. I don't think this is a coincidence. Contrary to what you may have heard about Mars, Earth was likely the only planet to have been completely flooded. The Bible tells us about this Flood in Genesis 6–8. And I think the creation of the plates and their movement was one of the keys to the Flood event. Don't get me wrong, the Flood was a miraculous judgment, but I believe God used the destruction of the original ocean floor as the mechanism to transport water onto the land. And by God's grace, He saved Noah and his family and at least two of every kind of land-dwelling animal on the Ark.

Some Christians have been hesitant to embrace the notion that the earth's outer surface is moving and had moved even more dramatically in the Flood year. Secular geologists had the same issue in the early 20th century, as we will discuss below. However, tremendous amounts of empirical data have suggested significant plate movement did occur thousands of years ago. And Global Positioning System satellites have confirmed the plates are still slightly moving today. Much of these data are independent of secular deep time and the geologic timescale. In addition, the catastrophic plate tectonics (CPT) model provides a mechanism for the flooding of the continents, the subsequent lowering and draining of the floodwaters, and a cause for the post-Flood Ice Age. Let's look first at the secular version of plate tectonics, then we'll look at catastrophic plate tectonics.[1]

Introduction to Plate Tectonics

Plate tectonics is a fairly new theory that has revolutionized geology since the 1960s. It is a theory about continually moving lithospheric plates that occasionally collide, scrape, and grind into each other as they move about the earth's surface. Prior

Thingvellir National Park, Iceland, showing an exposure of the Mid-Atlantic Ridge that marks the boundary between the North American plate and the Eurasian plate

to the modern theory, there were several other hypotheses, most notably continental drift, that hinted at a dynamic changing Earth.

The first written hint of moving continents is sometimes attributed to Sir Francis Bacon, a Bible-believing English scholar, who commented in 1620 on the similar outlines of Africa and South America. The apparent fit of these two continents was occasionally mentioned for many years thereafter, including references by French naturalist Georges de Buffon (1707-88) and German scientist Alexander von Humboldt (1769-1859). The splitting of Africa and South America was thought by some, including Antonio Snider-Pellegrini in 1858, to have occurred during the biblical Flood of Noah.[1] Between 1885 and 1909, an Austrian geologist named Eduard Suess proposed the fit of all of the Southern Hemisphere landmasses into a single continent he called Gondwanaland.

Georges de Buffon

> "Anytime a new hypothesis is proposed in science, it is often met with great skepticism, especially a 'wild' concept like continental drift."

Continental Drift

Between 1910 and 1930, German meteorologist Alfred Wegener proposed his theory of continental drift and backed it up with supporting scientific evidence. His data indicated the presence of a supercontinent he called Pangaea, which subsequently broke into the modern continental masses. Wegener believed the northern continental mass, which he called Laurasia, initially separated from the southern continents, which he called Gondwanaland, keeping with the name Suess suggested. Unfortunately,

Wegener died during an expedition in Greenland in 1930, ending his investigations.

For 20 years, Wegener was basically the solitary voice contradicting the geology establishment. Because he was not a formally trained geologist, the geologic community did not give much credit to his hypothesis. Wegener's ideas were also largely disregarded because his claim of moving continents was too radical for the times. The limited technology of the early 20th century prevented geologists from seeing any evidence that the continents were moving.

Anytime a new hypothesis is proposed in science, it is often met with great skepticism, especially a "wild" concept like continental drift. That is science's way of "weeding out" or testing new ideas. Unfortunately, many

Alfred Wegener

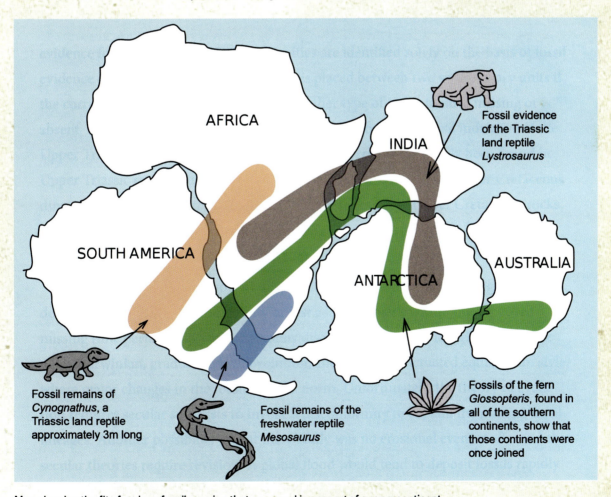

Map showing the fit of various fossil species that was used in support of a supercontinent

good ideas based on sound observations, like Wegener's, are often thrown out for the wrong reasons.

One American geologist, Frank Bursley Taylor, published several papers in 1928 on continental sliding, a concept similar to continental drift. Taylor believed continents could slide horizontally via some unknown "internal Earth tide," creating mountains as the continents collided with one another.

Wegener's ideas did cause concern for many geologists of the time. They knew he had some strong circumstantial geologic evidence and recognized enough value in his hypothesis to at least address it. You can still find references discussing the fallacies of continental drift in most geology textbooks published prior to the 1960s.

Basically, Wegener's evidence for continental drift came down to three observations: 1) the apparent fit of all continents into one supercontinent, Pangaea (in other words, the jigsaw pieces fit together); 2) fossil organisms found in similar sediments that matched up when the continents were reconstructed (in other words, the colors on the jigsaw pieces match); and 3) paleoclimatology data he claimed supported a common Ice Age prior to the breakup of Gondwanaland (again, the colors match when you reconstruct the pieces). However, this third point mostly consisted of striations on bedrock surfaces that can also form by debris flows. Young-earth creationists

Deep-sea drilling ship *Glomar Challenger*

do not feel there is compelling evidence for a pre-Flood Ice Age.

What destroyed the continental drift hypothesis was the lack of a viable mechanism for moving the continents. The geologists of the time went to physicists and mathematicians and had them prove, with sound established equations, that the continents could not drift about the globe. That was the end of the story, as far as they were concerned.

Seafloor Spreading

After Wegener and Taylor, there weren't a lot of advocates in support of continental drift until the 1960s, when the hypothesis of seafloor spreading was introduced. Harry Hess of Princeton University gets most of the credit for initiating this concept.[2] Because Hess was conscious of Wegener's failures, he introduced his paper by calling it "an essay in geopoetry" so that he did not alienate his geologic audience. The evidence

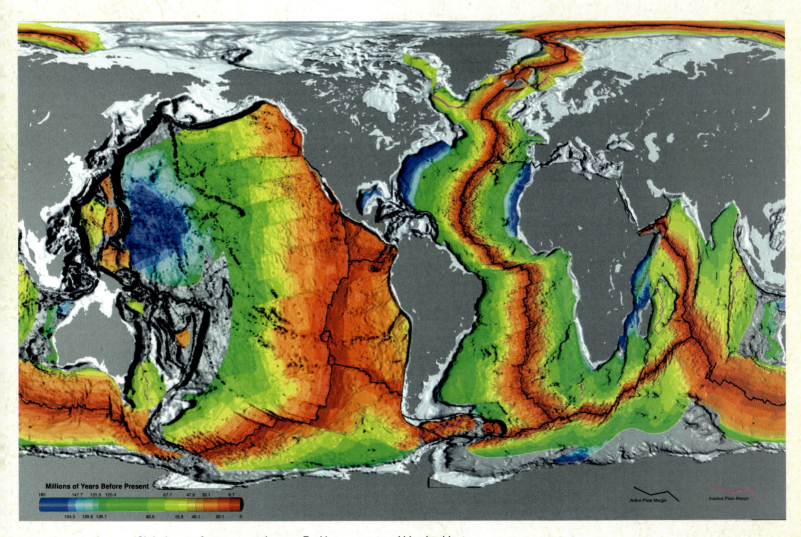

Global map of ocean crust by age. Red is youngest and blue is oldest.

for the spreading of the oceans came primarily from research related to the ocean ridges. Hess believed the ridges were where new seafloor was being created from mantle material. This, he argued, caused the seafloor to spread in a conveyor-like fashion.

Why didn't Wegener mention this in his continental drift hypothesis? Because he didn't have the knowledge of the ocean floor that science had accumulated by the 1960s. During the early Cold War era, the U.S. began an intensive study of the ocean as part of its submarine warfare agenda. War can cause great advances in science. Many research-funding organizations, like the Office of Naval Research and the National Science Foundation, were established to provide money for oceanographic research. Simultaneously, the Deep Sea Drilling Project began to drill cores in many locations, collecting sediments, crustal rocks, and temperature information. The first ocean floor maps were constructed using the latest depth information from sonar. These studies in the 1950s and 1960s provided all the necessary evidence for Hess to propose seafloor spreading.

The evidence for an expanding seafloor included: 1) the identification of the youngest sediments near the ridges and a progressive increase in age away from the ridges, 2) the identification of the youngest crust near the ridges and a progressive increase in age away from the ridges, 3) the identification of the highest heat flow near the ridges and a progressive cooling away from the ridges, and 4) the identification of symmetrical magnetic "stripes" across the ocean ridges.

Modern Plate Tectonics

By the late 1960s, the modern theory of plate tectonics was born. It is really just a combination of continental drift and seafloor spreading. It uses the geologic data from the continents gathered by Wegener and the oceanographic data from the seafloor as interpreted by Hess. One of the earliest papers on plate tectonics was formulated by Canadian geologist J. Tuzo Wilson. He had some trouble publishing his initial paper but finally was able to get his work printed in 1963.[3] He is now remembered as one of the founders of plate tectonics. Nearly 50 years after Wegener first proposed the concept of continental drift, the secular community somewhat reluctantly acknowledged the continents were in motion because they were overwhelmed with empirical data. And satellites were plotting the real-time motion of the plates from space.

There are about 8 to 10 major plates across the globe and a dozen or so smaller ones. Many plates contain both continental crust and oceanic crust. Most are delineated by the distribution of earthquakes along the edges of the plate boundaries. Prior to plate tectonics, the occurrence of earthquakes in linear patterns baffled scientists. Now, we believe that most earthquakes are the result of the grinding of one plate against another as they slowly continue to move.

Most volcanoes and mountain belts are also associated with current or past plate boundaries. It is also at the boundaries that magma is created for volcanic activity and where deformational forces are most active, creating mountain ranges. I will discuss this more later.

The plates are really the lithosphere, or the lithospheric plate. The lithosphere extends downward to about 100 kilometers (62 miles) in most places on Earth. The crust and the uppermost mantle are fused and behave as a single unit. It is the plate or lithosphere that moves across the earth. Plates are the coldest, outermost part of the earth and behave brittlely.

Oceanic crust is thin, about 6 to 10 kilometers (3.7 to 6.2 miles) and is composed

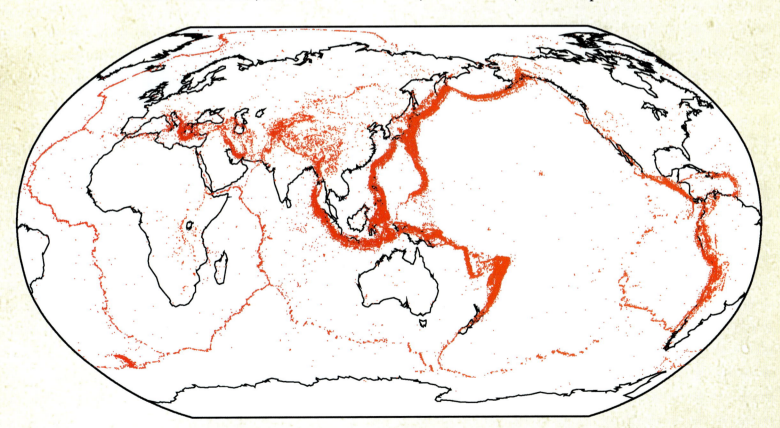

Earthquake epicenter map of events, 1960–2008. Note the close match of the earthquakes with the plate boundaries. The wider bands are caused by subduction at convergent boundaries.

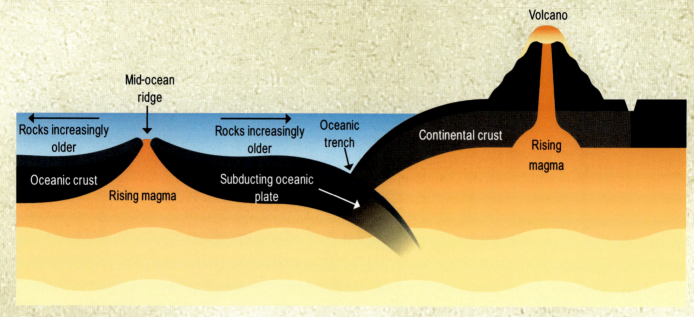

Seafloor spreading

of a basaltic material containing a lot of iron and magnesium-rich minerals. This makes it more dense than the continental crust at about 3 to 3.2 grams/cubic centimeter. New oceanic crust (really lithosphere) is produced at the spreading oceanic ridges and occasionally at continental rift zones. It is derived from the mantle below.

In contrast, the continental crust is much thicker than oceanic crust, about 20 to 70 kilometers (12.4 to 43.4 miles), and is composed of a completely different chemistry. It has a density less than oceanic crust at about 2.7 grams/cubic centimeter because it is composed of minerals that are more silica-rich, like quartz and feldspar.

Could the Pre-Flood Oceans Become Post-Flood Continents?

The modern oceans do not contain any early Flood sediments (Cambrian through Triassic), whereas the modern continents have both early- and late-Flood rocks spread across them. For this reason, some creationists have suggested that the original pre-Flood ocean crust rose during the Flood and became the modern continents, and the pre-Flood landmasses sank and became the modern oceans. However, one cannot easily turn one type of crust into the other. The major chemical and thickness differences described above make oceanic and continental crust unique and not interchangeable. The reason there are no early Flood sediments in the modern oceans is because of plate tectonic activity. Most of the pre-Flood oceanic lithosphere was subducted and recycled into the earth during the Flood, taking the sediments with them, or at least altering them severely to make them unrecognizable. Some remnants of these sediments may be found accreted to the coastal areas, as along the U.S. West Coast, but most were destroyed during subduction. Early, middle, and late-Flood sediments are preserved on the continents because they were not subducted. Continental lithosphere is too buoyant for subduction, except in isolated and/or limited instances.

The Low-Velocity Zone

Unbeknownst to Wegener, modern studies of seismic body waves have identified a low-velocity zone in the upper mantle that starts at the base of the lithosphere. This zone resides about 100 kilometers (62 miles) below the surface and extends down to about 350 kilometers (220 miles). Although this zone is a solid, it is a mushy solid and is estimated to be 1 to 2% melt. This causes the earthquake waves that penetrate it to slow as they pass through. And it allows the material above (the plates) to slide and move more easily across the surface of the layer. This low-velocity layer is termed the *asthenosphere*. Below 350 kilometers, the mantle again becomes more rigid.

Types of Plate Boundaries

There are three basic types of plate boundaries: 1) divergent, 2) convergent, and 3) transform. An example of divergent boundaries is Earth's ocean ridge systems. Here, new oceanic lithosphere is created from molten mantle material rising toward the surface, causing spreading and creation of new seafloor. Examples of convergent boundaries are subduction zones where older oceanic lithosphere is destroyed by subducting it beneath the less-dense continental lithosphere or younger and less-dense oceanic lithosphere. The volume of subducted (destroyed) oceanic lithosphere balances the vol-

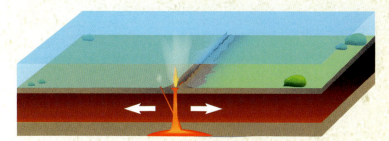
Divergent plate boundary

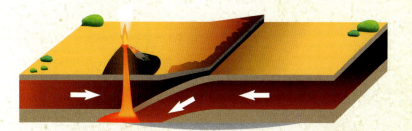
Convergent plate boundary

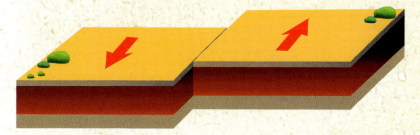
Transform plate boundary

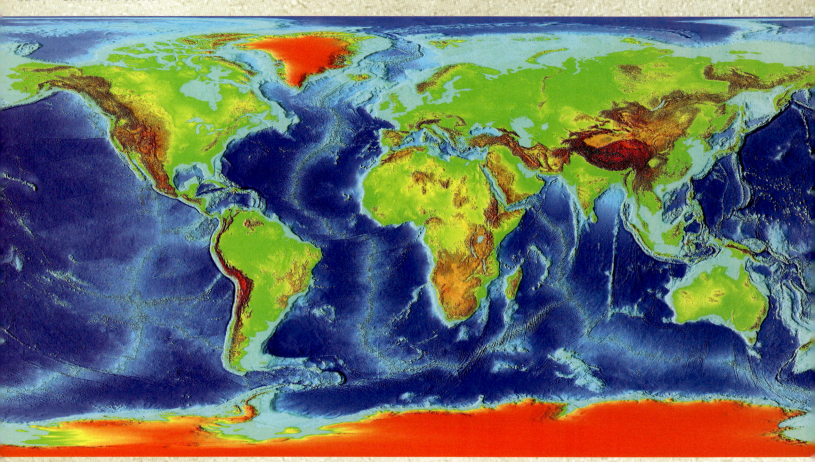
Bathymetry of the ocean floor showing shallow ridges as light blue lines through the oceans

ume of new oceanic lithosphere made at the ridges. Subduction also creates a source of magma from the partial remelting of the subducted slab. This type of magma causes silica-rich volcanoes to form along the subduction or convergent boundary. Transform boundaries form where the plates are merely sliding past one another horizontally and no lithosphere is created or destroyed. The San Andreas Fault in California is an example of this type of boundary.

Most divergent boundaries are ocean ridges. These are the spreading centers that Hess recognized where new seafloor is created. Every ocean has a ridge because the ridge is what produced the oceanic crust and lithosphere. The ridges are uplifted compared to the deep ocean basins due to high heat from below and form long linear mountain chains that connect and run from ocean to ocean. The classic example is the Mid-Ocean Ridge in the Atlantic Ocean. Collectively, these are the longest mountain chains in the world. The heat from

"Every ocean has a ridge because the ridge is what produced the oceanic crust and lithosphere."

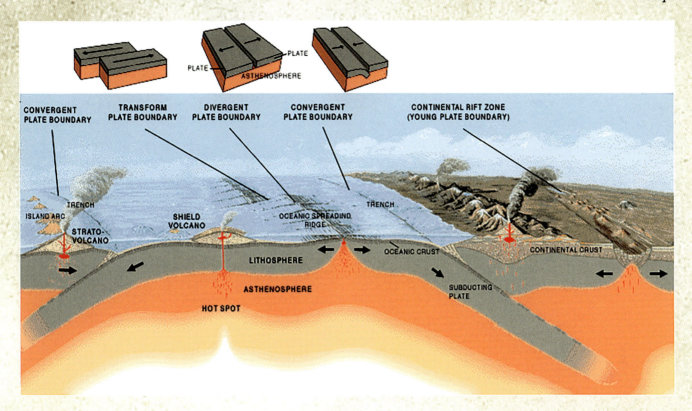

Diagram of the three types of plate tectonic boundaries and associated melt zones

the asthenosphere pushes the ridge upward for thousands of feet above the adjacent seafloor. The ocean deepens away from the ridges in both directions as the cooling and older crust becomes more dense and sinks down into the mantle. Only mild to moderate (< 6 M) and shallow (< 100 kilometers) earthquakes are associated with these types of boundaries.

Convergent boundaries are where two plates are colliding. If it is oceanic lithosphere against continental lithosphere, like along the west coast of South America, the more dense oceanic lithosphere is pushed under the continental plate, creating a subduction zone. This deformed the continent into mountains along the leading edge of the continental plate (Andes Mountains). The partial melt of the subducted plate creates volcanic activity. Many of the most dangerous geological hazards (lahars, pyroclastic flows, and tsunamis) are primarily associated with convergent boundaries. A lot of these hazards are due to the chemical differences in the magma that is generated at convergent boundaries. The volcanoes produced by the partial melt of the subducted lithosphere are very silica-rich and therefore explosive. Think for a moment about the hazards and loss of life from Mount St. Helens, Mt. Pinatubo, and Mt. Vesuvius. In addition, the earthquakes along the top and base of the subducted slab create the largest magnitude (> 9 M) and the deepest (up to 700 kilometers, or 430 miles) earthquakes.

Movements of the seafloor up or down during one of these massive earthquakes have created the largest and deadliest tsunamis.

In contrast, the Himalayas and the Alps formed from the collision of two continental plates. The oceanic lithosphere that was between the two continents was consumed by subduction until the continents themselves collided. At that point, subduction and volcanism ceased, but the mountains continued to be pushed upward and against one another. It is no surprise that the Himalayas are the highest mountain chain in the world today. They are the only active collisional zone between two continents, pushing India into Asia. Creation geologists believe this occurred late in the Flood event. Late uplift of these mountains also explains why early Flood fossils are found high up on Mt. Everest. The floodwaters never had to reach 29,000 feet above sea level to completely flood the earth. Instead, the early Flood sediments that were deposited across India were pushed upward late in the Flood as the Himalayas formed and after the

San Andreas Fault, California, is a transform boundary connecting two ocean ridges

water level had reached its peak. In the process, marine fossils were transported to the top of the Himalayas.

Transform boundaries or faults form as near-vertical faults with horizontal offset that cut completely through a tectonic plate (100 kilometers deep). Transform faults slide past one another and offset the ridges in hundreds of locations. They resemble a zipper in some ways. They are formed as ridges spread unevenly across a round earth. Most are small offsets (<100 kilometers) of the ridges and form at right angles to the spreading of the ridges system. They parallel the movement direction of the adjacent plates. Some of these are much greater offsets of the ridge system, like the San Andreas Fault. It connects to the ridge that is just offshore Oregon and Washington and back to the ridge in the Gulf of California. It is about 800 kilometers (500 miles) long and has lots of splaying faults associated with it as well. Because the San Andreas transform fault is so long, it can have fairly large-magnitude earthquakes that are much larger than most transform boundary earthquakes. Little volcanism is associated with transform boundaries, and no ocean lithosphere is created or destroyed by them.

Plate Tectonics Theory Explains the Geology We Observe Globally

Only plate tectonics theory can explain the systematic differences in volcanoes globally and offer a scientific reason for their differences in magma chemistry. Only plate tectonics can explain the linear pattern to earthquakes, mountain ranges, and volcanoes globally and tie them to a coherent theory of their origin. Only plate tectonics can offer a rational reason for the differences we observe in earthquake size, depth, and location along each type of plate boundary.

Mount St. Helens, 1980

Other proposed Flood models, like hydroplate theory, cannot account for the chemical and mineralogical differences in the magmas at convergent boundaries, like Mount St. Helens, and those that form elsewhere, like the less explosive and silica-poor Hawaiian-type magmas. To put it bluntly, plate tectonics provides a reason for these magma differences that other explanations cannot. That may be why so many creation geologists accept plate tectonics, or better, *catastrophic* plate tectonics. It is mostly non-geologists, who were never formally trained in igneous and metamorphic petrology, that gloss over and ignore the geologic details and develop alternative hypotheses like hydroplate theory.

No theory or hypothesis, other than plate tectonics, can explain so many global geologic observations.

However, plate tectonics cannot explain everything. There are still some unresolved issues. For example, plate tectonics can only explain the creation of new oceanic lithosphere and the complete destruction of older (pre-Flood) oceanic lithosphere. Plate tectonics does not explain the origin of massive amounts of continental crust, such as the supercontinent Pangaea. Day 3 of the creation week is still the best explanation for this apparent dilemma.

> Then God said, "Let the waters under the heavens be gathered together into one place, and let the dry land appear"; and it was so. (Genesis 1:9)

Secular Science Cannot Explain the Origin of Continents

The origin of the continental crust continues to baffle secular geologists, who often refer to this mystery as the "holy grail of geology."[4] Recall that Earth's plates are composed of two distinctly different types of crust: oceanic and continental. Explaining the reason for the unique crust and plates on Earth has been the subject of ongoing research and debate for decades.

We can observe basalt-rich oceanic crust pouring out of the mantle at ocean ridges today, and this provides us with a pretty good idea of how ocean crust forms. It origi-

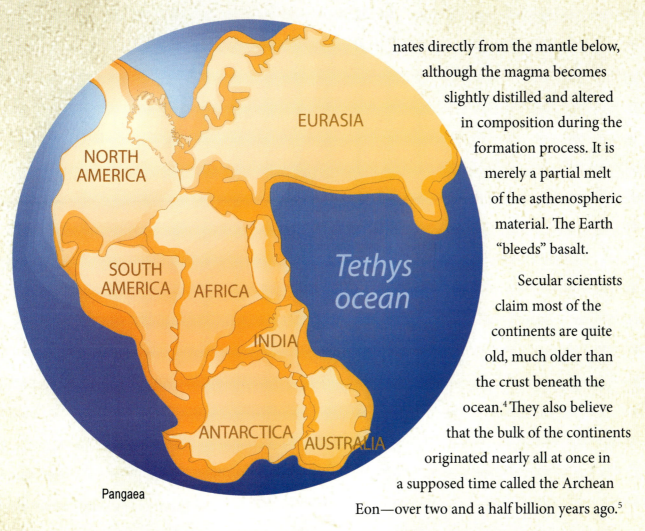

Pangaea

nates directly from the mantle below, although the magma becomes slightly distilled and altered in composition during the formation process. It is merely a partial melt of the asthenospheric material. The Earth "bleeds" basalt.

Secular scientists claim most of the continents are quite old, much older than the crust beneath the ocean.[4] They also believe that the bulk of the continents originated nearly all at once in a supposed time called the Archean Eon—over two and a half billion years ago.[5]

Recent research has attempted to shed light on the mystery of the continents, but it cannot account for the vast amounts of thick, granite-rich crust embedded in many tectonic plates on Earth.[4,5]

Secular geoscientists have demonstrated that slivers of continental crust can form during plate convergence when rising molten mantle material "mixes with fluid from subducted ocean crust and is distilled as it ascends, forming light continental crust, as well as a heavy slag that sinks back down [into the mantle]."[4]

Although secular geologists claim to have discovered clues to the way continents form, their research results only apply to a 120- to 250-mile belt of rock within the Himalayas.[4] They may be able to explain narrow bands of continental crust formation, but they still cannot explain the origin of continents thousands of miles wide. The bigger question remains, how did Earth's continents get here?

Dr. Esteban Gazel of Virginia Tech University and colleagues also concluded that continental crust can be made at subduction zones. But again, these authors were only able to explain the formation of narrow bands of new continental crust a mere 40 to 150 miles wide.[5]

Gazel and his coauthors further pointed out that the composition of these narrow and more recent bands of continental crust differ chemically from the supposed older Archean continental crust, adding another complication to the mystery and leaving the same questions unanswered.[5] Where did Earth's continents come from? And how did they form so rapidly and early in Earth's history?

Although we can learn much about the present Earth's crust through the study of empirical data like rocks and magmas, we can't go back in time and observe the occurrence of one-time events. The unique chemistry of the original continental crust and the size of the continents continue to perplex geoscientists. Our best recourse is to accept the Word of the One witness to the creation of the continents. Understanding *how* God formed the continents may be something we will better understand in the future through continued research.

God spoke the continents into existence on Day 3 of the creation week. A naturalistic explanation for such a supernatural event might never be found. Therefore, sec-

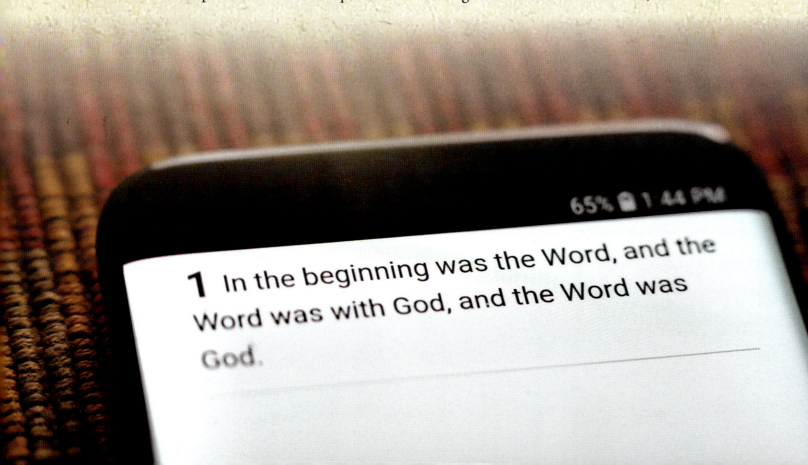

ular scientists will undoubtedly continue to struggle with the so-called "holy grail of geology." We can take heart knowing that the continents are visible evidence of God's eternal power of creation.

Modern Plate Movement Rates

You still may be wondering how we estimate the rates of plate movement or how we know which direction the plates are moving. With modern lasers and satellite data, scientists have demonstrated that the plates are still moving today at the approximate rates of 1 to 15 centimeters/year (0.5 to 6 inches/year).

In stark contrast to today's rates, many Flood geologists think the plates moved much more rapidly during the Flood event, at rates of *meters* per second. Complex computer models by Dr. John Baumgardner have shown that this type of movement is possible and that catastrophic plate tectonics is the likely cause of the world's

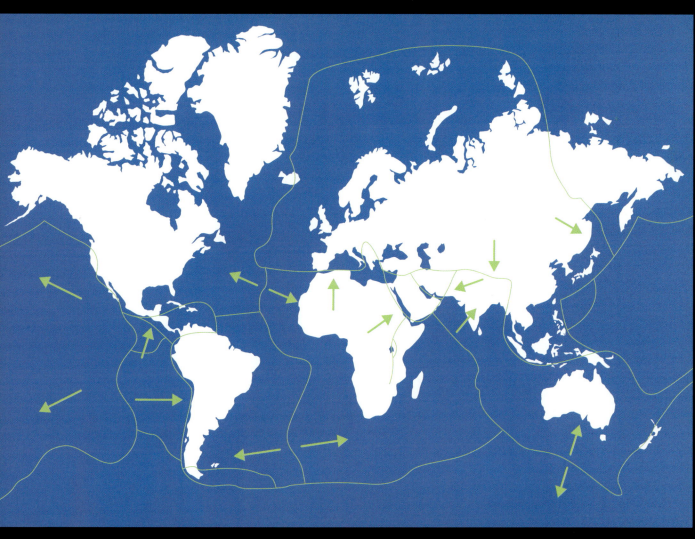

Map showing modern movement directions of the tectonic plates

continents separating from their pre-Flood configuration.[1,6] These discoveries led to a completely new theory called *catastrophic plate tectonics*.

Currently, secular and creation geoscientists agree that the likely mechanism for plate movement is *slab pull* from the high density of the subducted material as it enters the hotter, less-dense mantle. The resulting plate motion causes spreading and the opening of the ocean rifts. The gap in the crust is then filled by new melt from the asthenosphere below the ridges. Earlier speculation for the mechanism included mantle convection, but that seems less likely. There is, of course, movement in the mantle as the subducted plates drop in. Mantle material must move out of the way. And there must be mantle movement to fill the gaps created at the ridges, but there may not be the mantle convection as proposed by many earlier secular textbooks. Mantle convection or not, at least most geologists agree that convection is no longer the main driving force behind plate motion. Slab pull seems the better answer.

Catastrophic Plate Tectonics

Proponents of catastrophic plate tectonics believe that the plates moved rapidly during the Flood year and have since nearly ceased moving except for a few centimeters per year.[1] Creation geologist Dr. Steve Austin has even demonstrated an apparent decline in major earthquakes (magnitude 6 and greater) in the last century to support the claims of a slowing plate motion.[7] They contend that the plates are no longer moving rapidly and are merely settling in, and that the occurrence of major magnitude earthquakes will continue to decline in the next centuries as well. Their study was based on earthquake data provided by the National Earthquake Information Center, a branch of the United States Geological Survey.

Landslide resulting from the 1994 Northridge earthquake in California

Runaway Subduction and Superfaults

In 2013, a magnitude 8.3 earthquake struck deep below the Sea of Okhotsk in the Kuril-Kamchatka subduction zone just south of the Russian Kamchatka Peninsula and 950 miles north of Japan. It ruptured along a 110-mile-long fault about 378 miles below the surface where the Pacific Plate is being subducted, or pulled down, beneath the ocean crust.[8] Though it's been hailed as the largest deep earthquake ever recorded, the Okhotsk upheaval pales in comparison to the earthquake activity suspected during the great Flood. However, it does provide an excellent model of what may have transpired at the time of that catastrophic event long ago.

What perplexes scientists the most is the earthquake's great depth. Publishing in the journal *Science*, Lingling Ye and co-authors noted:

> The occurrence of earthquakes in the depth range from 400 to 720 km [250 to 450 miles] (the mantle transition zone) has long been enigmatic, given the immense pressure exerted by the overlying rock mass on any fault surface.[8]

The velocity of the rupture was also surprising, given that it transmitted the energy away from the focus at 4 to 4.5 kilometers per second—nearly 10,000 miles per hour! Geologists generally believe when dehydration reactions release water trapped in sediments, it reduces the stresses and pressure on rocks at shallower depths (30 to 250 miles), allowing for fluid-assisted earthquake movement.[8] However, at depths below 375 miles, scientists aren't confident as to what initiates incidents like the Okhotsk earthquake. They speculate that deep faults may shift as original fault-slip leads to frictional melting. This would create the runaway, or out-of-control, expansion of ruptures and subduction, the pulling under of plates. A cold-subducted plate may have been the key factor in allowing the Okhotsk earthquake to occur at such a great depth,

"Frigidity of the plates also indicates they had been subducted only a short time ago. If they had been subducted millions of years they would be much closer to the temperature of the surrounding mantle."

according to the authors. The cooler, more-brittle nature of the plate possibly allowed this event to behave like a shallower earthquake.[8] Frigidity of the plates also indicates they had been subducted only a short time ago. If they had been subducted millions of years, they would be much closer to the temperature of the surrounding mantle.

John Baumgardner has proposed a similar model of runaway subduction as the driving mechanism for the great Flood. He points out that the pre-Flood seafloor was evidently completely destroyed during the year-long event and rapidly replaced with today's young igneous ocean crust. He explains:

> In regard to the fate of the pre-Flood seafloor, there is strong observational support in global seismic tomography models for cold, dense material near the base of the lower mantle in a belt surrounding the present Pacific Ocean.[6]

This suggests that during the Flood cold plates were rapidly pulled down into the mantle, causing a thermal frictional envelope to develop around them by reducing viscosity (fluid-like thickness) in the mantle and "resulting in a sinking rate…higher than would occur otherwise."[6] Baumgardner found that once the older, colder, originally created oceanic crust and lithosphere began to subduct, it would speed up and drop into the less-dense hot mantle like a fishing weight in water. He referred to this as *runaway subduction*. He suggested rates of movement of meters per second, not centimeters per year as secular scientists like to suggest.

Although maximum displacement during this recent Russian earthquake was about 9.9 meters (32.5 feet), there is ample geological evidence for some superfaults showing offsets greater than 100 meters (328 feet).[9] Superfaults are rapid-moving slip events involving large offsets of the rock surfaces.

> Perhaps the concept of a "superfault" is best understood by contrast with the notion of a "regular fault." A modern magnitude 9 earthquake occurs on a "regular fault" with [a] displacement distance [of] less than 20 meters and a displacement rate of less than 0.1 meter per second. During the two-minute displacement event, friction on the fault generates heat. On the "regular fault" not enough heat is produced and retained within the interface to allow it to reach at least 1000°C [1800°F], a minimum temperature needed to melt rock. Instead, the thermal

Thick, black pseudotachylyte (PST) layers in a subduction zone on Kodiak Island, Alaska. The PST layers are nearly a foot thick, indicating extraordinary amounts of frictional heat during runaway subduction.

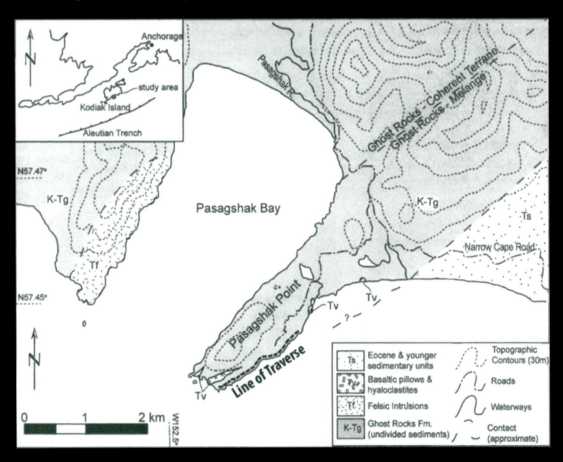

Map of Pasagshak Point, southeastern Kodiak Island, Alaska, showing line of traverse along the Pasagshak Thrust where pseudotachylyte is located.

conductivity of the rock around the fault allows the heat of friction to be transferred away from the surface. Rocks are converted to powder along the "regular fault" surface. However, a two-minute event on a "superfault" might have more than ten times the displacement distance and more than ten times the heat generation, allowing rocks to melt. The resulting "superquake" would be extraordinary, like none on Earth that humans have witnessed in historic times.[9]

The dark, glassy material formed by frictional melting upon the superfault surface is called *pseudotachylyte* (PST). The presence of PST is considered evidence of high-speed rock movement during superquakes, where displacements occur much faster than during modern, high-magnitude earthquakes. The existence today of superfaults, pseudotachylyte, and superquakes are proof of global catastrophic tectonic activity in Earth's present—and past.[9]

In our paper presented at the Seventh International Conference on Creationism in 2013, Dr. Steve Austin and I suggested that superfaults, and likewise this deep

Close-up of PST layers within the Pasagshak Thrust zone, Kodiak Island, Alaska. Note fracturing is commmonly orthogonal to the layering and has an overall resemblance to coal.

8.3-magnitude Okhotsk earthquake, may develop in subduction zones as trapped water becomes supercritical—i.e., undergoes a change in chemical properties, creating a liquefied slurry (thick mixture of water and solids).[9]

Subduction faults allow an entire oceanic plate to descend deep into the mantle of the earth. In a catastrophic superfault scenario, ocean sediment, groundwater, and ocean water descend rapidly. This places ocean sediment and water into extreme conditions, causing them to become supercritical at just 750°F and at pressures equivalent to a depth of only 3 kilometers (2 miles). As friction upon the fault generates melted rocks (PST), the supercritical seawater expands to three times its original volume, loses its viscosity, deposits salts, and reacts corrosively with the minerals that are present.[10] A slurry then develops along the fault surface. Perhaps the most significant element in fault dynamics is the expansion effect of supercritical water, which pressurizes and exerts tension on the fault surface. Buoyancy and pressure of the supercritical slurry could force it upward on the fault surface faster than the subducted rocks are descending. The melted silicate rock and supercritical water slurry, combined with the reduction in mantle viscosity around the subducted slab, make catastrophic plate subduction possible.[6,9]

Do we find any empirical evidence of runaway subduction preserved on Earth's surface today? The answer is yes. Superfaults and associated PST have been found and documented at many locations around the world, including several in subduction zone settings.[3] One of these is Pasagshak Point on Kodiak Island, Alaska, where the PST is the thickest in the world. Dr. Christen Rowe and her colleagues were the first to report PST found there in individual layers exceeding 12 inches, with each inch of PST thought to represent over 30 feet of fault movement.[11]

The PST at Pasagshak Point is found within an inactive subduction zone along the southeastern edge of Kodiak Island. Dr. Francesca Meneghini and co-authors refer to these PST layers as black fault rock (BFR). Some of the BFR is described as gray-black to blue-black rock devoid of foliation (i.e., mineral layering), possessing a conchoidal (glass-like) fracture pattern and exhibiting a satin-smooth luster. The Kodiak Island PST required "extraordinarily large slip" and occurred in "repeated seismic slip pulses," documenting catastrophic subduction.[11,12]

Superfaults containing PST, like the one at Kodiak Island, are some of the strongest

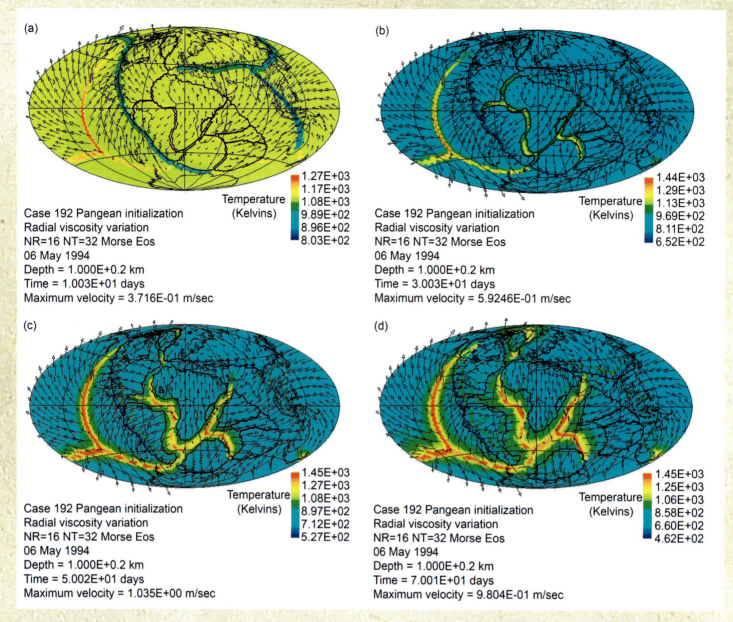

Dr. John Baumgardner's numerical model of the first 70 days of the Flood showing temperature contours, velocities, and continent boundaries at 100 km (62 mi) depth

preserved lines of evidence for the rapid subduction and catastrophic plate tectonic (CPT) activity that occurred during the great Flood. We can see the proof preserved in the rocks, in the seismic tomography, and in complex computer models. All testify of a real, global event that completely overturned the pre-Flood seafloor, creating a new world geography, separating the continents, and leaving behind billions of fossils as evidence of the catastrophic conditions that took place during the year-long biblical Flood. Today's large, deep earthquakes like the one near Okhotsk are small reminders of the watery catastrophe that took place just over 4,300 years ago.

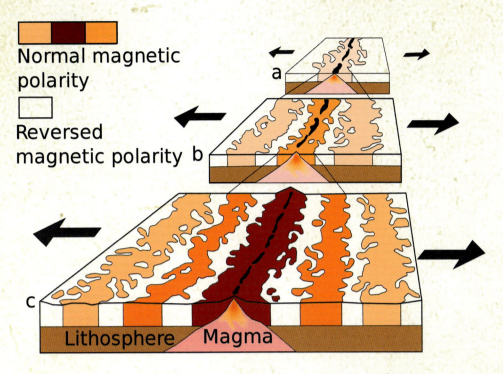

Magnetic reversal stripe pattern showing symmetry on either side of an ocean ridge

Sadly, most secular geologists reject the idea of runaway subduction. They insist that the plates have always moved at today's slow rates, employing their philosophy of uniformitarianism. It's not that they have found any mistakes in Dr. Baumgardner's math—on the contrary, his math is correct—or in his computer models, they just flat out don't believe it. So, they ignore his results and his powerful computer model and his math. They refuse to consider the validity of runaway subduction because it suggests a global catastrophe like the one described in the Bible. It is unfortunate that the secular world has been so willingly blinded to the truth of the global Flood.

Additional Empirical Evidence for CPT

Empirical data, independent of the chronostratigraphic timescale, demonstrate that the modern ocean lithosphere was completely created anew in conveyor belt fashion at the ridges during the Flood, causing systematic spreading in both directions. In the 1950s and 1960s, geologists discovered that the ocean crust is very young compared to many of the rocks on the continents. In fact, the oldest ocean crust only goes back to the Jurassic system, a point about midway through the Flood. Recall that at every ocean ridge, the crust gets systematically older in both directions. Although secular ocean floor maps claim ages in millions of years, they do seem to be correct in a relative sense. Older age dates usually indicate older rocks. In addition, a tremendous amount of data affirms seafloor spreading independent of absolute dating methods.

Consider, for example a few points. First, the temperatures recorded from wells drilled in the ocean crust and the heat flow measured near the ocean ridges show a systematic pattern of cooling with distance from the ridges in both directions. John Sclater and Jean Francheteau originally defined a relationship between heat flow and distance from the ocean ridge in 1970 that still holds today.[13] This empirical data set is not dependent on any dating methods, absolute or relative.

Second, the magnetic reversal stripe pattern shows symmetry on each side of the ocean ridges, supporting simultaneous seafloor spreading outward in both directions from the ridges. The overall symmetry to this data cannot be merely dismissed. The patterns initially observed by James Heirtzler and his colleagues for the ridge southwest of Iceland show a near-perfect symmetry for 200 kilometers in both directions about the ridge.[14] The raw, magnetic anomalies are based only on distance from the ridges and not on the secular ages of the rocks.

Third, the ubiquitous nature of ocean ridges suggests a common origin to the ocean crust (lithosphere) by seafloor spreading. Ocean ridges are found in every ocean of the world. The ridge system extends 45,000 miles, connecting all of the seas. They consist of huge, linear mountain chains rising 10,000 feet above the abyssal plains with a rift valley at the center, actively spewing out basaltic magma.

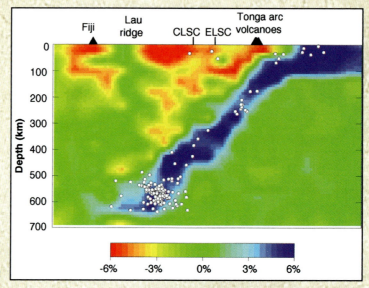

Fourth, modern seismic tomography data confirm runaway subduction happened just thousands of years ago. The internal images of the mantle (tomography) show visible lithospheric slabs of oceanic crust going down hundreds of miles beneath ocean trenches and into subduc-

P-wave tomography under the Tonga Trench, Pacific Ocean. The blue shows the colder ocean lithosphere descending into the mantle to a depth of nearly 700 km (435 miles). The white dots represent earthquake foci.

tion zones.¹⁵ Some of these now seem to extend to the top of the outer core. These are not merely faults, as some have proposed,¹⁶ but 62-mile-thick slabs of brittle, dense rock descending into the mantle. The cooler temperatures exhibited by these subducted slabs of rock create a thermal dilemma for the secular and old-earth geologists, who must demonstrate how these slabs remained cold for millions of years. Colder, subducted slabs are best explained by runaway subduction just thousands of years ago during the great Flood.⁶

Fifth, the correlation of oils from offshore Brazil and West Africa show demonstrable similarities when the continents are reunited.¹⁷ The unique chemistry in the oil families found on opposite sides of the Atlantic, when reunited, show an unmistakable match that can only be explained by later plate movement. The geochemical differences found in the oils from north to south along the coasts depend on the uniqueness of

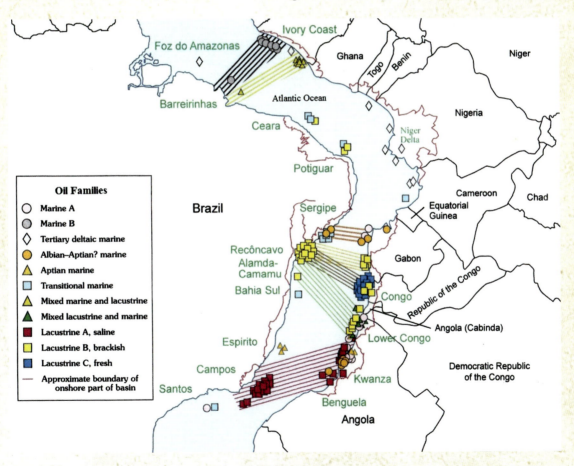

Map of the South Atlantic showing the correlation of families of oils and their similarities between Brazil and West Africa. The intervening Atlantic Ocean is reduced to illustrate the correlations. Oil chemical signatures are independent of perceived geologic age as they are based on chemical differences in liquid content regardless of rock unit. These data strongly support that plate movement has occurred to separate these source rocks and these oils.¹⁷

the source rocks themselves and not the perceived age of the rocks. These data indicate similar source rocks were deposited at different locations up and down the coasts of both continents that were later separated by plate motion.

The creation of new ocean lithosphere at the ridges is exactly what Harry Hess proposed in 1962.[2] John Baumgardner has further shown that this process was rapid and occurred just thousands of years ago—just like his computer models, the PST, and the cold, subducted slabs observed deep in the mantle demonstrate.

CPT Explains the Pattern of Earthquakes and Volcanoes

As noted above, maps of current earthquake epicenters define the boundaries of the majority of the plates, nearly 4,500 years after most of the plate movement ceased. This explains many of the world's largest and deepest earthquakes. Further support for these plate boundaries is shown by the linear chains of volcanoes found along the edge of the Pacific plate, associated with the Pacific Ocean's Ring of Fire. In addition, many of the major mountain ranges of the world also follow the edges of active plate boundaries, such as the Andes and Himalayas. These long, linear chains of mountains run

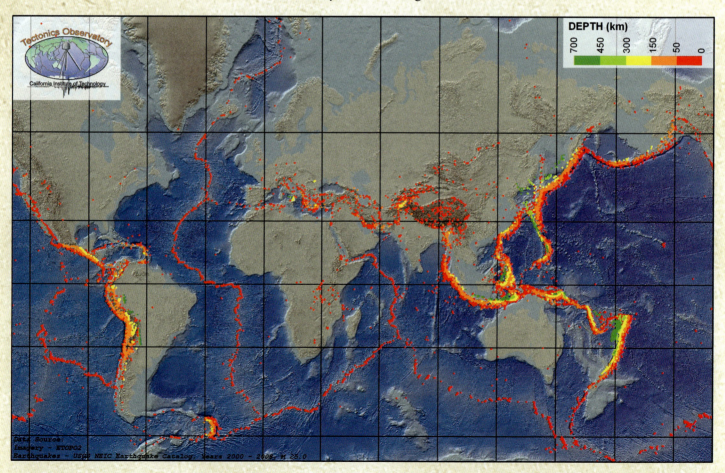

Most major earthquakes occur at convergent boundaries like subduction zones and collisional zones where earthquakes can extend to 700 km (435 mi) deep

parallel and in close proximity to many of the convergent-style plate boundaries.

CPT Explains the Flooding of the Continents

The Bible plainly states that the "fountains of the great deep were broken up, and the windows of heaven were opened" during the initiation of the Flood (Genesis 7:11). In terms of CPT, the breaking up of the fountains of the great deep may be a description of the initial rifting that took place at the ocean ridges and even rifting within continents.[18] It is even possible that this was the moment when the global tectonic plates first formed. Geoscientists today believe rifts are essential to the onset of subduction, although they are not sure exactly how this process begins. It appears that these long, linear rifts that formed as the fountains were broken up may have allowed the cold, dense, pre-Flood ocean crust to begin to subduct. Without a long, linear crack in the lithosphere, subduction cannot initiate.

A source of water for the fountains may have been in the upper mantle. Recent studies indicate that there are still massive quantities of water disseminated within the minerals of the upper mantle in a layer called the transition zone just below the asthenosphere (440 to 660 kilometers or 270 to 400 miles).[19] Secular scientists estimate that just as much water is trapped in the minerals at these depths as there is in all the oceans. Ringwoodite and wadesleyite, the two most common minerals at those depths, are estimated to contain 1-2% water by weight. It's possible that some of this water, and even limited shallower mantle water, was released as the fountains burst at the onset of the Flood. And today still, about 95% of the gases released by volcanoes are water and carbon dioxide, demonstrating that volcanoes do release water.

Obviously, the intense rainfall described as the opening of the "windows of heaven" contributed to the flooding of the pre-Flood landmasses. And some of this rainfall was likely from the water coming out of the volcanic eruptions as described above. But, because newly created oceanic lithosphere is hot, less dense, and more buoyant, the CPT model provides another source for water for the flooding of the continents. After its formation at the ridges, the freshly formed, low-density oceanic lithosphere rises and raises the top of the seafloor from below, displacing ocean water and forcing it on land. Creation geologist Dr. Andrew Snelling calculated that this elevated seafloor could have raised the global sea level by as much as 1.6 kilometers (1.0 mile), greatly helping flood the continents.[20] Think about the water in your bathtub. If the bottom of the tub is raised, the water will rise. If the bottom of the ocean is raised, sea level will rise. The more new ocean lithosphere, the more the ocean level will rise. This process is what seems to have caused the water to finally go over the top of the highest hills as the Flood approached the 150th day. More details of this process will be explained in chapters 12 and 13.

Rapid movement of the plates during runaway subduction also supplied innumerable tsunami-like waves to wash across the land, helping deposit blanket-type sediments across continents. Recent numerical modeling by Dr. Baumgardner has found that repetitive tsunami waves, caused by rapid plate movement, could result in water accumulation more than a kilometer (0.62 mile) deep on the continents, contributing to the flooding.[21] The runaway subduction model also provides a mechanism to lower the continental crust about two miles in the proximity of the subduction zones, causing more extensive flooding of the land and creating room for thousands of feet of sediment.[6]

In summary, plate motion provided two of the major potential sources of water to inundate the pre-Flood landmasses. First, the rapid creation of new seafloor during the Flood caused the ocean levels to rise up to a mile higher. Second, the tsunamis generated by plate motion (subduction especially) could have added half a mile or more of elevation to the water levels across the continents. Collectively, these two sources of water can account for the flooding of even the highest pre-Flood hills. The stratigraphic evidence that supports a progressive flooding of the continents is explained in detail in chapters 9–14.

Finally, subsequent cooling of the newly created ocean lithosphere later in the Flood year (after Day 150) offers an explanation for the lowering of the floodwaters. The 62-mile-thick, newly created ocean lithosphere slowly cooled and sank, lowering the bottom of the oceans and helping to draw the water off the continents and back into the ocean basins. What happened to the floodwaters? They are back in today's ocean basins. Remember, the Flood didn't have to cover pre-Flood land that was 29,000 feet above sea level. Those mountains and most others were pushed up toward the end of the Flood. The highest hills in the pre-Flood world were likely much less than people think, maybe only 5,000 feet above the pre-Flood ocean level.

CPT Explains Why the Plates Are Moving Slowly Today

It was the density contrast of the heavy, cold, original ocean crust (the lithosphere) that allowed the runaway subduction process to begin and continue. The

density difference served essentially as the fuel. Geophysicist John Baumgardner described it as "gravitational energy driving the motion" of the plates.[21] The runaway process continued until the original oceanic lithosphere was consumed. There was no geophysical means or reason to stop the rapid plate motion until the density contrast was fully alleviated. At that moment, the newer, more buoyant lithosphere ceased subducting, bringing plate motion to a virtual standstill. As a consequence, today we only witness small, residual plate motions of centimeters per year.

CPT Explains the Conditions Necessary for the Ice Age

Finally, CPT provides a mechanism for the Ice Age that occurred at the end of the Flood (chapter 15). A hot, newly formed ocean crust covering 70% of the world would have provided tremendous amounts of heat energy to the ocean waters above. This would have raised the overall temperature of the ocean and caused a greater amount of evaporation, resulting in staggering amounts of precipitation.[22] The increased volcanic activity from the subduction zone volcanoes and the unique chemistry of subduction zone magmas within the Ring of Fire and elsewhere late in the Flood would have placed huge volumes of ash and aerosols into the atmosphere, cooling the climate most noticeably in the higher latitudes.[22]

The distinctive chemistry of the magmas generated by the partial melt of subducted ocean lithosphere provides the perfect recipe for explosive, ash-rich eruptions.

Ice Age map of the extent of ice coverage across North America

These types of volcanoes (stratovolcanoes) are highest in silica, making them thicker and more explosive.[23] The net result of hotter oceans and tremendous silica-rich volcanic activity brought on from plate motion would be enough to start a widespread Ice Age. The hotter water provided higher evaporation, and the ash-rich volcanoes that erupted continually over many years provided the aerosols to cool the earth, especially in the higher latitudes.

In contrast, the most common type of volcanoes across the majority of the ocean basins have basalt-rich magmas (similar to shield volcanoes) and are less capable of producing the ash-rich explosions necessary to generate sun-blocking aerosols and ash.[23] This is another reason runaway subduction was an important part of the Flood mechanism. Only subduction provides these explosive ash-rich magmas. Finally, as the ocean water slowly cooled and volcanic activity diminished over the centuries after Flood, the Ice Age would have ended as abruptly as it began.[22]

Plate Movements During the Days of Peleg?

Many Christians have wondered if the plates could have moved during the days of Peleg. They read Genesis 10:25: "To Eber were born two sons: the name of one was Peleg, for in his days the earth was divided."

As a geologist, I find any significant division of the earth's plates and/or continents during the lifetime of Peleg very inconceivable. According to the Bible, Peleg lived about 100 to 340 years after the Flood.[24] If a major amount of plate movement did occur at this time, then hundreds of massive tsunamis would have been generated and possibly thousands of earthquakes greater than 8 M would have impacted the entire globe. I will concede that there was residual plate motion going on in the time of Peleg, and likely even a bit more movement than observed today. But, if the original ocean crust/lithosphere was already completely consumed, there would be an insufficient density difference between the newer ocean crust and the mantle to drive renewed or additional runaway subduction. Essentially, the driving gravitational energy would be gone. Plate movement would rapidly slow by several orders of magnitude to the rates we witness today.

In addition, plate motion during the lifetime of Peleg negates the reason for plate movement in the first place. It seems likely that CPT was the mechanism used to raise the sea level sufficiently high enough to Flood the entire Earth, through the creation

Mt. Pinatubo eruption of 1991

of an entirely new seafloor at the ocean ridges. Recall, the newer and hotter lithosphere would rise, pushing up the water from below. Second, the cooling and sinking of the new seafloor later in the Flood year would have lowered the ocean basins and assisted in draining the water back off the continents. These two processes must have been complete before Noah and his family stepped off the Ark. Any new or continued plate motion would have raised the water level again and caused renewed flooding. Therefore, any significant plate motion during Peleg's lifetime would have been highly limited and/or highly improbable.

Conclusion: The Reason for Catastrophic Plate Tectonics

Creation geologists who advocate CPT do not claim to understand all aspects of the theory, but they accept it as a sound working model steeped in empirical data. Secular and creation scientists alike debate how subduction is initiated[25] and how the major continents originated,[4] but most do not use this lack of understanding to question the overall validity of plate tectonics and/or the CPT model.

Catastrophic plate tectonics provides a Flood model and mechanism that explain much of the geology that scientists observe and measure. The overwhelming geological evidence supports the conclusion that catastrophic plate movement occurred just thousands of years ago and contributed to the flooding of the earth.

Catastrophic plate tectonics also provided a mechanism for the secession of plate movement. Once the original pre-Flood oceanic crust was completely consumed by subduction, plate motion would have ceased abruptly.

Between the runaway subduction and the secession of subduction, an entirely new seafloor of lava was created. This pushed up the water from below, causing the floodwaters to go over the top of the highest hills and prepared the conditions necessary for the Ice Age. We will see in chapter 15 that God had a plan and a reason for the Ice Age too.

Many well-meaning creationists have proposed alternative models to explain the global Flood. Sadly, most of these alternatives are being advocated by those who are not well-versed in the discoveries of modern geology. Some have argued that tens of thousands of meteorite impacts can explain the Flood and that massive plate motion is unnecessary. However, tens of thousands of impacts would have likely obliterated everything, leaving few fossils, organized strata, or even oceans today. Others have sug-

> "The overwhelming geological evidence supports the conclusion that catastrophic plate movement occurred just thousands of years ago and contributed to the flooding of the earth."

gested that the original created continents may have sunk and the pre-Flood oceans rose to explain the lack of early Flood sediments observed in the ocean basins today. And yet, knowledge of the chemistry and thicknesses of oceanic and continental crust precludes this idea. Ocean crust cannot just triple in thickness and completely change its chemistry, and vice versa. Hydroplate theory and the concept of vast water-filled caverns below the crust ignore vast amounts of modern seismic tomography data and the earthquake evidence for subduction (the Benioff zones). These ideas, although novel, all show a certain naïveté of the geologic data available today. Suggestions of thousands of impacts, flip-flops of the original ocean crust with the continental crust and/or thick, water-filled caverns below the crust are not necessary to explain the Flood. Catastrophic plate tectonics does a better job of explaining all of the available geologic data than any of these alternatives. And we will see that the stratigraphic data across the continents match nicely with the CPT model.

The more I studied the rocks, the more it all began to make sense. Catastrophic plate tectonics was the mechanism God used to implement the Flood. The result was a peak in volcanism at the right time and the heating of the oceans at the right time. Together these factors created the perfect conditions for the Ice Age after the Flood. Even the volcanism had to be the right chemistry to make explosive volcanoes, like at Mount St. Helens, to put ample aerosols into the atmosphere all at the right time. Not just any volcanism would do. It had to be the specific, volatile, silica-rich magmas generated by partial melts at subduction zones. Only these types of volcanoes would do the job. It was no coincidence.

References
1. Austin, S. A. et al. 1994. Catastrophic Plate Tectonics: A Global Flood Model of Earth History. In *Proceedings of the Third International Conference on Creationism.* R. E. Walsh, ed. Pittsburgh, PA: Creation Science Fellowship, 609-621.

2. Hess, H. 1962. History of Ocean Basins. In *Petrologic studies: a volume in honor of A. F. Buddington*. A. Engel, H. James, and B. Leonard, eds. Boulder, CO: Geological Society of America, 599-620.
3. Wilson, J. T. 1963. Hypothesis of Earth's behaviour. *Nature*. 198 (4884): 925-929.
4. Hecht, J. 2015. Rise of the upper crust. *New Scientist*. 226 (3017): 36-39.
5. Gazel, E. et al. 2105. Continental crust generated in oceanic arcs. *Nature Geoscience*. 8 (4): 321-327.
6. Baumgardner, J. 1994. Runaway Subduction as the Driving Mechanism for the Genesis Flood. In *Proceedings of the Third International Conference on Creationism*. R. Walsh, ed. Pittsburgh, PA: Creation Science Fellowship Inc., 63-75.
7. Austin, S. A. and M. L. Strauss. 1999. Are earthquakes signs of the end times?: a geological and Biblical response to an urban legend: *Christian Science Journal*. 21 (4): 30-39.
8. Ye, L. et al. 2013. Energy Release of the 2013 Mw 8.3 Sea of Okhotsk Earthquake and Deep Slab Stress Heterogeneity. *Science*. 341 (6152): 1380-1384.
9. Clarey, T. L. et al. 2013. Superfaults and Pseudotachylytes: Evidence of Catastrophic Earth Movements. In *Proceedings of the Seventh International Conference on Creationism*. M. Horstemeyer, ed. Pittsburgh, PA: Creation Science Fellowship, Inc.
10. Hovland, M. et al. 2006. Deep-rooted piercement structures in deep sedimentary basins— Manifestations of supercritical water generation at depth? *Journal of Geochemical Exploration*. 89 (1): 157-160.
11. Rowe, C. D. et al. 2005. Large-scale pseudotachylytes and fluidized cataclasites from an ancient subduction thrust fault. *Geology*. 33 (12): 937-940.
12. Meneghini, F. et al. 2010. Record of mega-earthquakes in subduction thrusts: The black fault rocks of Pasagshak Point (Kodiak Island, Alaska). *Geological Society of America Bulletin*. 122 (7/8): 1280-1297.
13. Sclater, J. and J. Francheteau. 1970. The implications of terrestrial heat flow observations on current tectonic and geochemical models of the crust and upper mantle of the Earth. *Geophysical Journal of the Royal Astronomical Society*. 20: 509-542.
14. Heirtzler, J., X. Le Pichon, and J. Baron. 1966. Magnetic anomalies over the Reykjanes Ridge. *Deep Sea Research*. 13: 427-433.
15. Schmandt, B. and F.-C. Lin. 2014. P and S wave tomography of the mantle beneath the United States. *Geophysical Research Letters*. 41: 6342-6349.
16. Brown Jr., W. 2008. *In the Beginning: Compelling Evidence for Creation and the Flood*, 9th ed. Phoenix, AZ: Center for Scientific Creation.
17. Brownfield, M. E. and R. R. Charpentier. 2006. Geology and total petroleum systems of the West-Central Coastal Province (7203), West Africa. *U.S. Geological Survey Bulletin 2207-B*, 1-52.
18. Reed, J. 2000. *The North American Midcontinent Rift System: An Interpretation Within the Biblical Worldview*. St. Joseph, MO: Creation Research Society Books.
19. Fei, H. et al. 2017. A nearly water-saturated mantle transition zone inferred from mineral viscosity. *Science Advances*. 3: e1603024.
20. Snelling, A. 2014. Geophysical issues: understanding the origin of the continents, their rock layers and mountains. In *Grappling with the Chronology of the Genesis Flood*. S. Boyd and A. Snelling, eds. Green Forest, AR: Master Books, 111-143.
21. Baumgardner, J. 2016. Numerical Modeling of the Large-Scale Erosion, Sediment Transport, and Deposition Processes of the Genesis Flood. *Answers Research Journal*. 9:1-24.
22. Oard, M. 2004. *Frozen in Time*. Green Forest, AR: Master Books.
23. Raymond, L. 1995. *Petrology: The Study of Igneous, Sedimentary, and Metamorphic Rocks*. Dubuque, IA: William C. Brown Communications.
24. Morris III, H. M. 2016. *The Book of Beginnings: A Practical Guide to Understanding Genesis*. Dallas, TX: The Institute for Creation Research.
25. Marques, F. et al. 2014. Subduction initiates at straight passive margins. *Geology*. 42 (4): 331-334.

7 The Pre-Flood Geologic Configuration

> **Summary:** Based on the geological data, it seems likely the pre-Flood continents were arranged much differently. The pre-Flood earth is assumed to have been differentiated into core, mantle, and crust. The oceanic crust was likely basaltic in composition, and the continental crust was likely granitic, similar to post-Flood compositions. From studying fossils, we can estimate creature growth rates and levels of atmospheric oxygen—and we think that oxygen levels may have been higher in Earth's past. Unfortunately, much of the information necessary to unlock the pre-Flood world is not preserved. The pre-Flood environments, oceans, and atmospheres were likely very different from today's, but we can only speculate as to how and why.

The Bible contains few details about the pre-Flood continental configuration. Genesis 3 tells us rivers were flowing out of Eden and precious metals like gold and minerals like onyx were readily available. Genesis 4:22 tells us the pre-Flood humans used Earth materials like iron, copper, and tin ores to make tools and possibly weapons of iron and bronze. This is about all the geology that is revealed for the pre-Flood earth, other than references to the hydrology system in Genesis 2:5-6.

Based on the geologic data discussed in the previous chapter on plate tectonics, it seems likely there was a supercontinent in the pre-Flood world. The continental margins fit nicely together into a supercontinent of sorts. This is especially obvious in the continents bordering the Atlantic Ocean. It's possible that the pre-Flood oceans were not as deep as today's oceans, and there were most likely no high mountains like we have today that tower tens of thousands of feet above sea level.

Genesis 2:5-6 tells us that there was no rain, "but a mist went up from the earth

Pre-Flood world model

and watered the whole face of the ground." Do these verses mean there was absolutely no rain after the creation week was over, or did this mean no rain just before God planted the vegetation on Day 3 of the creation week? I am not sure, but the ground was apparently well watered. It also seems likely there were no violent rainstorms like we witness on today's Earth. Therefore, the heavy rain for the first 40 days of the Flood would have been something new and unusual for all pre-Flood inhabitants.

Pre-Flood Earth Structure

The pre-Flood Earth is assumed to have already been differentiated into core, mantle, and crust.[1] This process was likely completed by the end of Day 3 of the creation week (Genesis 1:9-10). As noted by Steve Austin and his colleagues, the pre-Flood oceanic crust was likely basaltic or mafic in composition and the continental crust was granitic or felsic, similar to post-Flood compositions.[1] This was probably a consequence of God's command of "let the dry land appear" (Genesis 1:9). Granitic continental crust is less dense and able to maintain a higher topographic position due to buoyancy forces. This would tend to place most of the pre-Flood

> A mist went up from the earth and watered the whole face of the ground. (Genesis 2:6)

continents above sea level (Psalm 95:5). The more dense basaltic oceanic crust would tend to sink deeper into the mantle and allow the pre-Flood oceans to reside in deep basins.[1]

A pre-Flood basaltic crust is also essential to the initiation of plate tectonic movement and the destruction of nearly all original oceanic crust during the Flood. Recall, it is the uplift of the newly created ocean floor that formed during runaway subduction that raised the Flood's water level high enough to flood the highest hills. We will examine more of the details of the Flood's progression in chapters 9–14.

As the pre-Flood continents were brought into existence by God's command on creation Day 3, they were presumably derived from upper mantle material.[2] Some creation scientists refer to this event as the Great Upheaval. This extraction of the crust from the mantle would have left an upper mantle enriched in neodymium (Nd) and a crust depleted in Nd, as is observed. Overall, the crustal rocks are enriched in so-called incompatible elements.[3] Incompatible means the elements are too large to fit readily in common crystal lattices.

Geophysicist John Baumgardner reported that 40% of the heat-producing (radioactive) elements reside in the continental crust today, explaining today's observed heat flow.[3] The formation of continental crust on Day 3 of the creation week likely removed many of the incompatible elements (including the heat-generating radioactive elements like uranium, potassium, thorium) from the upper mantle during this differentiation. These elements were then emplaced in the continental crust, presumably during its formation. Baumgardner further reported that rock samples from magmas that were stripped from the lower crust

Interior of today's Earth showing the core, mantle, and crust

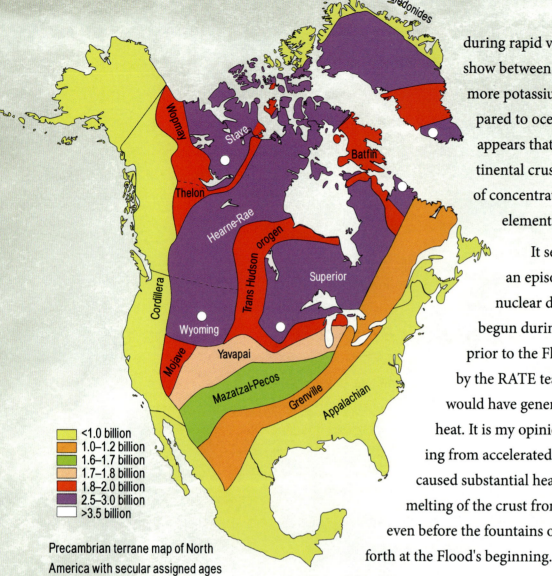

Precambrian terrane map of North America with secular assigned ages

during rapid volcanic ascent show between 10 and 30 times more potassium present compared to oceanic crust.[3] So, it appears that the lower continental crust became a zone of concentrated radioactive elements.

It seems possible that an episode of accelerated nuclear decay may have begun during the Flood or just prior to the Flood, as proposed by the RATE team.[4] This decay would have generated considerable heat. It is my opinion that the heating from accelerated decay may have caused substantial heating and even melting of the crust from below, possibly even before the fountains of the deep burst forth at the Flood's beginning.

It is also possible that this radioactivity-generated heat and subsequent rapid plate motion during the Flood may have altered many of the original continental rocks, creating the differences in the various terranes we observe. I think, as humans, we have a tendency to make the Flood event too small. We tend to give too much credit to naturalistic events and not enough to God's supernatural ability. It seems possible to me that the Flood affected the whole Earth right down to the base of the crust, heating and squeezing the continents differentially and leaving them altered so they look like separate terranes, or provinces. God might not have had to "piece together" the continents bit by bit on Day 3 of the creation week to explain these different Precambrian provinces. They may just be part of the Flood event. Admittedly, this is very speculative and much more research needs to be done on the Precambrian terranes.

Pre-Flood Continental Configuration

The model described herein begins with a pre-Flood world composed of essentially one major continental mass, namely a modified version of Pangaea. Secular geologists have invoked numerous pre-Pangaean configurations including Pannotia, Pangaea, Rodinia, and Nuna, increasing in age respectively.[5] Most of these earlier configurations are based on less and less actual rock data for each subsequent reconstruction, respectively. Secular scientists rely on paleomagnetic data to support these interpretations, but the fit of the various pieces becomes more and more subjective for older reconstructions due to metamorphism and erosion. And paleomagnetic studies have difficulty resolving longitude position. Cocks and Torsvik noted that there is at best an uncertainty of 300 to 500 kilometers in longitudinal estimates and at worst no constraint at all.[6]

Debate exists over the pre-Flood continental configuration, with some creation geologists advocating for an initial created supercontinent called Rodinia.[7] However, we chose a slightly modified Pangaea because it has the most empirical geological evidence supporting it, including the best fit of the continents.[8] We placed a narrow sea (300 to 500 kilometers) between North America and Africa/Europe, allowing for limited plate subduction, an early Flood closure of the pre-Atlantic, and the formation of the Appalachian/

Basalt dike in granite, Canadian Shield

Trilobite fossils

Caledonian mountains. The width of this pre-Atlantic is based on subducted plate remnants that diminish beneath the Appalachians below 300 kilometers, supporting this narrow-sea interpretation.[9]

The claims of a pre-Pangaean Rodinia are based on limited paleontological evidence. Cocks and Fortey found different groupings of trilobites along the North American seaboard compared to the European continent.[10] They point out that the distinct suite of trilobites and trace fossils found in each location indicates these continental masses were separated during the deposition of the

Secular reconstruction of Rodinia showing stable cratons and active mobile belts

Sauk Megasequence (Cambrian-Lower Ordovician systems). However, by the end of the deposition of the Tippecanoe Megasequence (Middle Ordovician-Silurian systems), these faunal differences nearly disappear and more similar suites of marine fossils are found on both continents. A narrow sea, about 300 to 500 kilometers wide between North America and Africa/Europe would seem to satisfy this needed separation. The difference in these trilobite suites may merely amount to breeding. A narrow sea, as suggested, would most likely provide sufficient separation to amass these differences.

Another reason we favored Pangaea over Rodinia is that the current ocean floor was evidently created when the original creation week seafloor was consumed by subduction during the Flood event. It was the density contrast of the heavy, cold, original ocean crust (the lithosphere) that allowed the runaway subduction process to begin and to continue. The density difference served essentially as the fuel. Dr. John Baumgardner described it as "gravitational energy driving the motion" of the plates.[11] The runaway process continued until the original oceanic lithosphere was consumed. There was no geophysical means or reason to stop the rapid plate motion until the density contrast was fully alleviated. At that moment, the newer, more buoyant lithosphere ceased subducting, bringing plate motion to a virtual standstill. As a consequence, today we only witness small, residual plate motions of centimeters per year.

A pre-Flood world that resembled Rodinia would require the consumption of nearly all the pre-Flood ocean crust twice. The first time would be while the continents from Rodinia moved into the configuration of Pangaea, and then a second time when Pangaea split into the present global configuration. Geophysically, the first break-up of Rodinia and reconfiguration into Pangaea would be possible, but it would also consume all of the dense pre-Flood ocean crust. A second move would then be rendered impossible since any significant

The Precambrian section of the geologic column as currently understood by the U.S. Geological Survey, reflecting the uniformitarian belief in long ages of Earth history

EONOTHEIM / EON	ERATHEM / ERA	SYSTEM / PERIOD **	Age estimates of boundaries in mega-annum (Ma) unless otherwise noted
Proterozoic (P)	Neoproterozoic (Z)	Ediacaran	635*
		Cryogenian	850
		Tonian	1000
	Mesoproterozoic (Y)	Stenian	1200
		Ectasian	1400
		Calymmian	1600
	Paleoproterozoic (X)	Statherian	1800
		Orosirian	2050
		Rhyacian	2300
		Siderian	2500
Archean (A)	Neoarchean		2800
	Mesoarchean		3200
	Paleoarchean		3600
	Eoarchean		-4000
Hadean (pA)			-4600*

Map of Precambrian banded iron formations, courtesy of J. D. Dieterle

amount of new ocean crust created while splitting up Rodinia would not have enough density contrast to fuel a second episode of subduction. As mentioned above, it is the consumption of the cold, more dense pre-Flood ocean crust that caused runaway subduction in the first place.[11] Therefore, if there had been a Rodinia, we would still be in a Pangaea continental configuration today.

Pre-Flood Sediments

After studying the megasequences for over five years, I have to admit I still don't know for sure where the pre-Flood/Flood boundary is in the rock record in every location. It's easy to find in most places, but there are locations with extensive pre-Flood and/or early-Flood sedimentation that cause some confusion.

Austin et al believe significant thicknesses of pre-Flood sediments existed. These sediments were either created on Day 3 as direct fiat creation, or sediments that formed in the approximate 1,700 years between creation and the Flood.[1] They concluded that substantial quantities of clastic and carbonate sediment must have existed in the pre-Flood ocean and were redistributed in the Flood. Just how much pre-Flood sediment actually existed is unknown.

Map of Precambrian stromatolite occurrences, courtesy of J. D. Dieterle

Secular geologists claim that Archean rocks are the oldest segments of the continents. These include the crystalline greenstone belts and the granite and gneissic belts of the Canadian Shield and equivalent shield areas on most other continents. Many creation geologists interpret these rocks as part of the original continental crust that was formed on Day 3 of the creation week when God separated the waters from the land.[12] Some Archean rocks also contain unique sedimentary deposits called *banded iron formations*, or BIFs.[13] Archean sedimentary rocks, many of which were later metamorphosed, are presumed to be pre-Flood,[1] but some may have been formed during creation. Exactly which portions are from Day 3 of the creation week and which formed in the intervening 1,700 years before the Flood remains unclear.

Most Paleoproterozoic and some Mesoproterozoic sedimentary rocks are also probably pre-Flood, and some may even be from the creation week. These include many of the BIFs on many continents around the globe.[13] These iron-rich layers apparently formed in conditions that no longer exist on Earth today.[14] Some of these pre-Flood rocks also contain stromatolites and stromatolite mounds, a type of algal deposit that still exists today in selective environments.

The Kona Dolomite in Michigan is an example of these pre-Flood stromatolite-rich rocks and is conventionally dated as 2.2 billion years old.[15] Stromatolites and BIFs have been found fairly close to one another. Both have been tied to similar depositional environments, and they have also been tied directly or indirectly to hydrothermal activity.[16-18] Stromatolites and the nearby BIFs possibly represent a unique environment that may have existed in shallow marine locations in great abundance in the pre-Flood world. Dr. Kurt Wise has proposed such an environment that may have extended across many sections of the pre-Flood continent.[18]

Scientists believed stromatolites were extinct—until they found them living in Shark Bay, Australia, in 1956.[19] Stromatolites keep popping up in newer and more diverse environments.

Living stromatolites have been found in highly saline marine environments in the Bahamas and in atolls in the Central Pacific.[19] Stromatolites have even been found in freshwater lakes and streams in Spain, Canada, Germany, France, Australia, and

Stromatolites in Hamelin Pool, Denham, Western Australia

Japan. Although these are freshwater bodies, they all have an unusual water chemistry, allowing the stromatolites to thrive in both saltwater and freshwater environments.[19,20]

A 2017 report in the journal *Scientific Reports* identified stromatolites in Australia again, but this time on land in what has been termed a peat-bound wetland.[20] Bernadette Proemse and her colleagues from the University of Tasmania, Australia, were the first to identify stromatolites living as "smooth mats of yellowish and greenish, globular structures growing on the wetted surface of tufa barriers."[20] Unlike typical stromatolites, these were not submerged in water but rose above the surface of a calcium-rich, spring-fed ecosystem.

Polished stromatolite

The scientists further speculate that the calcium-rich groundwater prevented predation from snails, allowing the stromatolites to thrive. "This discovery, the scientists say, means that stromatolites may be more common than realized, because people have not been looking for them in freshwater springs."[19]

Ancient and modern stromatolites consist of finely laminated biomats formed by colonial cyanobacteria. Fossil stromatolites are found in mounded structures similar to modern stromatolites. The fossils are not composed of the bacteria themselves but a sediment-trapping mat formed by "biologically mediated mineral precipitation."[20]

Evolutionary scientists claim stromatolites are perhaps the oldest evidence of life on Earth and date some to be 3.7 billion years.[21] Fossil stromatolites are found all over the world in Archean and Proterozoic rocks and to a lesser extent in Cambrian and later rocks. Evolutionary scientists have tried to explain the decline in stromatolites in post-Cambrian rocks due to the sudden appearance of grazing organisms that presumably eat the cyanobacteria.[20]

However, creation scientists have a far better explanation. We believe God created stromatolites as part of His original created biome. It's no surprise we find numerous stromatolite fossils still preserved and buried beneath Flood sediments in many of the remnants of pre-Flood rocks of the world—Archean and many Proterozoic rocks.

Furthermore, creation scientists Drs. Georgia Purdom and Andrew Snelling proposed that the catastrophic nature of the Flood reshaped the earth's surface sufficiently to destroy the pre-Flood environments where the stromatolites formerly thrived.[22] Today, it is only in specialized environments that stromatolites are able to re-establish their mat-making abilities and grow, whether on land or in the sea.

Finally, modern stromatolites are another example of a living fossil. Although secular science claims these organisms are around 3.7 billion years old, they show essentially no evidence of evolution. The saga of the stromatolites fits best with a recent creation and a recent Flood, as described in the Bible.

Flood Alteration of Pre-Flood Sediments

Many of the pre-Flood sedimentary rocks were subsequently heated, deformed, and metamorphosed, sometimes distorting the original layering and likely also changing their radioisotope ages. If the banded iron formations represent a pre-Flood environment that existed between the creation week and the Flood, alteration of these rocks must have taken place during the Flood itself. There must have been tremendous heat transferred to these areas. Many of the pre-Flood rocks in and around the BIFs in Michigan do exhibit high-grade metamorphic minerals such as garnet, indicating temperatures above 500°C were attained. This may be where the heat generated by the aforementioned accelerated radioactive decay came into play. However, much more research on this topic is needed before we can be conclusive.

> "Creation scientists Drs. Georgia Purdom and Andrew Snelling proposed that the catastrophic nature of the Flood reshaped the earth's surface sufficiently to destroy the pre-Flood environments where the stromatolites formerly thrived."

Location of the Pre-Flood/Flood Boundary

The exact location of the pre-Flood/Flood boundary everywhere is still uncertain. In many places it is fairly obvious, as noted by Austin et al,[1] and resides at the Precambrian/Cambrian boundary, particularly where Cambrian system deposits are in direct contact with Archean crystalline rocks. In these instances, the boundary coincides with the Great Unconformity, as observed in the bottom of Grand Canyon.

In other places, the sedimentary structures and grain size of the sediments may help make a determination of the pre-Flood/Flood boundary, as in the Sixtymile Formation of eastern Grand Canyon. Creation scientists Drs. Steve Austin and Kurt Wise formulated their interpretation that the Sixtymile Formation is the bottom unit of the Sauk Megasequence in Grand Canyon in 1994 using observable sedimentological evidence within the strata, noting that the formation contains large angular clasts indicative of high-energy deposition at the start of the Flood.[23] The formation is composed

Great Unconformity in Grand Canyon showing dark brown Tapeats Sandstone (Cambrian) on top of pink Zoraster Granite (Precambrian)

Chief Mountain (Ninaistako), Montana, showing Precambrian Belt Supergroup sediments thrust over Mesozoic sediments

primarily of sandstones and breccias and occasional mudstones and is approximately 200 feet at its maximum.

In 2018, the secular geologic community arrived at a similar conclusion, finding that the Sixtymile Formation was much younger than originally thought.[24] Prior to this study, the secular community insisted that the formation was Precambrian and 650 million years old. Dr. Karl Karlstrom and his co-authors concluded that the Sauk Megasequence includes the Sixtymile Formation based on their age-dating of detrital zircons. However, they believe this unit marks the beginning of the first of several flooding events, not the beginning of the great Flood.[24]

Even though creation and secular scientists agree that the Sixtymile Formation is part of the Sauk Megasequence, they have different worldviews. For this reason, secular scientists struggle to explain the sudden appearance of fully formed, shelled fossils in Cambrian strata, the Sauk Megasequence, and the significance of the Sixtymile Formation.

Other Late Precambrian Rocks

However, where substantial Neoproterozoic and even Mesoproterozoic sedimentary rocks exist, it is difficult to locate an exact, universal boundary. For example, where the Belt Supergroup resides in Idaho and Montana, the location of the pre-Flood/Flood boundary has been difficult to establish. These units are reportedly over 15,000 meters (50,000 feet) in places and are conventionally dated between 1.5 to 1.07 billion years old, but there are plenty of conflicting age determinations in the secular geologic literature.[25-27] An additional complication in the location of this boundary is the structural deformation that has occurred in this area. Numerous thrust faults have placed many of these sediments on top of one another, probably thickening the sequence considerably. For this reason, the lack of fossils within the sedimentary package, and the repetitive nature of the sediments, it is difficult to know the original thicknesses of the Belt Supergroup.

Thrust faults involving the Belt Supergroup demonstrate that tectonic transport of some of these sediments occurred later in the Flood, even after the Zuni Megasequence (Cretaceous system) was deposited. The Lewis Overthrust places Belt Supergroup sediments on top of late Zuni Flood sediments. The secular mechanism for overthrusting generally requires high-fluid pressures to support the thrusted sheet so that it moves almost weightless. I have proposed a Flood explanation for overthrusts that overcomes many of the mechanical difficulties that encumber secular geologists.[28] Chapter 16 contains a more thorough discussion of overthrust faulting.

Pre-Flood Oceans

The chemistry of the pre-Flood ocean was probably considerably different from our modern ocean. The early oceans were likely less salty. Substantial volumes of salt and gypsum appear to have been added to the oceans during rifting events and/or volcanic activity of the fountains of the great deep. We will see in chapters 9–13 that there are vast salt and gypsum layers deposited globally as part of the Flood record. These precipitate rocks will be addressed further in the discussions of individual megasequences.

Trilobite fossil

Also, the special environments responsible for the banded iron formations, abundant stromatolites, and even the biological niches for trilobites, many corals, most brachiopods, and many other extinct marine invertebrates were no longer present in the post-Flood oceans. This likely resulted in a rapid population decline and extinction of these animals in a few years or less after the Flood.

Pre-Flood Atmosphere

Much of the information for the pre-Flood atmosphere is from indirect evidence, including fossils and geochemical signatures in the rocks. And neither is very exact. Studies by uniformitarian scientists have suggested there was a very different set of environmental conditions in the geologic past, which creationists apply to the pre-Flood world. Computer models based on geochemical signatures in the rocks have shown conflicting results, but most indicate there were higher carbon dioxide levels in the pre-Flood world (Paleozoic and Mesozoic Eras), as much as three to seven times today's levels. Some models and many paleontologists also suggest the oxygen levels were higher during much of this time, as much as 30% of the atmosphere. Not only do geochemical models indicate higher oxygen levels but the gigantism exhibited by pre-Flood insects (dragonflies with wingspans of two feet) and dinosaurs (up to 160 feet long) supports this conclusion. Gigantism in various animals has been tied again and again to higher oxygen levels. Along with higher concentrations of carbon dioxide, this would have allowed plants to flourish and animals to grow to tremendous sizes as exhibited by the fossils we find. There was most likely a worldwide greenhouse effect due to these higher oxygen and carbon dioxide levels, with a warm and moist climate everywhere.

Conclusion: The Pre-Flood Environment Was Very Different

The pre-Flood was vastly different from the modern world. The atmosphere likely changed dramatically during the Flood. These changes would have had major effects on the animals and the plants. The extinction of dinosaurs after the Flood may have been partially caused by rapid climatic changes and even changes to the post-Flood atmospheric composition. A significant drop in oxygen concentration to the current 21%, combined with cooler conditions in the north and south latitudes, would have limited the range and possibly slowed the activity levels of dinosaurs, especially if they were cold-blooded.

Many plants were different in the pre-Flood world. The warmer climate and atmospheric compositional differences allowed the pteridophytes (seedless vascular plants like ferns) and gymnosperms (naked seeded vascular plants like conifers) to

Dragonfly fossil

Cycad tree

dominate the landscape, at least in the lower elevations. There were many angiosperms (flowering plants), but they were much less dominant than today, and many may only have resided in higher elevations. We will discuss the plant fossil record more as we go through the megasequences.

References
1. Austin, S. A. et al. 1994. Catastrophic Plate Tectonics: A Global Flood Model of Earth History. In *Proceedings of the Third International Conference on Creationism*. R. E. Walsh, ed. Pittsburgh, PA: Creation Science Fellowship, 609-621.
2. Snelling, A. A. 2000. Geochemical processes in the mantle and crust. In *Radioisotopes and the Age of the Earth: A Young-Earth Creationist Research Initiative*. L. Vardiman, A. A. Snelling, and E. F. Chaffin, eds. El Cajon, CA: Institute for Creation Research and St. Joseph, MO: Creation Research Society, 123-304.
3. Baumgardner, J. R., Distribution of radioactive isotopes in the Earth, *Radioisotopes and the Age of the Earth: A Young-Earth Creationist Research Initiative*, 49-94.
4. Vardiman, L., A. A. Snelling, and E. F. Chaffin, eds. 2005. *Radioisotopes and the Age of the Earth: Results of a Young-Earth Creationist Research Initiative*. San Diego, CA: Institute for Creation Research and Chino Valley, AZ: Creation Research Society.

5. Wicander, R. and J. S. Monroe. 2013. *Historical Geology*, 7th ed. Belmont, CA: Brooks/Cole.
6. Cocks, L. R. M. and T. H. Torsvik. 2002. Earth geography from 500 to 400 million years ago: a faunal and palaeomagnetic review. *Journal of the Geological Society*. 159: 631-644.
7. Snelling, A. A. 2014. Geological Issues: Charting a scheme for correlating the rock layers with the Biblical record. In *Grappling with the Chronology of the Genesis Flood*. S. W. Boyd and A. A. Snelling, eds. Green Forest, AR: Master Books, 77-109.
8. Clarey, T. L. 2016. Empirical data support seafloor spreading and catastrophic plate tectonics. *Journal of Creation*. 30 (1): 76-82.
9. Schmandt, B. and F.-C. Lin. 2014. *P* and *S* wave tomography of the mantle beneath the United States. *Geophysical Research Letters*. 41 (18): 6342-6349.
10. Cocks, L. R. M. and R. A. Fortey. 1982. Faunal evidence for oceanic separations in the Palaeozoic of Britain. *Journal of the Geological Society*. 139: 465-478.
11. Baumgardner, J. R. 2016. Numerical Modeling of the Large-Scale Erosion, Sediment Transport, and Deposition Processes of the Genesis Flood. *Answers Research Journal*. 9: 1-24.
12. Snelling, A. A. 2009. *Earth's Catastrophic Past: Geology, Creation & the Flood*, vol. 2. Dallas, TX: Institute for Creation Research.
13. Klein, C. 2005. Some Precambrian banded iron-formations (BIFs) from around the world: their age, geologic setting, mineralogy, metamorphism, geochemistry, and origin. *American Mineralogist*. 90: 1473-1499.
14. Boggs Jr., S. 2006. *Principles of Sedimentology and Stratigraphy*, 4th ed. Upper Saddle River, NJ: Pearson/Prentice Hall.
15. LaBerge, G. L. 1994. *Geology of the Lake Superior Region*. Phoenix, AZ: Geoscience Press, Inc.
16. Isley, A. E. 1995. Hydrothermal plumes and the delivery of iron to banded iron formations. *The Journal of Geology*. 103: 169-185.
17. Powell, C. M. et al. 1999. Synorogenic hydrothermal origin for giant Hamersley iron oxide ore bodies. *Geology*. 27 (2): 175-178.
18. Wise, K. P. 2003. The hydrothermal biome: a pre-Flood environment. In *Proceedings of the Fifth International Conference on Creationism*. R. L. Ivey, Jr., ed. Pittsburgh, PA: Creation Science Fellowship, 359-370.
19. Frazer, J. 2017. Stromatolites Defy Odds by A) Living B) on Land. *Scientific American*. Posted on scientificamerican.com December 21, 2017, accessed January 4, 2018.
20. Proemse, B. C. et al. 2017. Stromatolites on the rise in peat-bound karstic wetlands. *Scientific Reports*. 7: 15384.
21. Mueller, P. A. and A. P. Nutman. 2017. The Archean-Hadean Earth: Modern paradigms and ancient processes. In *The Web of Geological Sciences: Advances, Impacts, and Interactions II*. M. E. Bickford, ed. Geological Society of America Special Paper. 523: 75-237.
22. Purdom, G. and A. A. Snelling. 2013. Survey of microbial composition and mechanisms of living stromatolites of the Bahamas and Australia: Developing criteria to determine the biogenicity of fossil stromatolites. In *Proceedings of the Seventh International Conference on Creationism*. M. Horstemeyer, ed. Pittsburgh, PA: Creation Science Fellowship.
23. Austin, S. A. and K. P. Wise. 1994. The Pre-Flood/Flood Boundary: As Defined in Grand Canyon, Arizona and Eastern Mojave Desert, California. In *Proceedings of the Third International Conference on Creationism*. R. E. Walsh, ed. Pittsburgh, PA: Creation Science Fellowship, 37-47.
24. Karlstrom, K. et al. 2018. Cambrian Sauk transgression in the Grand Canyon redefined by detrital zircons. *Nature Geoscience*. 11: 438-443.
25. Lageson, D. R. 1994. Metallogeny of the Belt-Purcell Basin, southern British Columbia and northern U.S. Rocky Mountains. *Northwest Geology*. 23.
26. Lageson, D. R. 2003. Tobacco Root Geological Society Field Conference. Symposium IV. *Northwest Geology*. 32
27. Link, P. K. 1993. *Geologic Guidebook to the Belt-Purcell Supergroups, Glacier National Park and Vicinity, Montana and Adjacent Canada*. Whitefish, MT: Belt Association, Inc.
28. Clarey, T. L. 2013. South Fork and Heart Mountain Faults: examples of catastrophic, gravity-driven "overthrusts," Northwest Wyoming, USA. In *Proceedings of the Seventh International Conference on Creationism*. M. Horstemeyer, ed. Pittsburgh, PA: Creation Science fellowship.

8 The Flood Begins: The Fountains of the Great Deep

Summary: The Flood began with the breaking open of the "fountains of the great deep" (Genesis 7:11). It seems likely these fountains initiated the tectonic plates because geologic evidence supports the simultaneous formation of multiple rift zones. As the continents moved apart, magma pushed up between them and rapidly formed new ocean crust. There is also evidence of a spike in volcanic activity just prior to the Cambrian, the geologic layer that creation scientists typically believe is the first in the Flood. This corroborates the opening of the catastrophic fountains of Genesis. Their breaking up may be a description of the great rifting that took place at the ocean ridges and within continents. This initial phase of the Flood was likely when the pre-Sauk Megasequence was laid.

> In the six hundredth year of Noah's life, in the second month, on that day all the fountains of the great deep were broken up, and the windows of heaven were opened. (Genesis 7:11)

On Day 1 of the Flood, the Bible tells us the fountains of the great deep opened all across the earth. Exactly what the fountains entailed is unclear. We know that today's volcanoes release a tremendous amount of gases, like water and carbon dioxide, along with lava. Water trapped in mantle minerals, like wadsleyite, can produce water when they melt. It seems likely that the fountains produced a lot of water/steam as discussed in chapter 6. Intensive pressure would have been released as the magma rose upward in the earth, but exactly how high this water/steam shot up into the atmosphere is unclear.

It seems likely the breaking of these fountains also initiated the tectonic plates. There is a lot of geologic evidence for the simultaneous formation of multiple rift zones. These may be the fountains described in Genesis. And several recent discoveries have suggested certain conditions were vastly different just prior to the deposition of the Sauk Megasequence.

Mantle Hotter at the Beginning of the Flood

In a paper published in *Nature Geoscience*, German scientists found evidence suggesting the earth's mantle was up to 300°F hotter during the initial, formative stages of the Atlantic Ocean—when the continents began to violently pull apart to create it—compared to today.[1] Over time, the mantle cooled to current levels.

El Tatio Geysers, Chile

The scientists studied the composition of oceanic crust using deep-sea drilling core samples and found a systematic change in chemistry from the shoreline—the edge of the continents—to that of the middle of the ocean. The shifts in the core sample's geologic chemistry were linked to changes in the temperature of the underlying mantle that generated the oceanic crust.

These findings suggest that the initiation of the great Flood began with an anomalously high-temperature mantle beneath the pre-Flood continents. As the continents rifted apart, new ocean crust formed rapidly between them, and molten mantle filled the ever-widening gap, supporting the concept of catastrophic plate tectonics as postulated by creation scientists.[2]

> "The ridge system extends for over 37,000 miles and accounts for about 75% of the present volcanic activity on Earth."

Today, new ocean crust forms along ocean ridges where tectonic plates continue to slowly pull apart. The ridges, a system of subsea mountains that run through every ocean, are elevated because the spreading seafloor is hot and rises due to its lower density. The ridge system extends for over 37,000 miles and accounts for about 75% of the present volcanic activity on Earth.[1]

The German scientists also noted that the average ocean ridge today resides at a depth of 1.8 miles below sea level. In contrast, they calculated that the ridges above the hotter mantle in the past would have only been about 0.6 miles below sea level—well over a mile higher![1]

What effect would this have had? Shallower ridges from higher heat flow would have raised global sea levels, at least partially accounting for the inundation of the continents during the Flood event. Later, as the mantle progressively cooled, as confirmed by this paper, the ocean ridges would have sunk, dramatically dropping sea level and draining the water off the continents to end the Flood.

In a related article summarizing these findings, it was pointed out that "much of the ancient oceanic crust seems to have been generated under conditions that are rare beneath present-day ridges."[3]

Rare indeed. The global Flood was a one-time unimaginable event, never to be repeated (Genesis 9:15). The high-mantle heat that apparently initiated the breakup of the pre-Flood continents only occurred once in the past. This heat flow also raised the seafloor ridges and helped flood the landmasses. The subsequent cooling of the mantle dropped the ridges and provided a way for the water to drain off the land and back to the deepening ocean basins. Present-day volcanic activity at the ocean ridges is minute compared to the catastrophic formation of oceanic crust during the year-long Flood. It's amazing how science again and again confirms biblical truth.

Peak in Volcanism and Massive Release of CO_2 in the Pre-Sauk

Two additional discoveries may provide important validations of the "fountains of the great deep" that started the deluge. Timothy Paulsen from the University of Wisconsin Oshkosh and co-authors from Michigan Technological University and ETH Zurich found evidence of a spike in volcanic activity and a rapid release of massive amounts of carbon dioxide just prior to the deposition of Cambrian rock layers.[4]

Many creation geologists consider the Cambrian rocks to represent the first extensive Flood deposits. Cambrian sediments are the bottommost layer in the Sauk Mega-

sequence and contain fossils of the so-called Cambrian Explosion—the first sediments with prolific numbers of hard-shelled organisms.[5]

Paulsen and his colleagues conducted trace element analysis on zircon crystals from Antarctica and compared them to previous studies of global magmatism. They determined there is evidence of a massive outpouring of carbon dioxide and associated volcanic activity just before the deposition of the Sauk Megasequence in a system called the Ediacaran. The Ediacaran is what secularists call the latest Precambrian or Neoproterozoic.

Sascha Brune and co-authors report:

> CO_2 behaves as an incompatible element and is readily transported to shallow crustal levels during melt migration. Although a certain fraction is released to the atmosphere during volcanic eruptions, recent studies indicate that much higher CO_2 release occurs due to circulation of hydrothermal fluids and degassing along normal faults without eruptive volcanism.[6]

These scientists claim that spikes in CO_2 are associated with continental rifting that produces normal faults and conduits for hydrothermal minerals like copper.[6]

So, whether by increased volcanic activity or increased hydrothermal activity, there appears to be significant evidence of a spike in global volcanic action and associated rifting prior to the deposition of the Sauk Megasequence. Could this be evidence of the breaking up of the fountains of the great deep mentioned in Genesis?

The Bible offers a solution to this newly identified spike in volcanism and rifting. Genesis 7:11 plainly states that "all the fountains of the great deep were broken up, and the windows of heaven were opened" during the initiation of the Flood. The breaking up of the fountains of the great deep may be a description of the great rifting that took place at the ocean ridges and within continents.[7]

Today, we merely see the remnants of this activity in the ridges of our modern oceans. No longer are they spewing out tremendous volumes of lava and massive amounts of CO_2. The Flood was a one-time event like no other. It is no wonder scientists find evidence of its catastrophic history in the rocks. The geologic history of the Flood is clearly seen, including the billions of fossils entombed in the rocks on every continent. Time and again, science confirms the historical accuracy of the Bible.

Chapter 8 ✦ 177

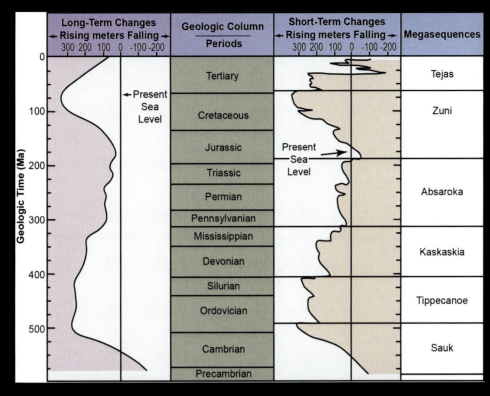

Secular chart showing presumed geologic time, global sea level, and the six megasequences (after Vail and Mitchum[8])

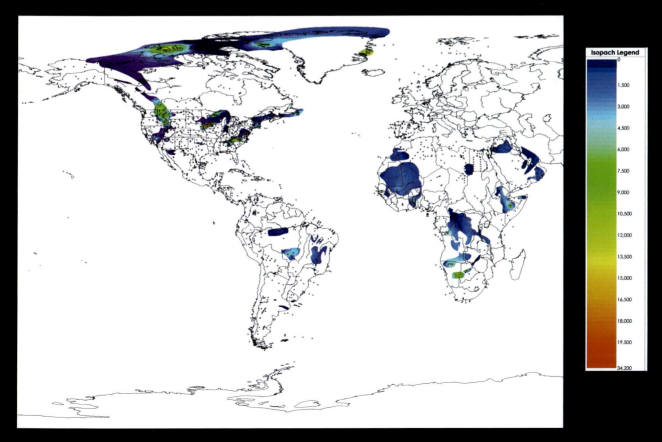

Figure 8.1. Pre-Sauk (Precambrian) thickness map for the three continents in this study. Measurement is in meters.

The Seventh Megasequence: The Pre-Sauk

Our results demonstrate the presence of a seventh megasequence below the six common fossil-bearing megasequences. There is no official name for this sequence, but it is found just below the Sauk Megasequence in what secularists call the latest Pre-cambrian, or Proterozoic Era. However, this seventh sequence was likely instrumental to the onset of the Flood.

Figure 8.1 shows the extent and thickness of the pre-Sauk across the world where it has been mapped (North America, South America, and Africa). Some of these layers are volcanic rocks and some are sedimentary units. We will examine each of the continents individually, and then we will discuss the more global implications.

Plate Configuration for the Pre-Sauk

Figure 8.2 shows an approximate pre-Sauk continental configuration, based partly on secular models (i.e., Christopher Scotese's work) and also on interpretations using our own global stratigraphic data set. The result was a slightly modified Pangaea. I will discuss more about this configuration later. Keep in mind that this map shows modern

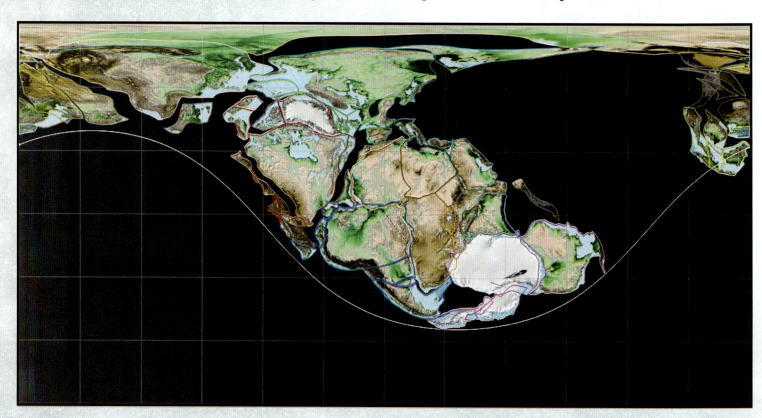

Figure 8.2. Pre-Flood configuration of the continents. Modern terrains, mountains, and ice sheets are shown for identification only.

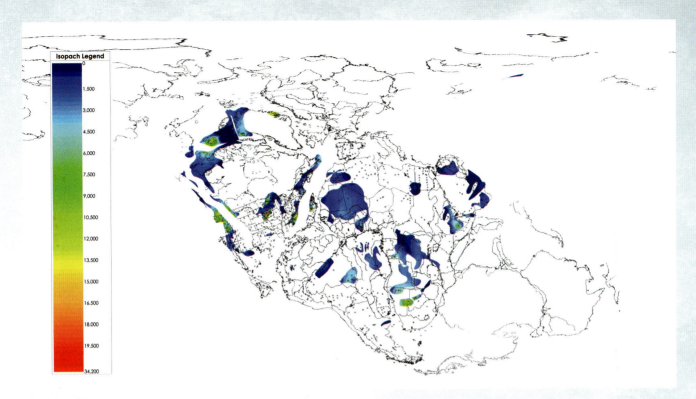

Figure 8.3. Pre-Sauk thickness map projected on the pre-Flood configuration of Pangaea. Measurement is in meters.

mountain ranges and ice sheets that did not exist during the deposition of the pre-Sauk. We include them to allow better identification of the various landmasses.

Figure 8.3 is the same continental configuration but shown with the thicknesses and exposures of the pre-Sauk sediments and volcanic rocks described above. Note that the only continents we have completed to date are North America, South America, and Africa, along with the Middle East.

Pre-Sauk Across North America

Figure 8.4 shows the thickness of the pre-Sauk across North America. Figure 8.5 shows the same map but contoured to differentiate the volcanic rocks and sedimentary rocks. Much of the pre-Sauk in North America is likely a combination of pre-Flood (most likely post-creation week) and early material derived from the fountains of the great

Pre-Sauk thickness map projected onto a globe

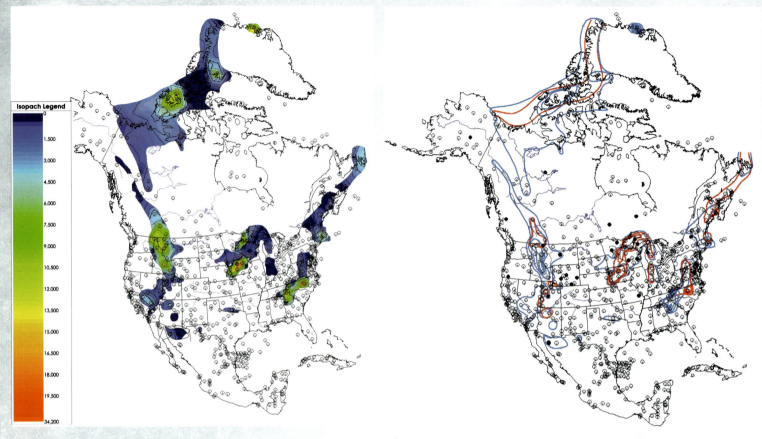

Figure 8.4. Pre-Sauk thickness map for North America. Measurement is in meters.

Figure 8.5. Contoured thickness map of pre-Sauk rocks for North America. Blue is sedimentary rocks and red is volcanic rocks. Contour interval is 5 km (3.1 mi).

deep at the onset of the Flood. In many locations, including the Belt Supergroup of Montana, it is difficult to determine the exact boundary between the pre-Flood material and any early Flood-related material from the activity of the fountains. There are likely more volcanic and sedimentary rocks than shown in Figure 8.5 since much of the pre-Sauk rock remains unexposed or unpenetrated by boreholes.

Note the three major areas with thick pre-Sauk material in this map. Most of the pre-Sauk Megasequence across North Americas exhibits linear patterns, indicative of plate boundaries. It seems likely these three areas were also affected by the Flood fountains because of this linear pattern and the evidence of extensive volcanic activity in each of the three locations. There were likely more extensive pre-Flood soils and sediments across much of North America, but the erosive power of the early Flood event seems to have stripped the land down to the basement rocks (possibly the creation week rocks) between the three primary pre-Sauk locations shown in Figure 8.4.

Across the Upper Midwest of the U.S., running from Lake Superior to Nebraska and across the Lower Peninsula of Michigan to Indiana, is an area known as the Midcontinent Rift. A second location is a linear pattern of the pre-Sauk that parallels the East Coast of the North American continent. And the final location is a linear trend that extends up the American West and continues upward to northernmost Canada and Greenland.

The Grand Canyon Supergroup in the American Southwest, for example, is tilted and deformed compared to the Sauk Megasequence rocks deposited above, which are still quite horizontal. This tilting likely occurred at the onset of the Flood event as this area was near the edge of the pre-Flood continent of North America. It seems likely the deformation occurred rapidly, was quickly eroded, and then the Sauk Megasequence was deposited shortly thereafter over a fairly extensive planation surface. Again, like the Belt rocks, the Grand Canyon Supergroup rocks are most likely post-creation week and pre-Flood sediments. The

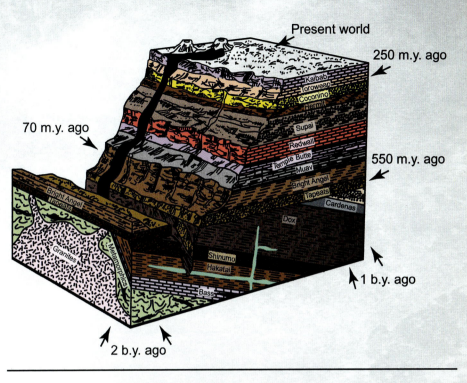

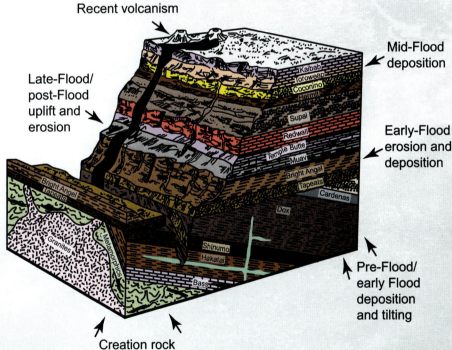

Secular view of Grand Canyon (top) and a Flood view (bottom). Most of Grand Canyon was carved in the receding phase of the Flood prior to Pleistocene lava flows pouring over the side.

lack of hard-shelled fossils like trilobites and brachiopods in either the Grand Canyon or the Belt Supergroups also indicates these strata are mostly pre-Flood. However, volcanic rocks (Cardenas Basalt) in the Grand Canyon Supergroup may be a product of the earliest Flood fountains, indicating that some of these units formed after the onset of the Flood, including the Sixtymile Formation. Furthermore, as both the Belt and Grand Canyon Supergroup rocks are near the pre-Flood continental margin, it is probable that these areas were pre-Flood depositional centers and possibly even river deposits along the edge of the continent.

There is ample evidence of pre-Sauk volcanism and rifting and sedimentation down the middle of the North American continent—the Midcontinent Rift. Here, the pre-Sauk produced a tremendous outpouring of basaltic lava that split open central North America.[9] This north-south fracture extends over 1,800 miles across what is now Lake Superior all the way to Kansas. It produced nearly 500,000 cubic miles of

Fuego Volcano Lava Flow, Antigua, Guatemala

lava up to 18 miles deep and yet was less than the width of the Red Sea.[9,10]

Kenneth Klewin and Steven Shirey found that the earliest basalts that poured from the Midcontinent Rift were enriched in the element neodymium (Nd), commonly produced as a byproduct of radioactive decay of samarium (Sm).[11] They interpreted that the lavas enriched in Nd were sourced from the melting of the uppermost mantle beneath the continent. They also concluded that later lava flows, those that were more depleted in neodymium, likely originated from deeper in the mantle. These findings support the notion that radioactive isotopes like Sm-147 were concentrated in great abundance at the base of the continental crust prior to the Flood. The decay of these elements may have even contributed to the hotter temperatures in the upper mantle as described above.

The Midcontinent Rift contains more than lava flows. It also holds the world's largest unmined copper-nickel deposit within its confines in Minnesota.[9] And between 1845 and 1968, over 11 billion pounds of copper were mined from Michigan's Keweenaw Peninsula.[12] Tremendous volumes of superheated groundwater, estimated

Pahoehoe lava flow within the Midcontinent Rift, Canada

Basaltic lava flow from the Midcontinent Rift eruptions, Lake Superior region

at temperatures of 430°F, are interpreted to have followed the eruption of lavas and assisted in emplacement of the copper deposits within the lava flows.[12] Could this be the fountains of the great deep mentioned in Genesis 7:11? Dr. John Reed believed this.[13] His synopsis of the Midcontinent Rift led to the conclusion that this was one of the fountains.

Following the catastrophic extrusion of lava within the Rift, partial subsidence of the center of the Rift system allowed accumulation of up to several miles of sedimentary strata locally (mostly shale, sandstone, and conglomerate) in a linear basin centered along the Rift valley.[10] These sedimentary layers, although devoid of visible fossils, are shown by physical contact to be a pre-Sauk sequence, as are the volcanic rocks.

Further collapse of the apex of the Rift system caused an inversion of the normal faulting (caused by extension) along the margins of the Rift valley near Lake Superior, producing steep reverse faults with vertical offset of over a mile (Keweenaw Fault) and creating steeply folded lava beds.[14] For lack of a better explanation, secular scientists have proposed that these reverse or compressional faults within an extensional system formed because of the collision of North America with another continent to the east, terming this event the Grenville Orogeny.[15] They speculate this collision prevented the Midcontinent Rift from opening further and activity eventually waned.

Some creation scientists interpret the Midcontinent Rift as part of the creation week events of Day 3, when God created the continents and separated the ocean from the land.[16] However, the Midcontinent Rift stands in stark contrast to the older Precambrian terranes that are likely from the creation week. John Reed has pointed out that the Midcontinent Rift cuts across at least four, and possibly five, earlier Precambrian terranes, demonstrating that it is the most recent event and trends contrary to all earlier rock fabrics.[13]

This research also demonstrates that the Midcontinent Rift played a primary role in the topography of the North American continent during the Flood year. The volcanic rift valley and pre-Sauk sediments formed merely at the apex of a larger, uplifted area, which I have termed Dinosaur Peninsula.[17] Flood deposition for the next six megasequences had to detour around this uplifted area, as demonstrated by thickening of the various megasequences on either side of Dinosaur Peninsula.

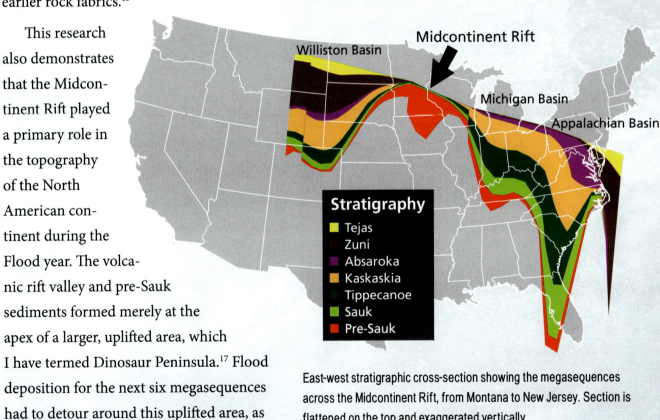

East-west stratigraphic cross-section showing the megasequences across the Midcontinent Rift, from Montana to New Jersey. Section is flattened on the top and exaggerated vertically.

In addition to the Midcontinent Rift, we also observe evidence for massive rifting along the edge of northernmost Canada and Greenland, up the East Coast, and also along parts of the American West. Note that these deposits, both volcanic and sedimentary, mark the likely eastern and western edges of the pre-Flood North American continent. There was apparently no pre-Flood Washington and Oregon or northern California. The continent ended where the isopach map shows the pre-Sauk along the American West. The modern states of Washington, Oregon, and much of California were added or accreted during the runaway subduction process and the formation of the Pacific Ocean. There will be more about this in chapters 12 and 13.

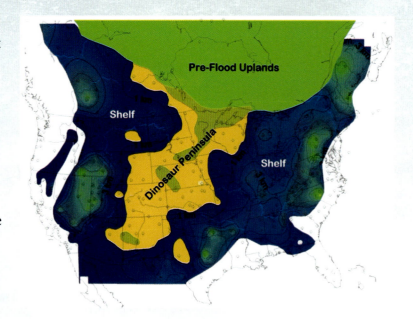

Pre-Flood geography of the United States

It is also possible that Florida and the southeasternmost section of the U.S. were not originally part of North America. This piece of continental crust may have been part of westernmost Africa that was added to the U.S. during the closure of a pre-Flood ocean, forming the Appalachian Mountains. Later, when North America was wrenched away from Africa during the Absaroka Megasequence (chapter 12), this piece stayed adhered to North America.

Numerous secular geologists have also identified a Late Precambrian-Early Cambrian (pre-Sauk) rift system along what we believe to be the pre-Flood West Coast of North America.[18,19] Their evidence entails relatively thin mafic basalt flows within the Neoproterozoic section in western Utah and eastern Nevada and north into the Canadian Rockies, along with overlying thick Cambrian (Sauk) sediments (Figure 8.5).

Pre-Sauk Across South America

Figure 8.6 is a thickness map of the pre-Sauk across South America. In contrast to North America, there is very little volcanic activity evident in the pre-Sauk Megasequence in South America. Most of the pre-Sauk is exposed across the Brazilian Shield,

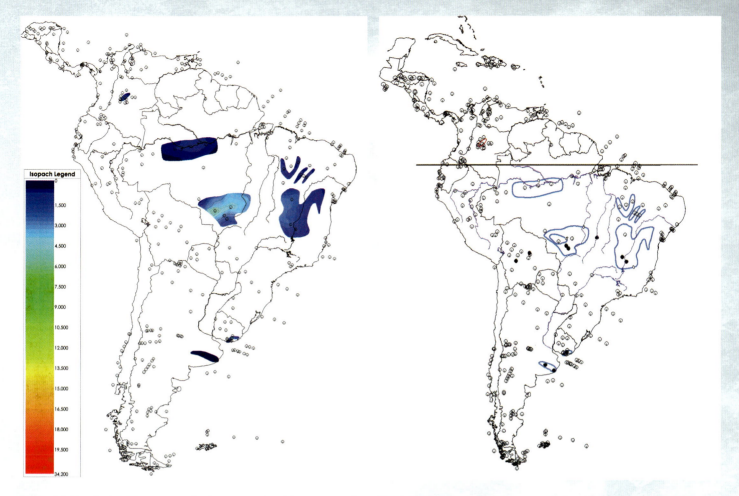

Figure 8.6. Pre-Sauk thickness map for South America. Measurement is in meters.

Figure 8.7. Contoured thickness map of pre-Sauk rocks for South America. Blue is sedimentary rocks and red is volcanic rocks. Contour interval is 5 km (3.1 mi).

including the Amazonian Craton of Central Brazil and the São Francisco Craton of easternmost Brazil. Undoubtedly there is more unexposed pre-Sauk in areas that are now covered by Flood sediments and that were not penetrated by well bores, but the basic pattern is visible in Figure 8.6.

The lack of any extensive volcanism in the pre-Sauk across South America indicates that early Flood fountains and/or the formation of plate boundaries did not directly affect the continent until later in the Flood (Figure 8.7). However, we do know that the South American and African continents eventually split during the deposition of the Zuni Megasequence. Also note there are no pre-Sauk sediments or volcanic rocks in southern Central America since this narrow landmass did not exist until later in the Flood, forming as part of the volcanic activity associated with subduction. This will be discussed later in chapter 13.

Some of the preserved pre-Sauk material in South America is found in small rift valleys, as shown in eastern Argentina and eastern Brazil. These areas likely are remnants of more extensive pre-Flood sediments that blanketed much of the continent. The erosive power of the early Flood probably stripped off the soils and thin sediments that had built up across much of the pre-Flood landmasses, similar to what apparently occurred across North America. This erosion was likely a global phenomenon, removing and destroying much of the pre-Flood soil and many sedimentary deposits across every continent. It is only in rift valleys and areas that were somehow protected from the onslaught that any pre-Flood material is preserved.

Pre-Sauk Across Africa and the Middle East

Figure 8.8 is a thickness map of the pre-Sauk Megasequence across Africa and the Middle East. Again, similar to South America, and in contrast to North America, there is very little pre-Sauk volcanism across Africa. The pre-Sauk is composed of nearly all sedimentary material. This implies that the earliest Flood fountains and the formation of plate boundaries did not directly impact the continent of Africa. However, there are indications of a plate boundary and possible rifting on the northeast edge of the Saudi Arabian Peninsula. Here we find thick deposits of pre-Sauk salt. Unfortunately, at the

Landscape of the Great Rift Valley, Tanzania and Kenya

time of writing I have not finished compiling the column data across Asia to know the full extent of this deposit. Nonetheless, thick salt deposits are usually a sign of rifting. The salt was possibly generated by hydrothermal activity associated with the rifting.[16] We observe rift-related salt deposition in many locations globally, including the deposition of several thousand feet of Louann Salt during the rift-opening of the Gulf of Mexico (Zuni Megasequence) and over 10,000 feet of salt in the rift-bounded Red Sea (Tejas Megasequence).

Fairly extensive pre-Sauk deposits are preserved in Taoudeni Basin of West Africa (West African Craton) and in the Congo Basin area in Central Africa (Kasai Craton) (Figure 8.8). Many of these sedimentary rocks contain extensive stromatolites and are likely post-creation week and pre-Flood. Other areas where pre-Sauk deposits are preserved are in rift valleys that formed during the Flood event, causing the pre-Sauk to be down-dropped along faults and protected from erosion. Some of these are illustrated by the linear patterns in the pre-Sauk exposures in southernmost Africa in Figure 8.8. Figure 8.9 shows there was some volcanic activity during the pre-Sauk in eastern Africa.

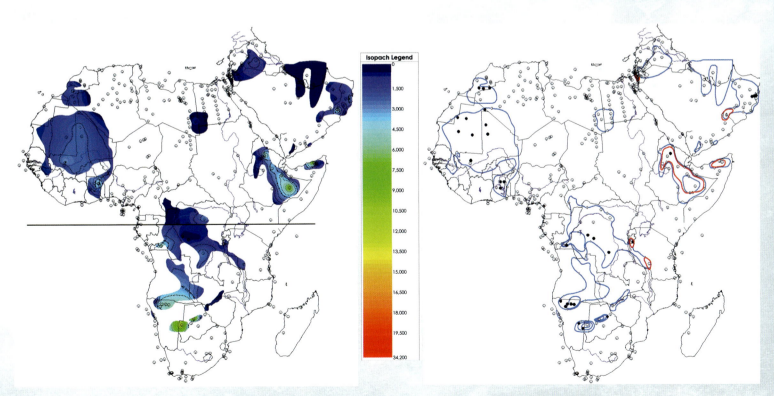

Figure 8.8. Pre-Sauk thickness map for Africa. Measurement is in meters.

Figure 8.9. Contoured thickness map of pre-Sauk rocks for Africa. Blue is sedimentary rocks and red is volcanic rocks. Contour interval is 5 km (3.1 mi).

Again, like North America and South America, it seems probable there were much more extensive pre-Flood soils and sedimentary deposits across Africa. Erosion during the Flood event likely removed much of this pre-Flood material.

How Did Subduction Begin?

To move the tectonic plates and split apart continents like Pangaea requires a combination of an equal amount of seafloor spreading and subduction. I do not think the earth changed size appreciably during the Flood event. The volume of new seafloor created at the ridges must therefore balance with the volume that was subducted. One of the great mysteries in geology is trying to explain how subduction actually begins. Sometimes this is referred to as *subduction initiation*. Any answer must begin with a plate boundary of some sort. Unfortunately, the secular community has reached no consensus as to which type of boundary. But without the formation of global plate boundaries, subduction cannot initiate.

Some have suggested it begins with rifting and the opening of long linear cracks in the crust, allowing more dense oceanic plate to drop by gravity into the rift. Pre-Sauk volcanism and rifting along each coast of North America may have caused the pre-Flood ocean crust and continental crust to rift, buckle, and split just prior to subduction initiation. Many of the Proterozoic sediments in these rift valleys along each coast may be evidence of this earliest, deformational phase of the Flood.

Recent studies are suggesting that transform boundaries may also be a way to initiate subduction.[20,21] Either way, there must be a large crack of sufficient length to allow denser and colder oceanic lithosphere at the surface to begin sliding down into the less dense and hotter mantle. However the process of subduction initiation began, the first day of the Flood event likely provided the cracks necessary for initiation.

Human Perspective

If you were a human at this point in the Flood who was not on the Ark, you would likely have noticed some major things happening to the world. One, you would have probably felt your first earthquakes as portions of the crust rifted open all over the earth. Some of these quakes would have been much larger in magnitude than any we experience today. Second, you would have noticed a lot of volcanic activity in the newly formed rifts if you were close enough to observe one. There was probably not much in the way of ash being produced at this point, as would be expected in a subduction

zone volcanic eruption. Instead, there was likely a lot of flowing lava and gases shooting out of the ground similar to geysers. Finally, an intense rainfall event began to occur globally. This rain may have been tied directly or indirectly to the water and gases shooting out of the rifts (the fountains). The Bible doesn't provide details on this, but it is clear that there was an intense rain that began at the same time that the fountains erupted (Genesis 7:11-12). Other than feeling and witnessing the effects of these new and devastating disasters, life would have gone on uninterrupted for most humans at this point. However, I would assume that people who had heard Noah preach about the coming Flood were getting gravely concerned.

Conclusion: Explaining the Purpose of the Fountains

The bottom line is that there can be no plate motion unless the plates are first formed. That was apparently one of the purposes of the first day of the Flood: to make the initial plate boundaries. The fountains most likely released a lot of gases and water at the onset of the Flood event. Some of this water obviously contributed to the flooding of the earth. But a secondary purpose of the fountains was to create the plate boundaries for what would initiate the runaway subduction process. It was subduction and the simultaneous formation of an entirely new seafloor during the Flood year that pushed the water higher and higher as described in chapter 6.

References
1. Brandl, P. A. et al. 2013. High mantle temperatures following rifting caused by continental insulation. *Nature Geoscience*. 6 (5): 391-394.
2. Austin, S. A. et al. 1994. Catastrophic Plate Tectonics: A Global Flood Model of Earth History. In *Proceedings of the Third International Conference on Creationism*. R. E. Walsh, ed. Pittsburg, PA: Creation Science Fellowship, 609-621.
3. Langmuir, C. 2013. Older and hotter. *Nature Geoscience*. 6 (5): 332-333.
4. Paulsen, T. et al. 2017. Evidence of a spike in mantle carbon outgassing during the Ediacaran period. *Nature Geoscience*. 10 (12): 930-934.
5. Clarey, T. 2015. Grappling with Megasequences. *Acts & Facts*. 44 (4): 18-19.
6. Brune, S., S. E. Williams, and R. D. Müller. 2017. Potential links between continental rifting, CO_2 degassing and climate change through time. *Nature Geoscience*. 10 (12): 941.
7. Clarey, T. 2016. Embracing Catastrophic Plate Tectonics. *Acts & Facts*. 45 (5): 8-11.
8. Vail, P. R. and R. M. Mitchum Jr. 1979. Global cycles of relative changes of sea level from seismic stratigraphy. *American Association of Petroleum Geologists Memoir*. 29: 469-472.
9. Marshall, J. 2013. North America's broken heart. *Nature*. 504 (7478): 24-26.
10. Cannon, W. F. 1992. The Midcontinent rift in the Lake Superior region with emphasis on its geodynamic evolution. *Tectonophysics*. 213 (1-2): 41-48.
11. Klewin, K. W. and S. B. Shirey. 1992. The igneous petrology and magmatic evolution of the Midcontinent rift system. *Tectonophysics*. 213 (1-2): 33-40.
12. Bornhorst, T. J. and L. D. Lankton. 2009. Copper Mining: A Billion Years of Geologic and Human History. In *Michigan Geography and Geology*. R. Schaetzl, J. Darden, and D. Brandt, eds. New York: Pearson Custom Publishing.
13. Reed, J. K. 2000. *The North American Midcontinent Rift System: An interpretation within the Biblical worldview*. Creation Research Monograph Series No. 9. St Joseph, MO: Creation Research Society Books, 154.
14. Hinze, W. J. et al. 1992. Geophysical investigations and the crustal structure of the North American Midcontinent Rift system. *Tectonophysics*. 213 (1-2): 17-32.
15. Cannon, W. F. 1992. Speculations on the origin of the North American Midcontinent rift. *Tectonophysics*. 213 (1-2): 49-55.
16. Snelling, A. A. 2009. *Earth's Catastrophic Past: Geology, Creation & the Flood*, vol. 2. Dallas, TX: Institute for Creation Research, 493-499.
17. Clarey, T. 2015. Dinosaur Fossils in Late-Flood rocks. *Acts & Facts* 44 (2): 16.
18. Bond, G. C. et al. 1985. An early Cambrian rift to post-rift transition in the Cordillera of western North America. *Nature*. 315: 742-746.
19. Colpron, M. et al. 2002. U-Pb zircon age constraint for late Neoproterozoic rifting and initiation of the lower Paleozoic passive margin of western Laurentia. *Canadian Journal of Earth Sciences*. 39: 133-143.
20. Kim, G. et al. 2018. Transition from buckling to subduction on strike-slip continental margins: Evidence from the East Sea (Japan Sea). *Geology*. 46 (7): 603-606.
21. Zhou, X. et al. 2018. Subduction initiation dynamics along a transform fault control trench curvature and ophiolite ages. *Geology*. 46 (7): 607-610.

9 Rising Water: Sauk Megasequence

> **Summary:** In the first 40 days of the Flood, the fountains of the deep formed the tectonic plates, which began to subduct into the mantle. These subductions caused tsunamis over the continents that laid down the first major megasequence: the Sauk, which includes the Great Unconformity and the Cambrian Explosion. Since most of the fossils in the Sauk are marine, these early waves likely only buried the shallow seas of the pre-Flood world.

Initial Flooding of the Continents

Sometime in the first 40 days of the Flood, after God caused the fountains of the great deep to burst open, some of the newly formed plates began to fold and buckle near their edges, and dense oceanic lithosphere began to subduct into the mantle. This subduction process, possibly limited to just a couple of locations at first, began to generate tsunamis. These high-velocity water waves came crashing across the continents, bringing sediment and debris and burying many hard-shelled organisms in great numbers. However, it appears that the earliest wave pulses were limited to locations that may have been pre-Flood shallow seas because nearly all of the fossils found in this megasequence are marine organisms. There will be a more thorough discussion on this and the rationale behind it in chapter 17.

For now, we will examine the global distribution of the Sauk Megasequence and its rock types. The Sauk consists of strata that begin at the base of the Cambrian system and extend upward through the Lower Ordovician system. We will first describe the actual rocks in place, including the rock types, thicknesses, the total volume of strata observed, and the total surface areas covered for each continent.

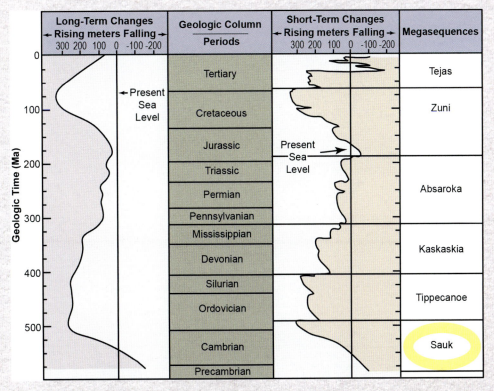

Secular chart showing presumed geologic time, global sea level, and the six megasequences (after Vail and Mitchum[1])

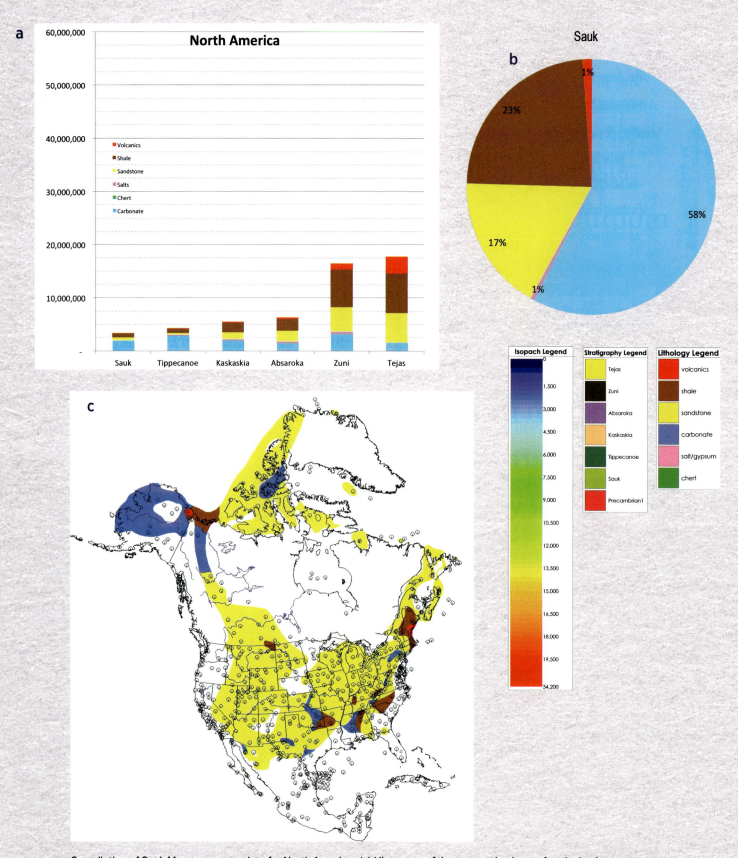

Compilation of Sauk Megasequence data for North America: (a) Histogram of the type and volume of rocks in place across North America for each megasequence; (b) pie chart of the total Sauk Megasequence by rock type; (c) map of the bottommost (basal) Sauk rock type; (d) thickness map of the total Sauk Megasequence; (e) stratigraphic cross-sections showing the relative thickness of the Sauk Megasequence. Histogram measurement is in cubic kilometers, isopach map measurement is in meters.

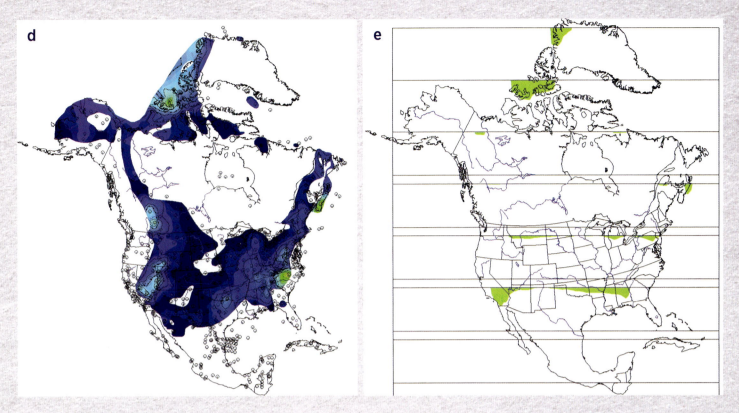

What Do the Rocks Show?

North America

Across North America, this megasequence has the most extensive sandstone layer at its base compared to all subsequent megasequences. However, much of this sandstone layer is very thin, often less than 100 meters. This is especially true along the NE-SW-trending Transcontinental Arch (which I have termed Dinosaur Peninsula) that runs from Minnesota to New Mexico. Here, the Sauk Megasequence thins to just a few tens of meters in many places or is nonexistent. The thickest deposits of the Sauk Megasequence are found along the pre-Flood eastern and western edges of the North American continent and in northernmost Canada, with thicknesses exceeding 3 kilometers. Although there is an extensive blanket sandstone (Tapeats equivalent) at the base of the megasequence, sandstones only make up about 17% of the entire Sauk sequence. The shale and limestone layers on top of the basal sandstone unit are much thicker, making up 23% and 58% of the Sauk, respectively. Clearly, carbonate rocks like limestone and dolomite dominated the Sauk Megasequence across North America. The Sauk Megasequence across North America covers approximately 12 million square kilometers and averages only about 275 meters thick. This is the thinnest (and least volume) of all subsequent megasequences. The total volume of sediment within the Sauk Megasequence is about 3.3 million cubic kilometers.

198 ⋆ *Carved in Stone*

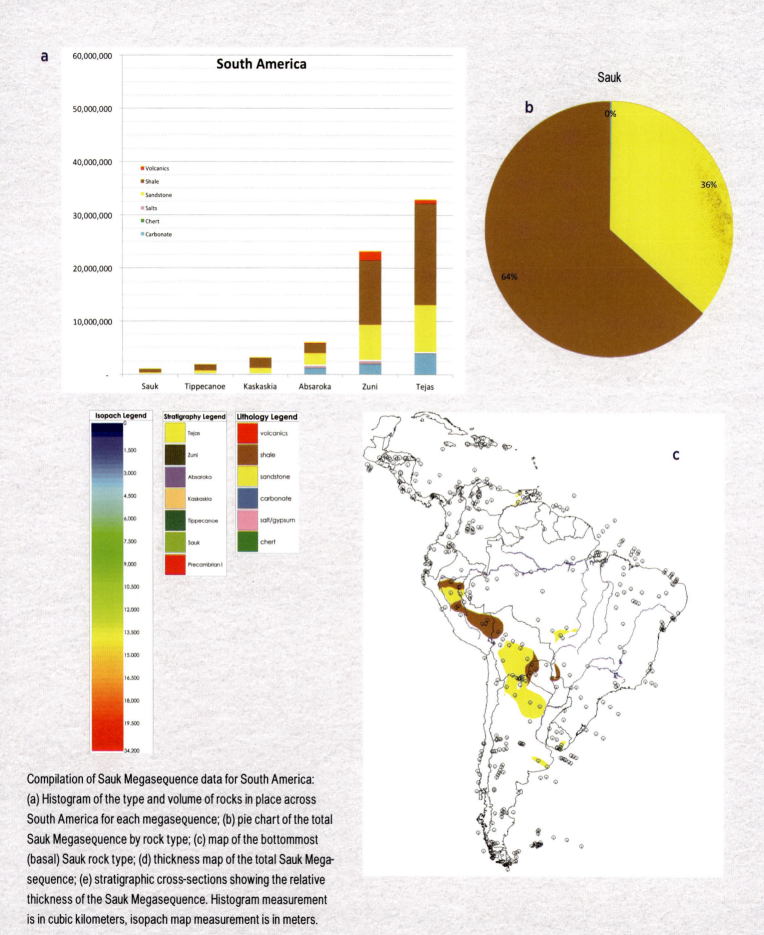

Compilation of Sauk Megasequence data for South America: (a) Histogram of the type and volume of rocks in place across South America for each megasequence; (b) pie chart of the total Sauk Megasequence by rock type; (c) map of the bottommost (basal) Sauk rock type; (d) thickness map of the total Sauk Megasequence; (e) stratigraphic cross-sections showing the relative thickness of the Sauk Megasequence. Histogram measurement is in cubic kilometers, isopach map measurement is in meters.

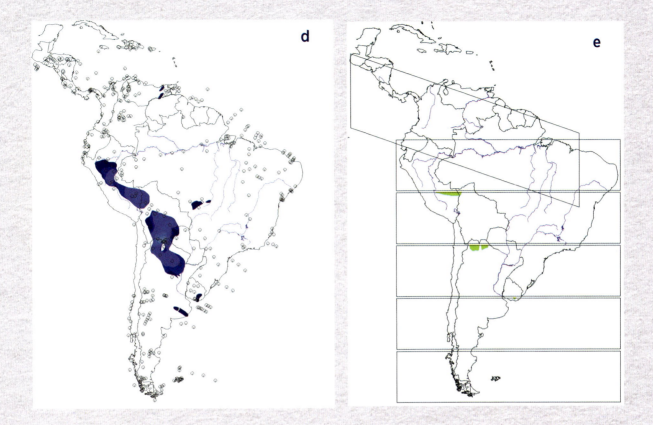

South America

The Sauk Megasequence across South America is limited in areal extent to the central-western part of the continent. The base of this megasequence consists of a mixed sand and shale unit that covers only a small part of South America just to the east of the present-day central Andes Mountains. In addition, there are a few small, isolated Sauk outliers at other locations across the continent. However, much of South America shows no evidence of Sauk deposition. Siliciclastics (sand and shale) dominate the strata of the Sauk Megasequence, with less than 1% composed of non-clastic material like carbonates. Approximately 64% of the total rock volume is shale and about 36% is sandstone. The Sauk Megasequence covers an area of only 1.4 million square kilometers across South America, the least surface area of any megasequence. Where present, the Sauk Megasequence is spread relatively thin, averaging about 700 meters thick. In addition, the volume of Sauk sediment is also the least of any megasequence at a little over 1 million cubic kilometers.

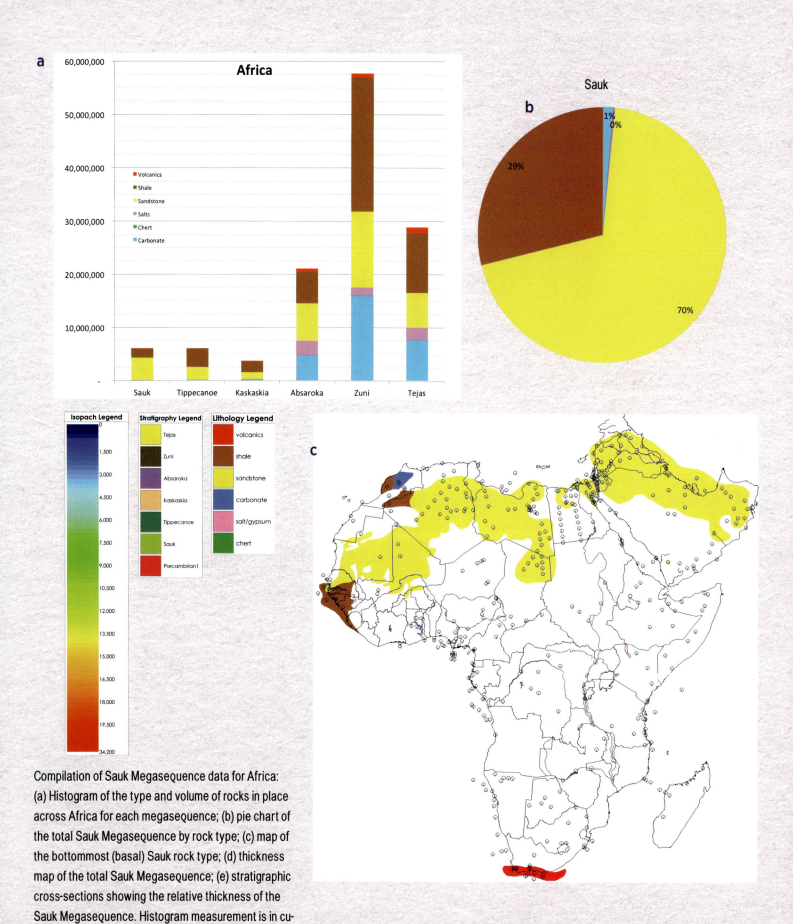

Compilation of Sauk Megasequence data for Africa: (a) Histogram of the type and volume of rocks in place across Africa for each megasequence; (b) pie chart of the total Sauk Megasequence by rock type; (c) map of the bottommost (basal) Sauk rock type; (d) thickness map of the total Sauk Megasequence; (e) stratigraphic cross-sections showing the relative thickness of the Sauk Megasequence. Histogram measurement is in cubic kilometers, isopach map measurement is in meters.

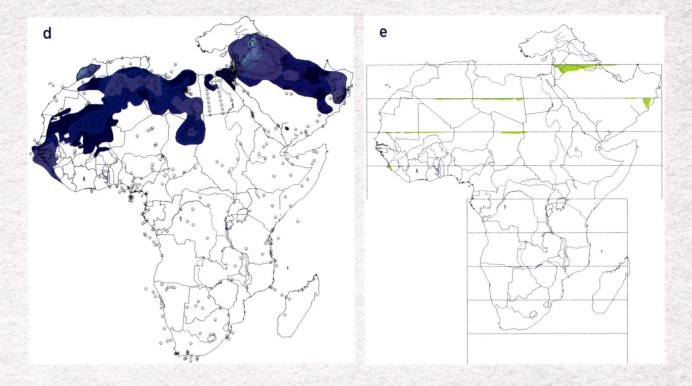

Africa

Across Africa, this megasequence consists of an extensive blanket sandstone that covers much of North Africa and the Saudi Arabian Peninsula. However, much of the rest of Africa shows no evidence of Sauk deposition. Most of the Sauk is spread relatively thin, only exceeding a thickness of a kilometer in isolated locations. The thickest depo-center is in the northwestern Saudi Arabian Peninsula, where the Sauk exceeds 3 kilometers thick. Sandstones dominate the strata of the Sauk sequence, with 69.6% of the total rock volume. Shale is the next-most common rock in the Sauk sequence, at 28.9%. Carbonate layers are rare and thin by comparison to clastic layers. The Sauk Megasequence is spread across nearly 9 million square kilometers, mainly in North Africa and the Middle East. It averages about 675 meters in thickness and has a total volume of just over 6 million cubic kilometers.

Current Data for the Sauk Megasequence

Dr. Art Chadwick and his students at Southwestern Adventist University have compiled current direction data for many years across the world.[2] They plotted 500,000 cross-beds, scour marks, and ripples from 15,000 locations globally that preserve the paleocurrent directions in the sedimentary rocks. These data were then represented by arrows indicating current directions on global maps. Figure 9.1 shows plots of the dominant current directions for Cambrian system rocks across North America, South America, and Africa, which is the bulk of the Sauk Megasequence.

The North American data show a dominant southwesterly direction across much of the continent. Little data were acquired across South America in the Sauk Megasequence, partly because there is so little Sauk deposited in South America. Whereas in Africa, the Cambrian rocks show a dominant northerly current direction across the north half of the continent.

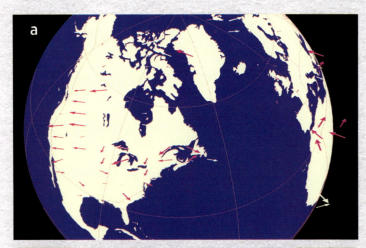

Figure 9.1. Current directions compiled across (a) North America, (b) South America, and (c) Africa for the Cambrian system (Lower Sauk). Courtesy of Art Chadwick.

Summary of the Rock Data

The Sauk includes the Great Unconformity and the Cambrian Explosion. The Great Unconformity is as mysterious to secular geologists as the Cambrian Explosion, which was discussed previously in chapter 5. Recall, the Cambrian Explosion is the location in the rock record where prolific life of all sorts suddenly shows up as fossils. This occurred in the Sauk Megasequence, particularly in the Cambrian system in most locations. Fossils exploded onto the scene and in rocks for the first time.

The Great Unconformity is the erosional contact at the base of the Sauk Megasequence in many locations globally, including Grand Canyon. In fact, the term was first used to describe the contact between the Cambrian Tapeats Sandstone and the underlying Precambrian metamorphic rocks of the Vishnu Schist and/or the tilted

The Great Unconformity, Grand Canyon

Precambrian sediments of the Grand Canyon Supergroup.[3]

The Great Unconformity has been traced globally across every continental mass, "making it the most widely recognized and distinctive stratigraphic surface in the rock record. Thus, the great Unconformity marks the termination on an extended period of continental denudation [erosion] that exhumed and exposed large areas of igneous and metamorphic rocks to subaerial weathering before marine trans-

> "The Great Unconformity has been traced globally across every continental mass, 'making it the most widely recognized and distinctive stratigraphic surface in the rock record.'"

Cambrian brachiopod

gression and subsequent sedimentation [deposition of the Sauk Megasequence]."[4]

In other words, to make the Great Unconformity, much of the pre-Sauk surface material (and some of the pre-Flood sedimentary rocks) must be stripped off large segments of a majority of the world's continents simultaneously, before the deposition of the Sauk sequence. However, as described above, the extent of the Sauk Megasequence was limited to selected locations on each of the continents, such as across North Africa. In addition, most of the fossils found in the Sauk are shallow marine organisms. This implies that the Sauk transgression likely eroded and scoured areas that were shallow seas in the pre-Flood world. I do not think the 40 days of intense rain described at the beginning of the Flood removed all soil/vegetation in the pre-Flood land areas. If it had, we would see more plants and coal layers in the Sauk Megasequence, and probably more land animals too. Instead, we see virtually no land plants or land animals in the Sauk rocks. There will be more on this in chapter 17.

Continental Configuration for the Sauk Megasequence

The continents in the Sauk are essentially the same as the configuration in the preceding pre-Sauk Megasequence. Any subduction and plate motion that had begun had not yet moved the continents appreciably. Figure 9.2 shows the Pangaea-like configuration we have developed over the course of this study for the onset of the Flood.

Figure 9.3 shows the extent and thickness of the Sauk Megasequence across the three continents completed in this study and in the proper Pangaea-like configuration of Figure 9.2.

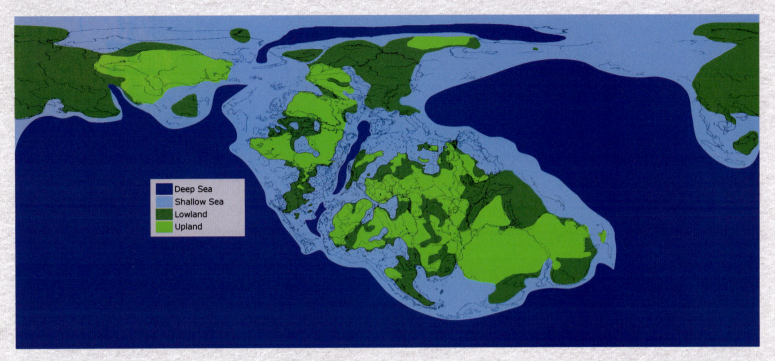

Figure 9.2. The global continental configuration for pre-Flood Pangaea

Solving the Mystery of the Paleozoic Intracratonic Basins

There are numerous Paleozoic-age basins (massive geologic synclines) on every continent that all seemed to have formed at once, globally. Many of these began to form during the Sauk Megasequence since we see thickening of some of the Sauk Megasequence rocks in the center of these basins. Most of these basins are not near plate boundaries, so conventional plate tectonic theory fails to explain their origin. They remain a geologic enigma. Let's examine what the secular geologists have concluded, and then I will speculate on a possible Flood origin for these huge basins.

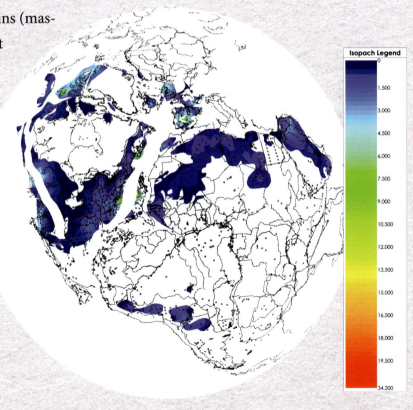

Figure 9.3. Continental configuration during the deposition of the Sauk Megasequence and its sedimentary thickness. Measurement is in meters.

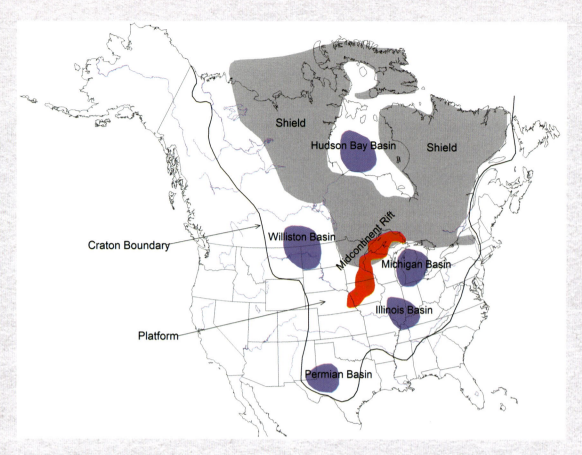

Figure 9.4. Map of North America showing the extent of the craton boundary (black line), the five intracratonic basins in blue, the sedimentary platform (colorless), and the exposed shield area (shaded). In red is the western and most active limb of the Midcontinent Rift.

The secular literature explains the Sauk Megasequence as a marine transgression caused as a consequence of crustal subsidence.[5-7] However, this explanation calls upon subsidence of the crust due to thermal contraction and cooling following simultaneous rifting events on the eastern and western margins of the continent in the Neoproterozoic (Latest Precambrian).[7] Price explains this subsidence as a response to loads imposed upon the margins that allowed the sea to flood (transgress) the continent.[6] He explained that these loads include thermal contraction and the weight of the sediments deposited along the continental margins. Secular geologists believe this Sauk transgression took 30 to 40 million years to go from the edges of the continent to the interior.[7]

To explain the later megasequences (i.e., Tippecanoe and Kaskaskia), secular science has had to call on the synchronous subsidence of the continental margins for a second and third time, along with mantle downwelling in the continental interior to create the subsidence of the intracratonic basins and flooding of the continents.[8] Unfortunately, these secular explanations fail to address how thermal subsidence along the continental margins could happen a second and third time without renewed heating and rifting along the margins to create the subsidence.

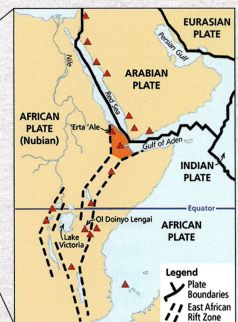

Secular science has a plate tectonic model to explain long linear rifts but not the formation of more circular intracratonic basins that are not near plate boundaries. The map above shows the geometry of the East African Rift.

Other secular explanations include flexure-related buckling from distant orogenic (mountain-building) belts and/or extensional and thermal subsidence.[9] All of these hypotheses have fallen short in their explanation of how subsidence within the normally stable craton could occur without plate tectonic influences to provide either heating or rifting to create the basins. And why do these large basins all begin to subside at nearly the same time across the entire globe?

Geologists define a geologic basin as a folded depression filled with sediment that thickens in the center, forming a roughly three-dimensional, circular structure called a *syncline*. Cratons are the relatively undeformed and more stable parts of the continental crust. In North America, the craton falls in the area between the Appalachian Mountains and the Rocky Mountains (Figure 9.4). Many creationists consider the cratonic part of the continent as forming on Day 3 of the creation week.[10] Intracratonic basins are ubiquitous to every major craton, and all seem to have formed at about the same time, beginning in the early Paleozoic Era, or during earliest episodes of Flood sedimentation.[9]

It is likely that the intracratonic basins formed by a different method altogether. Figure 9.4 shows five of the Paleozoic basins across North America. It's possible that these basins sank downward not due to thermal subsidence but to the removal of material underneath the crust. Recall, tremendous quantities of lava and water and other gases were expelled catastrophically as the fountains exploded at the onset of the Flood (i.e., Midcontinent Rift).

Any explanation has to explain the near-simultaneous occurrence of at least five major intracratonic basins in North America and more around the globe (Figure 9.4). These five basins all formed in the so-called Early Paleozoic Era (early in the Flood) during the deposition of the Sauk, Tippecanoe, and Kaskaskia Megasequences. The sedimentary fill in each basin varies from about 1.3 kilometers (1 mile) in the Hudson Bay Basin to over 3 kilometers (2 miles) in the Michigan and Illinois Basins, and up to 5 kilometers (3.1 miles) in the Permian Basin.[11]

If the Midcontinent Rift is taken as an example of the fountains of the deep, then rapid expulsion of nearly 2 million cubic kilometers of magma and unknown volumes of superheated water from the upper mantle would surely have created voids below the crust or deep in the crust, possibly influencing the subsidence of these sedimentary basins.

Parameter	Hudson Bay	Michigan	Williston	Illinois	Permian
Sediment Thickness	1.3	3.4	3.4	3.0	5.0
Basin Radius	250	250	250	200	240
Sediment Vol. (km^3)	255,000	668,000	668,000	377,000	905,000
Total Sediment Volume = 2,873,000 km^3					

Table 9.1. Data for five intracratonic basins of North America. All values are listed in km. Modified from Kaminski and Jaupart.[11]

Table 9.1 lists data from the five intracratonic basins of North America. The thickness data were taken from Kaminski and Jaupart,[11] with the exception of the Permian Basin, which was estimated from the megasequence isopach maps in this study. The size of the basins was estimated from various maps, including Wicander and Monroe.[12]

These data provide an estimated total sediment volume within the five intracratonic basins of 2.9 million cubic kilometers. Therefore, the previously estimated 2 million cubic kilometers of lava extruded from the Midcontinent Rift could easily account for a large portion of the cumulative subsidence of the five basins. It seems reasonable to conclude that the additional amount of basin subsidence (0.9 million cubic kilometers) could be accounted for by the expulsion of massive amounts of hydrothermal water following and during the eruption of the lava flows. The extent and distribution

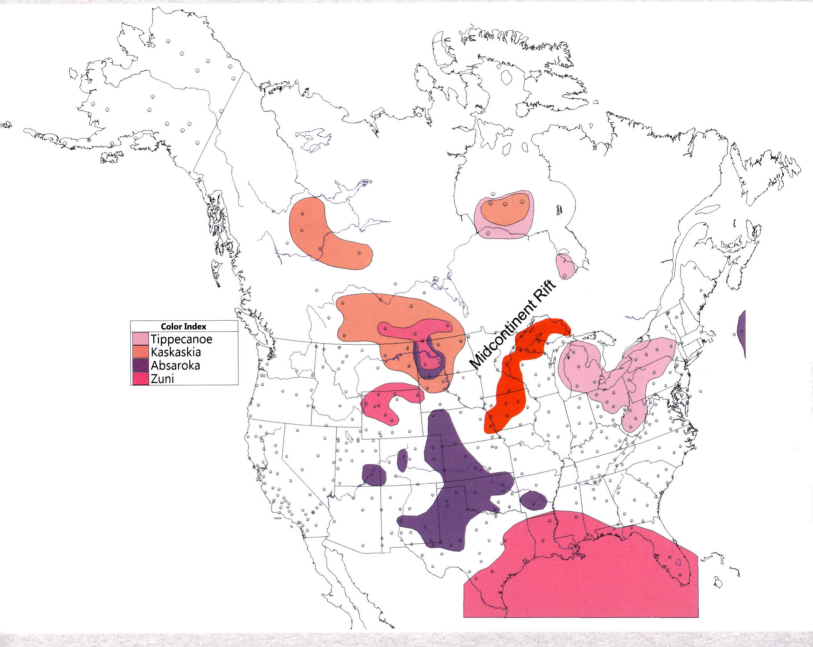

Figure 9.5. Extent of salt and/or gypsum deposits by megasequence. Midcontinent Rift is shown in red.

of the vast salt and gypsum deposits in the intracratonic basins support this hydrothermal water interpretation (Figure 9.5). Midcontinent Rift volcanism and hydrothermal activity, as one of the fountains of the great deep, seems to be a likely cause of the observed crustal subsidence. The result was the formation of the five intracratonic basins in a rough ring around the Midcontinent Rift (Figure 9.4). Admittedly, this is speculative, but it does offer a potential Flood-related explanation for the basins. Much additional research on this topic is still needed.

Resolving the Sheet Sand Enigma

Sheet sands are widespread, thin sandstones that blanket large regions of the continents. Most are composed of extremely pure quartz of uniform, well-rounded grains that contain almost no shale. Secular geologists have tried to explain their presence for decades and have failed to develop a satisfactory answer.[13] Their best models invoke "atypical depositional conditions unique to shallow epeiric seas" and "are viewed as sufficiently different from other modern and ancient sedimentary successions that some textbooks treat them as a separate category of stratigraphic unit."[14]

In other words, not only are the sands hard to explain, they fail to follow uniformitarian expectations. Many of these sheet sandstones extend for hundreds of miles and are just a few tens of feet thick. The so-called Tapeats Sandstone that blankets much of North America is an excellent example (Figure 9.6) and is found at the base of the Sauk Megasequence (Cambrian through Lower Ordovician systems).[15]

The continuity of the basal Sauk sandstone layer across North America is a testament to the Flood, specifically to the extent and uniformity of the first marine transgression of the continents. Most creationists recognize this sandstone layer as the Flood's first extensive deposit.

This same layer also extends across North Africa and the Middle East, including Jordan and the city of Petra (Figure 9.7). It can even be found across parts of South America, demonstrating that the basal Sauk sandstone layer (the Tapeats equivalent) extends across multiple continents.

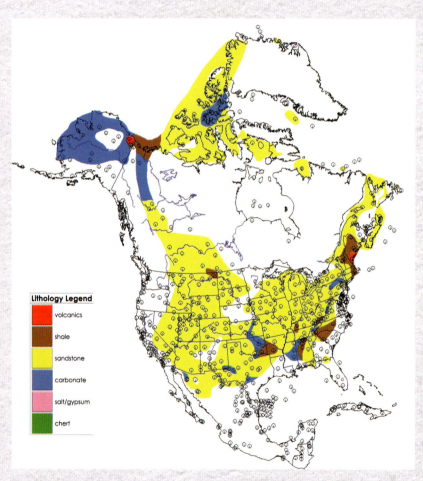

Figure 9.6. Basal Sauk Megasequence map for North America

> **"The continuity of the basal Sauk sandstone layer across North America is a testament to the Flood."**

The Tippecanoe Megasequence (Ordovician to Silurian systems) just above the Sauk also exhibits a large sheet sand at its base called the St. Peter Sandstone. This thin sandstone spreads across the midsection of North America. It can be correlated from Canada to Texas and Montana to West Virginia. A St. Peter-equivalent Tippecanoe sandstone is also found across North Africa and the Middle East in a similar location and extent as the Sauk basal sandstone. And this same sandstone is again found across parts of South America.

Extensive, thin sheet sandstone layers continue to baffle secular geologists. They have failed to develop an acceptable model to explain these widespread deposits, and yet there they are, stacked one on top of the other across multiple continents.

It appears that these geologists' refusal to take into account the history recorded in the Word of God is blinding them to the real explanation for the vast sandstone layers. Genesis clearly describes a global flood event. The Flood offers the only reasonable explanation for the thin, uniform sandstones that were deposited at the same time across multiple continents. God's Word can resolve many mysteries if we simply choose to believe it.

Human Perspective

Humans during this time would have witnessed continuous massive earthquakes caused by the rifting and the initiation of subduction and plate movement. These quakes created the devastating tsunamis that rushed across the continents. However, at this point in the Flood, the tsunamis only

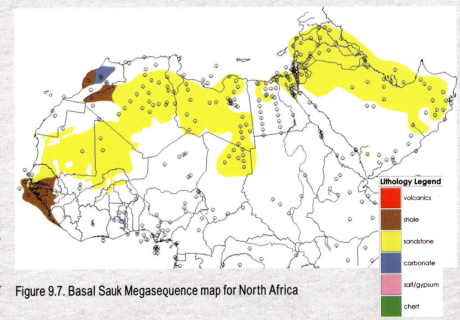

Figure 9.7. Basal Sauk Megasequence map for North Africa

affected the surrounding shallow seas and did not impact the higher land areas where people were most likely living. Humans would have also noticed that the intense rain was not letting up. Volcanism was still occurring in the rifts, along with the gases and water being expelled. Geologic disasters were getting more and more severe every day. More humans were most likely seriously regretting not listening to Noah.

> "The Flood offers the only reasonable explanation for the thin, uniform sandstones that were deposited at the same time across multiple continents. God's Word can resolve many mysteries if we simply choose to believe it."

Conclusion: The Sauk Marks the Start of the Flooding

The stratigraphic data show that a significant rise in sea level occurred across all three continents simultaneously during the Sauk Megasequence. However, the sedimentary strata of the Sauk are consistently one of the least extensive megasequences in terms of volume, surface area, and average thickness (Table 9.2). North America differs from the other two continents because the Sauk Megasequence does cover a high percentage of the continent. The basal sandstone (Tapeats and equivalent) blankets North America but is quite thin, often just a few tens of meters or less. In fact, the entire Sauk Megasequence averages the least thickness of any subsequent megasequence, at 275 meters thick.

North America is also different because the Sauk Megasequence (and the next two megasequences also) is dominated by carbonate rocks like limestone and dolomite. These are purely marine rocks. Africa and South America have about 1% or less carbonate rocks in the Sauk Megasequence. They are dominated by sandstones and shales instead. All of the continents have similar shallow marine fossils in the Sauk strata, regardless of rock type.

It seems likely that more of North America was a shallow sea in the pre-Flood world, thereby explaining the higher percentage of the carbonate rocks in the Sauk (see chapter 17). Africa and South America had less extensive pre-Flood shallow seas and therefore exhibited less surface coverage in the Sauk.

The Sauk transgression cycle was probably caused by the initiation of runaway subduction during the first 40 days of the Flood. This tectonic activity generated a series of tsunamis that propagated across the shallow seas of all three continents, scouring and depositing as they went. This scouring created the Great Unconformity. The subsequent depositional cycle created the Cambrian Explosion.

It appears that runaway subduction only began in selected locations at first, like along the edge of the pre-Atlantic Ocean and possibly in the western Pacific. Later in the Flood, we see evidence that runaway subduction was operating on a more massive scale. The later and more extensive formation of hot, new seafloor pushed the tsunami waves and sea level higher and higher until a peak was reached at Day 150 in the Zuni Megasequence. But for now, the Sauk Megasequence seems to have only affected the pre-Flood shallow seas.

Surface Area (km²)	North America	South America	Africa	Total
Sauk	12,157,200	1,448,100	8,989,300	22,594,600
Tippecanoe	10,250,400	4,270,600	9,167,200	23,688,200
Kaskaskia	11,035,000	4,392,600	7,417,500	22,845,100
Absaroka	11,540,300	6,169,000	17,859,900	35,569,200
Zuni	16,012,900	14,221,900	26,626,900	56,861,700
Tejas	14,827,400	15,815,200	24,375,100	55,017,700
Volume (km³)	**North America**	**South America**	**Africa**	**Total**
Sauk	3,347,690	1,017,910	6,070,490	10,436,090
Tippecanoe	4,273,080	1,834,940	6,114,910	12,222,930
Kaskaskia	5,482,040	3,154,390	3,725,900	12,362,330
Absaroka	6,312,620	6,073,710	21,075,040	33,461,370
Zuni	16,446,210	23,198,970	57,729,600	97,374,780
Tejas	17,758,530	32,908,080	28,855,530	79,522,140
Average Thickness (km)	**North America**	**South America**	**Africa**	**Total**
Sauk	0.275	0.703	0.675	0.462
Tippecanoe	0.417	0.430	0.667	0.516
Kaskaskia	0.497	0.718	0.502	0.541
Absaroka	0.547	0.985	1.180	0.941
Zuni	1.027	1.631	2.168	1.712
Tejas	1.198	2.081	1.184	1.445

Table 9.2. Surface area, sediment volume, and average thicknesses for North America, South America, and Africa for each of the six megasequences

Bridal Veil Falls, Pictured Rocks National Lakeshore, Michigan

Next, we will examine the Tippecanoe Megasequence, where we see a similar pattern of deposition and even the same general types of marine fossils that we observe in the Sauk.

References
1. Vail, P. R. and R. M. Mitchum Jr. 1979. Global cycles of relative changes of sea level from seismic stratigraphy. *American Association of Petroleum Geologists Memoir.* 29: 469-472.
2. Chadwick, A. V. 1993. Megatrends in North American Paleocurrents. *Society of Economic Paleontologists and Mineralogists Abstracts.* 8: 58.
3. Peters, S. E. and R. R. Gaines. 2012. Formation of the 'Great Unconformity' as a trigger for the Cambrian explosion. *Nature.* 484 (7394): 363-366.
4. Ibid, 363.
5. Bond, G. C. and M. A. Kominz. 1984. Construction of tectonic subsidence curves for the early Paleozoic miogeocline, southern Canadian Rocky Mountains–implications for subsidence mechanisms, age of breakup, and crustal thinning. *Geological Society of America Bulletin.* 95: 155-173.
6. Price, R. A. 1994. Cordilleran tectonics and the evolution of the Western Canada Sedimentary Basin. *In Geological Atlas of the Western Canada Sedimentary Basin.* G. D. Mossop and I. Shetsen, comp. Canadian Society of Petroleum Geologists and Alberta Research Council.
7. Spencer, C. J. et al. 2014. Detrital zircon geochronology of the Grenville/Llano foreland and basal Sauk Sequence in west Texas, USA. *Geological Society of America Bulletin.* 127 (7/8): 1117-1128.
8. Kominz, M. A. and G. C. Bond. 1991. Unusually large subsidence and sea level events during middle Paleozoic time: new evidence supporting mantle convection models for supercontinent assembly. *Geology.* 19: 56-60.
9. An, M. and M. Assumpcao. 2006. Crustal and upper mantle structure in the Parana Basin, SE Brazil, from surface wave dispersion using genetic algorithms. *Journal of South American Earth Sciences.* 21: 173-184.
10. Snelling, A. A. 2009. *Earth's Catastrophic Past: Geology, Creation & the Flood.* Dallas, TX: Institute for Creation Research.
11. Kaminski, E. and C. Jaupart. 2000. Lithosphere structure beneath the Phanerozoic intracratonic basins of North America. *Earth and Planetary Letters.* 178: 139-149.
12. Wicander, R. and J. S. Monroe. 2013. *Historical Geology,* 7th ed. Belmont, CA: Brooks/Cole.
13. Runkel, A. C. et al. 2007. High-resolution sequence stratigraphy of lower Paleozoic sheet sandstones in central North America: The role of special conditions of cratonic interiors in development of stratal architecture. *Geological Society of America Bulletin.* 119 (7-8): 860-881.
14. Ibid, 861.
15. Clarey, T. 2015. Grappling with Megasequences. *Acts & Facts.* 44 (4): 18-19.

10 More Rising Water: Tippecanoe Megasequence

> **Summary:** Tectonic plate subduction created tsunamis that continued throughout the first 40 days of the Flood. They formed the Sauk Megasequence and then the second megasequence, the Tippecanoe. The two megasequences are roughly equal in thickness and composition, but the Tippecanoe rose slightly higher on the continents than the Sauk. At this point in the Flood, it's possible that some human settlements noticed the increasing pulses of water and began retreating to higher ground.

Flooding Level Increases

The deposition of the Tippecanoe Megasequence possibly began just days after the deposition of the Sauk Megasequence. This was also likely within the first 40 days of the Flood event. The runaway subduction process that created tsunami waves for the Sauk Megasequence continued to operate through the Tippecanoe cycle.

Secular geologists refer to the initial closing of the pre-Atlantic Ocean, called the Iapetus Ocean, as the Taconic Orogeny. It involved the subduction of the Iapetus Ocean beneath North America and affected the states of New York, Massachusetts, and Vermont.[1] In effect, this helped to close up the northern section of the pre-Flood narrow sea between Europe and North America. It also created a thick wedge of clastic sediment several kilometers thick in the subduction zone, referred to as the Queenston Delta Clastic Wedge.[1] Dr. John Baumgardner predicted that the

> *And the rain was on the earth forty days and forty nights.*
> (Genesis 7:12)

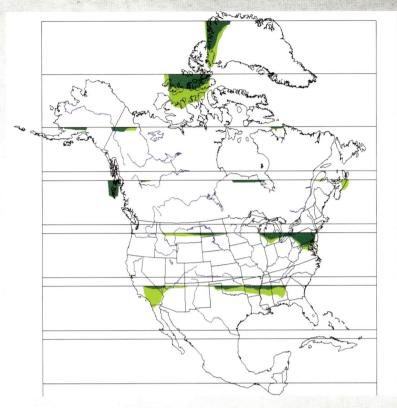

Figure 10.1. East-west stratigraphic cross-sections showing the relative thickness of the Sauk (light green) and the overlying Tippecanoe (dark green). Note the thick Queenston Delta Clastic Wedge in Pennsylvania.

runaway subduction process, because it was so rapid, would pull down the continental crust about 3 kilometers and fill with Flood sediments.[2] The observed clastic wedge fulfills his prediction perfectly. This wedge is visible in Figure 10.1, the east-west cross-sections across North America. The discovery of this clastic wedge should be no surprise to those who accept catastrophic plate tectonics. Finally, there is also evidence for simultaneous subduction on the eastern side of the Iapetus Ocean, termed the Caledonian Orogeny, which affected western Europe.[1]

All of this runaway subduction caused the creation of new ocean lithosphere. Recall that new ocean lithosphere is hotter and rises, pushing up the seafloor and raising global sea levels slightly (see chapter 6). For this reason, the tsunami waves that generated the Tippecanoe Megasequence were able to reach slightly higher elevations on some of the continents compared to the Sauk.

What Do the Rocks Show?

North America

The Tippecanoe sequence extends from the Middle Ordovician to the top of the Silurian system. This megasequence contains the highest percentage of carbonate rock of any of the megasequences in North America. It has a fairly extensive basal sandstone layer in the Midcontinent region of the U.S. (St. Peter Sandstone and equivalent), including an incursion into Hudson Bay. This sandstone layer is also quite thin, commonly less than 100 meters. A large part of the basal Tippecanoe consists of an extensive carbonate layer that was deposited across northern Canada and along the eastern and western margins of the pre-Flood continental U.S. The uplifted Transcontinental Arch (Dinosaur Peninsula) still caused thinning of this sequence across the center of the U.S., and in many places prevented any Tippecanoe deposition. The thickest Tippecanoe sections are found in northernmost Canada, Alaska, and up the East Coast, including the Queenston Delta Clastic Wedge. Sandstone only makes up about 8% of the entire Tippecanoe sequence, with shale making up an additional 18%. Carbonate rock dominates, however, comprising about 67% of the rock volume of this sequence. The Tippecanoe across North America averages about 420 meters thick and covers an area of over 10 million square kilometers. The total volume of Tippecanoe strata across North America is 4.3 million cubic kilometers. The surface area for the Tippecanoe is a little less than the Sauk, but the volume deposited was significantly greater.

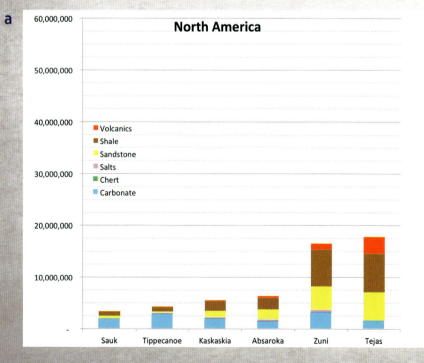
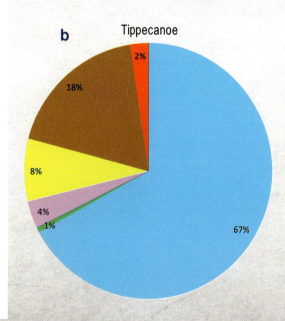

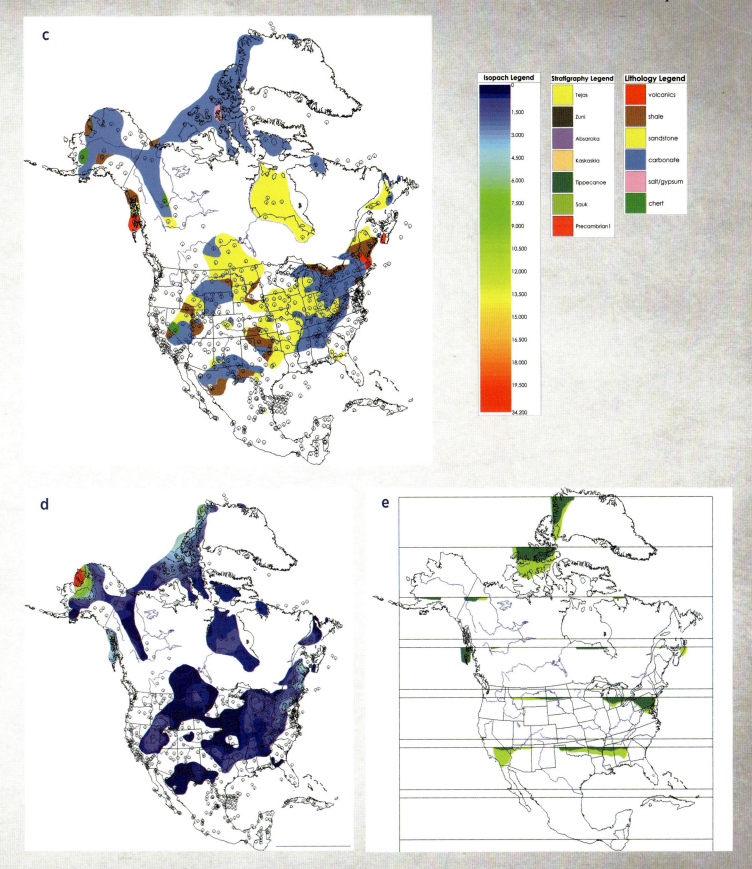

Compilation of Tippecanoe Megasequence data for North America: (a) Histogram of the type and volume of rocks in place across North America for each megasequence; (b) pie chart of the total Tippecanoe Megasequence by rock type; (c) map of the bottommost (basal) Tippecanoe rock type; (d) thickness map of the total Tippecanoe Megasequence; (e) stratigraphic cross-sections showing the relative thickness of the Tippecanoe (dark green) and Sauk (light green) Megasequences. Histogram measurement is in cubic kilometers, isopach map measurement is in meters.

South America

The base of this megasequence consists of a slightly more extensive sandstone layer compared to the earlier Sauk Megasequence across South America. Like the Sauk, most of South America shows little evidence of Tippecanoe deposition. The Tippecanoe includes new deposits in the Amazonas and Parnaíba Basins of Brazil. Most of the Tippecanoe is also spread relatively thin, only averaging about 430 meters where present. Siliciclastics are the dominant sedimentary rock type. Shale rocks compose about 58% of the total rock volume of the Tippecanoe Megasequence. Sandstone is a close second at nearly 38%. Carbonate layers are rare, only making up 4% of the total megasequence volume. The Tippecanoe covers nearly 4.3 million square kilometers of South America and has a total volume of over 1.8 million cubic kilometers. Both of these values are greater than the preceding Sauk Megasequence.

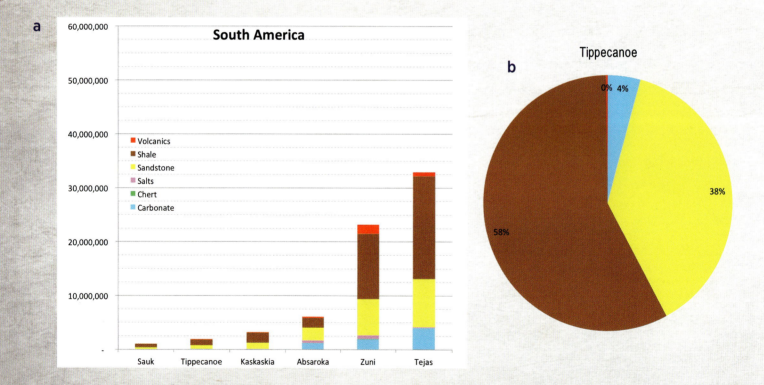

Compilation of Tippecanoe Megasequence data for South America: (a) Histogram of the type and volume of rocks in place across South America for each megasequence; (b) pie chart of the total Tippecanoe Megasequence by rock type; (c) map of the bottommost (basal) Tippecanoe rock type; (d) thickness map of the total Tippecanoe Megasequence; (e) stratigraphic cross-sections showing the relative thickness of the Tippecanoe (dark green) and Sauk (light green) Megasequences. Histogram measurement is in cubic kilometers, isopach map measurement is in meters.

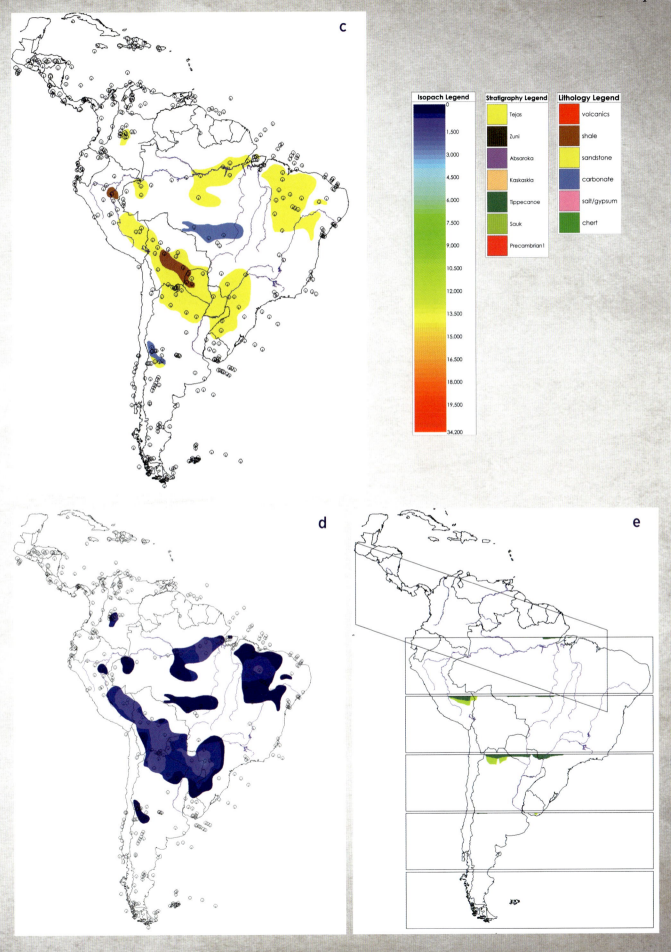

Africa

The base of this megasequence also consists of an extensive blanket sandstone that covers much of North Africa and the Saudi Arabian Peninsula, nearly mirroring the location of the Sauk Megasequence. Like the Sauk, much of the rest of Africa shows no evidence of Tippecanoe deposition other than the Cape Fold Belt in South Africa. Most of the Tippecanoe is also spread relatively thin, only exceeding a thickness of one kilometer in isolated locations. Similar to the Sauk Megasequence, the thickest depo-center is in the northwestern Saudi Arabian Peninsula, where the Tippecanoe exceeds 3 kilometers thick. The average thickness of the Tippecanoe is 667 meters. Shales dominate the strata of the Tippecanoe Megasequence with 57.7% of the total rock volume. Sandstone is a close second at 40.3%. Carbonate layers are rare, only making up 2% of the total sequence volume. The Tippecanoe covers close to 9.2 million square kilometers of Africa and the Middle East and has a total volume exceeding 6.1 million cubic kilometers. Both of these values are slightly greater than the preceding Sauk Megasequence.

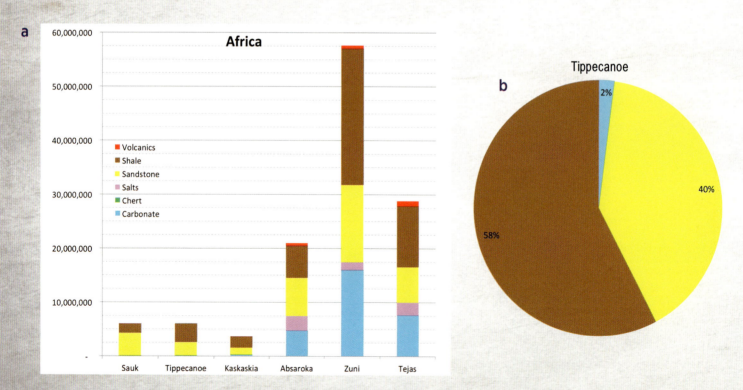

Compilation of Tippecanoe Megasequence data for Africa: (a) Histogram of the type and volume of rocks in place across Africa for each megasequence; (b) pie chart of the total Tippecanoe Megasequence by rock type; (c) map of the bottommost (basal) Tippecanoe rock type; (d) thickness map of the total Tippecanoe Megasequence; (e) stratigraphic cross-sections showing the relative thickness of the Tippecanoe (dark green) and Sauk (light green) Megasequences. Histogram measurement is in cubic kilometers, isopach map measurement is in meters.

Chapter 10 • 223

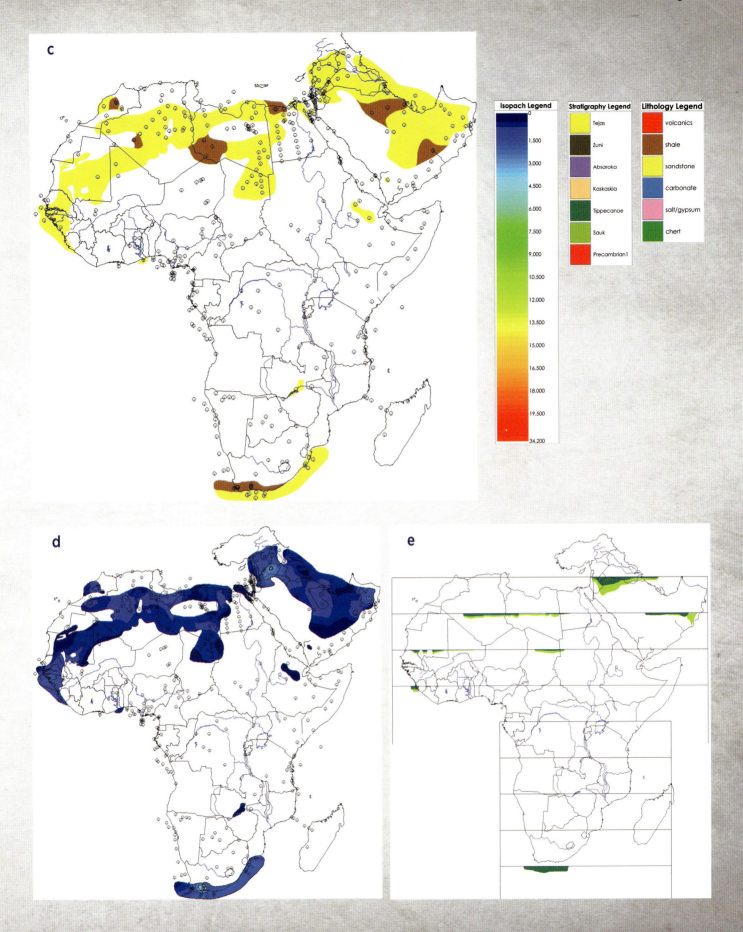

Current Data for the Tippecanoe Megasequence

Dr. Art Chadwick and his students also created maps showing the paleocurrent directions during the Ordovician and Silurian systems.³ Figure 10.2 shows plots of the dominant current directions for the Ordovician system rocks across North America, South America, and Africa. The Ordovician is the lower part of the Tippecanoe Megasequence.

The current directions during the Tippecanoe are very similar to the Sauk Megasequence, with each maintaining a strong westerly or southwesterly direction. South America shows some paleocurrent data in the Ordovician section, indicating a weak westerly direction. The Tippecanoe data in Africa are also very similar to the current directions in the Sauk, showing a northerly, and possibly westerly, flow direction across North Africa.

Summary of the Rock Data

The Tippecanoe Megasequence generally increases in volume and extent across most continents, North America being the only exception in terms of extent. Most of the fossils are still shallow marine fossils, similar to the Sauk Megasequence but with many new varieties. Overall, the sedimentary rocks deposited were very similar to those

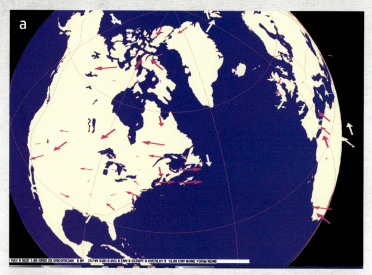

Figure 10.2. Current directions compiled across (a) North America, (b) South America, and (c) Africa for the Ordovician system (Lower Tippecanoe). Courtesy of Art Chadwick.

deposited during the Sauk. South America and Africa were again dominated by clastic material, mostly shales and sandstones and very few carbonates. In contrast, North America continued to be dominated by carbonate deposition throughout most of the Tippecanoe. In fact, the Tippecanoe across North America has the highest carbonate percentage (67%) by volume of any megasequence on any continent.

The average thickness of the Tippecanoe was not much different from the earlier Sauk Megasequence. The Tippecanoe averaged about 516 meters globally, whereas the Sauk averaged about 462 meters. Overall, the patterns of deposition for the Tippecanoe were a repeat of the Sauk in terms of lithology and extent, with only a slight increase in the overall extent and volume globally. This increase was likely due to the small increase in new oceanic lithosphere, rising sea level, and the height of the tsunami waves slightly above Sauk levels.

Many previous maps of the St. Peter Sandstone across North America showed too much extent. Figure 10.3 shows the basal lithology across North America. The St. Peter is the basal sandstone shown in yellow. Note that it is still massive, covering the central part of the continent, but it does not extend to the coastal regions. The basal Tippecanoe in those areas was limestone and dolomite (carbonate) deposits.

The reason for the high percentage of carbonate rocks may be that a great deal of the flanks of North America was covered by large shallow seas in the pre-Flood world (see chapter 17). If the regression of the Sauk Megasequence was incomplete prior to the onset of the Tippecanoe transgression, then the observed carbonate deposition around the edges of North America could merely represent deeper and less energetic water and therefore continuous carbonate deposition from one megasequence to the next. As a consequence, the highest energy and resulting sandstone

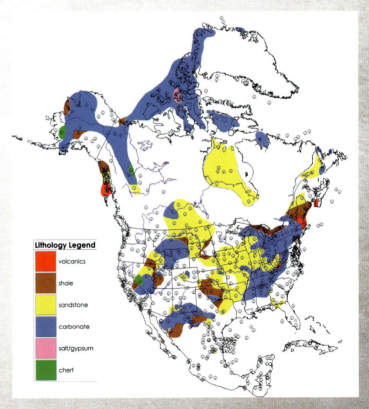

Figure 10.3. Map of the bottommost (basal) Tippecanoe rock type for North America

deposition for the basal Tippecanoe were centered only in the middle of the continent, producing the St. Peter Sandstone.

Continental Configuration for the Tippecanoe Megasequence

The continents in the Tippecanoe are essentially the same as the configuration in the preceding Sauk Megasequence, except for the partial closing of the Iapetus (pre-Atlantic) Ocean in the north (Taconic Orogeny). Subduction and runaway plate motion still had not yet appreciably moved the majority of the continents. Neither had there been substantial new oceanic lithosphere created since the runaway subduction process was apparently limited to just a few locations globally. The slight increase in new seafloor, however, did seem to raise sea level slightly, resulting in a slightly greater volume and extent for the Tippecanoe Megasequence. Figure 10.4 shows the extent and thickness of the Tippecanoe Megasequence across the three continents completed in this study and in the proper Pangaea-like configuration.

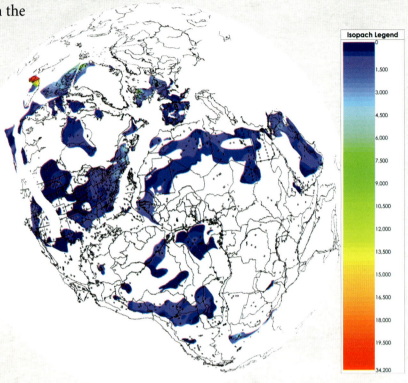

Figure 10.4. Continental configuration during the deposition of the Tippecanoe Megasequence and its sedimentary thickness. Measurement is in meters.

Limited Flooding of the Land During the Early Flood (First 40 Days)

Numerous authors have speculated on the extent of the early floodwaters and on when the Flood peaked. Many questions remain unanswered. For example, did the Flood cover the continents early in the Flood year, recede, and then rise again? Did the waters rise once and peak around Day 150? Or was it some combination? Our research provides some answers.[4]

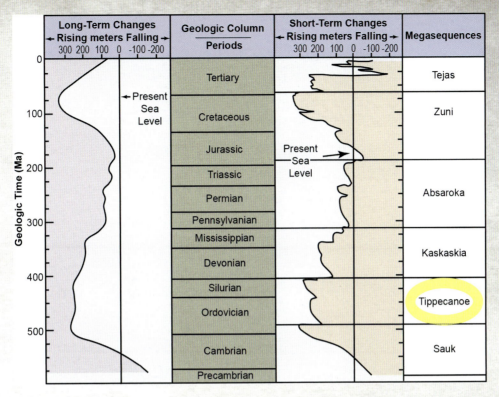

Figure 10.5. Secular chart showing presumed geologic time, global sea level, and the six megasequences (after Vail and Mitchum[10])

Surface Area (km²)	North America	South America	Africa	Total
Sauk	12,157,200	1,448,100	8,989,300	22,594,600
Tippecanoe	10,250,400	4,270,600	9,167,200	23,688,200
Kaskaskia	11,035,000	4,392,600	7,417,500	22,845,100
Absaroka	11,540,300	6,169,000	17,859,900	35,569,200
Zuni	16,012,900	14,221,900	26,626,900	56,861,700
Tejas	14,827,400	15,815,200	24,375,100	55,017,700
Volume (km³)	North America	South America	Africa	Total
Sauk	3,347,690	1,017,910	6,070,490	10,436,090
Tippecanoe	4,273,080	1,834,940	6,114,910	12,222,930
Kaskaskia	5,482,040	3,154,390	3,725,900	12,362,330
Absaroka	6,312,620	6,073,710	21,075,040	33,461,370
Zuni	16,446,210	23,198,970	57,729,600	97,374,780
Tejas	17,758,530	32,908,080	28,855,530	79,522,140
Average Thickness (km)	North America	South America	Africa	Total
Sauk	0.275	0.703	0.675	0.462
Tippecanoe	0.417	0.430	0.667	0.516
Kaskaskia	0.497	0.718	0.502	0.541
Absaroka	0.547	0.985	1.180	0.941
Zuni	1.027	1.631	2.168	1.712
Tejas	1.198	2.081	1.184	1.445

Table 10.1. Surface area, sediment volume, and average thicknesses for North America, South America, and Africa for each of the six megasequences

Figure 10.5 shows the megasequences defined by geologist L. L. Sloss and the secular sea level curve during the time of their deposition.[5] Although Sloss initially defined his megasequences across only the interior of North America, oil industry geologists quickly extended these boundaries to the offshore regions and adjacent continents using oil well logs and seismic data.[6] Secular geologists claimed global sea level reached an extreme high during the Sauk Megasequence (Figure 10.5).

ICR's research is one of the first attempts to map the true extent of the sedimentary rocks across the continents and to test the validity of the published secular sea level curve model (Figure 10.5).[4] The area covered by the Sauk Megasequence across North America is over 12 million square kilometers (Table 10.1 and Figure 10.6a). Sauk deposits are several kilometers thick along the east and west coasts of the North American continent and yet are very thin (only a few tens of meters) to nonexistent across the central part of the continental United States.

In stark contrast to North America, the sedimentary rocks of the Sauk Megasequence show very little coverage across South America (Figure 10.6b) and Africa

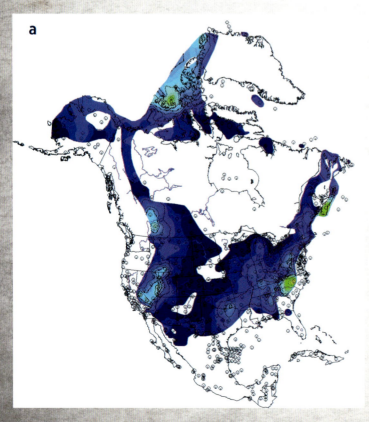 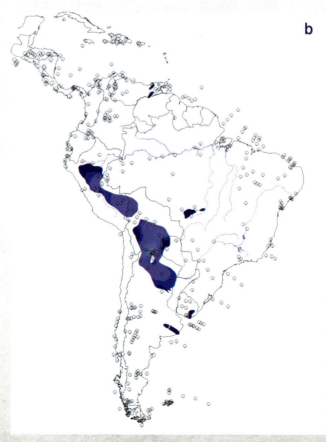

Figure 10.6. Thickness maps of the Sauk Megasequence across (a) North America, (b) South America, and (c) Africa

(Figure 10.6c). Only the northernmost part of Africa and the west-central portion of South America show any Sauk sediments.

The rock data suggest the floodwaters rose progressively as described in Genesis 7 with only limited flooding during the Sauk event. The Sauk Megasequence was only the violent beginning of the Flood, creating the Great Unconformity at its base and encasing prolific numbers of the hard-shelled marine fossils that define the Cambrian Explosion. The peak height of the floodwaters seems to have occurred later, possibly around Day 150 during the Zuni Megasequence, as that is when the coverage and volume of sediment also peak on most of the continents (Table 10.1).

After reviewing the vast oil well data, we concluded that the published secular global sea level curve is inaccurate. In addition, the amount of pre-Flood dry land inundated during the Sauk, Tippecanoe, and Kaskaskia Megasequence deposition seems to have also been limited. As we suggested in 2015, the Sauk, Tippecanoe, and Kaskaskia Megasequences probably represent the effects of tsumani-like waves that transported sediment across pre-Flood shallow seas and not across landmasses.[8,9] This could also possibly explain why so few, if any, terrestrial animals and plants are found as fossils in Sauk, Tippecanoe, and Kaskaskia Megasequence rocks.

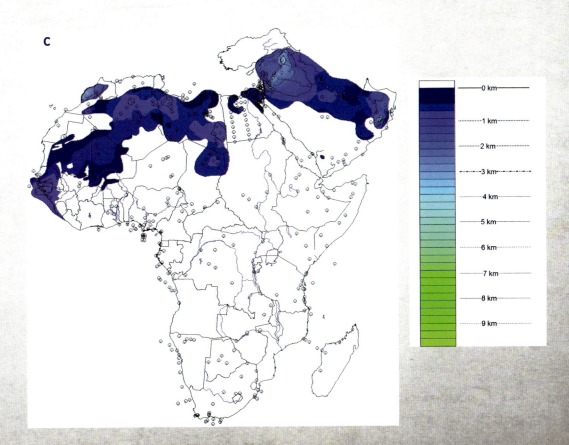

Inside an underground salt mine

Human Perspective

At this point in the Flood, the relentless rain and the numerous massive earthquakes were likely taking a physical and psychological toll on the humans who were not on the Ark. Many were likely thinking that Noah was right after all. Although the tsunami waves and transported sediment were probably not reaching the elevation of most human settlements, they may have been able to see the activity from afar. They might have noticed that the waves were washing a bit higher with each pulse of the tsunami waves. And with the continuous rainfall and volcanic activity, they were likely getting much more concerned for their own safety. Some may even have begun to travel to higher ground.

Conclusion: Flooding Continued in the Tippecanoe

The Tippecanoe Megasequence is similar to the Sauk in a lot of ways. The stratigraphic data show a significant rise in sea level advanced across all three continents simultaneously, similar to the Sauk but higher. The Tippecanoe covers the same general areas as the preceding Sauk Megasequence, with a few exceptions. It averages about the same general thickness and contains much of the same lithologies and even similar shallow marine types of fossils. However, there are enough differences in the specific types of fossils to be able to differentiate the two megasequences. Exactly why the fossils differ slightly from megasequence to megasequence is unclear. It could be because the tsunami waves for the Tippecanoe reached a bit higher in elevation and/or were more energetic compared to the Sauk, resulting in transportation of different types of invertebrates within each megasequence.

It appears that the secular global sea level curve is wrong for the earliest megasequences. The high flooding level suggested by the secular sea level curve for the Sauk/Tippecanoe, and even by some creationists, seems unjustified by the limited extent of the Sauk and the Tippecanoe Megasequences. South America and Africa, in particular, show only limited amount of surface coverage of both megasequences. Evidence for a significant rise in sea level early in the Flood event is weak at best.

Note that the surface extent of both the Sauk and the Tippecanoe Megasequences mirror one another quite closely. Simple erosion of the megasequences during their regressive cycle (and later) cannot explain this consistent pattern. Erosional remnants would be more random and not leave megasequences that stack almost on top of one another as observed. This is particularly evident across Africa where the surface extent of the first three megasequences covers essentially the same areas of North Africa.

Runaway subduction seems to have remained limited during the deposition of the Tippecanoe Megasequence. It was likely still confined to the pre-Atlantic and possibly the western Pacific. There was just not enough new ocean lithosphere created at this point in the Flood to push water levels much higher than during the preceding Sauk Megasequence. It appears very little of the dry land portions of the continents were being affected by the tsunami waves and the rising floodwaters at this point. The fossils are still overwhelmingly marine, with virtually zero land animals and land plants buried in the Tippecanoe Megasequence.

We will examine the Kaskaskia Megasequence next. This sequence is also very similar to the Sauk and Tippecanoe Megasequences in terms of fossils, rock types, volume, and surface extent. This is why I believe the first three megasequences fall within the first 40 days of the Flood. God tells us in Genesis 7:17 that it wasn't until after 40 days that the "waters increased and lifted up the ark." After this point in the Flood year, when the Ark began to float, we should expect to see a sudden influx of fossils of land flora and fauna. This does not occur until the Absaroka Megasequence.

> "They might have noticed that the waves were washing a bit higher with each pulse of the tsunami waves. And with the continuous rainfall and volcanic activity, they were likely getting much more concerned for their own safety."

References

1. Wicander, R. and J. S. Monroe. 2013. *Historical Geology,* 7th ed. Belmont, CA: Brooks/Cole.
2. Baumgardner, J. 2005. Recent Rapid Uplift of Today's Mountains. *Acts & Facts.* 34 (3).
3. Chadwick, A. V. 1993. Megatrends in North American Paleocurrents. *Society of Economic Paleontologists and Mineralogists Abstracts.* 8: 58.
4. Clarey, T. L. and D. J. Werner. 2017. The Sedimentary Record Demonstrates Minimal Flooding of the Continents During Sauk Deposition. *Answers Research Journal.* 10: 271-283.
5. Sloss, L. L. 1963. Sequences in the Cratonic Interior of North America. *Geological Society of America Bulletin.* 74 (2): 93-114.
6. Haq, B. U., J. Hardenbol, and P. R. Vail. 1988. Mesozoic and Cenozoic chronostratigraphy and cycles of sea-level change. In *Sea-Level Changes: An Integrated Approach.* Tulsa, OK: Society of Economic Paleontologists and Mineralogists, 71-108.
7. Peters, S. E. and R. R. Gaines. 2012. Formation of the 'Great Unconformity' as a trigger for the Cambrian explosion. *Nature.* 484: 363-366.
8. Clarey, T. L. 2015. Examining the floating forest hypothesis: a geological perspective. *Journal of Creation.* 29 (3): 50-55.
9. Clarey, T. 2015. Dinosaur Fossils in Late-Flood Rocks. *Acts & Facts.* 44 (2): 16.
10. Vail, P. R. and R. M. Mitchum Jr. 1979. Global cycles of relative changes of sea level from seismic stratigraphy. *American Association of Petroleum Geologists Memoir.* 29: 469-472.

11 Water Continues to Rise: Kaskaskia Megasequence

> **Summary:** As the tectonic activity continued, the Kaskaskia Megasequence was being deposited—still within the first 40 days of the Flood. This third megasequence rose slightly higher and shows a slight increase in average thickness over the two previous megasequences. At this point in the Flood, the waters still did not cover much land. The Sauk, Tippecanoe, and Kaskaskia Megasequences contain almost 100% marine fossils. It's possible that creatures such as dinosaurs retreated to higher ground or were still living at high enough elevation during these early parts of the Flood. That's why so few dinosaur fossils exist in these megasequences.

Flooding Again Increases

The Kaskaskia cycle began possibly just a few days after the deposition of the Tippecanoe Megasequence. This was also likely within the first 40 days of the Flood event as discussed at the end of the last chapter. Runaway subduction was still apparently limited to just a few locations globally, including the Atlantic seaboard of North America and Europe. In addition, the Kaskaskia shows evidence of the closure of the southern portion of the Iapetus Ocean (pre-Atlantic) extending southward to the collision with Africa. These are marked by the Acadian Orogeny (mountain-building event) and the continuation of the Caledonian Orogeny.[1]

What Do the Rocks Show?

North America

The Kaskaskia Megasequence extends from the Devonian to the top of the Mississippian system. This interval contains the most extensive basal layer of carbonate rock of any sequence across North America. However, some basal sandstone was

Appalachian Mountains

deposited in western Canada and along the East Coast of the U.S. and in northwestern Mexico. The uplifted Transcontinental Arch (Dinosaur Peninsula) still caused thinning of this megasequence across the center of the U.S., and in many places prevented any Kaskaskia deposition. The thickest Kaskaskia stratigraphic columns are found in the western U.S. and Canada and along the East Coast. Kaskaskia deposits also show thickening into the Michigan, Illinois, and Williston Basins, indicating their continued subsidence. Some basal chert-rich layers are also found across Arkansas, Illinois, southwest Texas, and Alaska (shown in green on map "c" on page 237). Sandstones make up about 24% of the entire Kaskaskia Megasequence, while shale makes up 33%. Carbonate rock, although a lesser percentage than previous megasequences, still dominates, comprising about 36% of the total rock volume.

The Kaskaskia shows a slight increase in average thickness over the two previous megasequences, averaging nearly 500 meters thick. The Kaskaskia covers an area greater than 11 million square kilometers and has a total volume of nearly 5.5 million cubic kilometers.

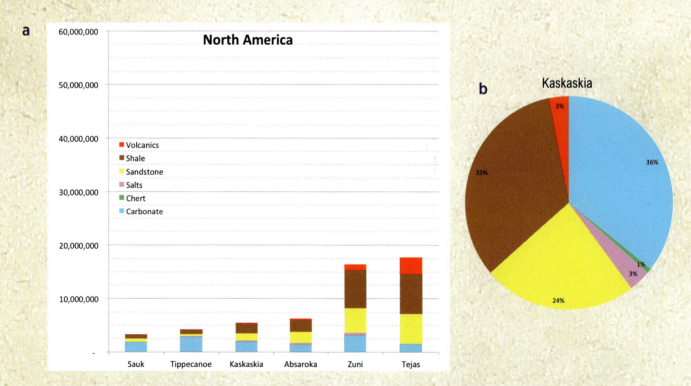

Compilation of Kaskaskia Megasequence data for North America: (a) Histogram of the type and volume of rocks in place across North America for each megasequence; (b) pie chart of the total Kaskaskia Megasequence by rock type; (c) map of the bottommost (basal) Kaskaskia rock type; (d) thickness map of the total Kaskaskia Megasequence; (e) stratigraphic cross-sections showing the relative thickness of the Kaskaskia Megasequence in tan on top of the earlier megasequences. Histogram measurement is in cubic kilometers, isopach map measurement is in meters.

Chapter 11 * 237

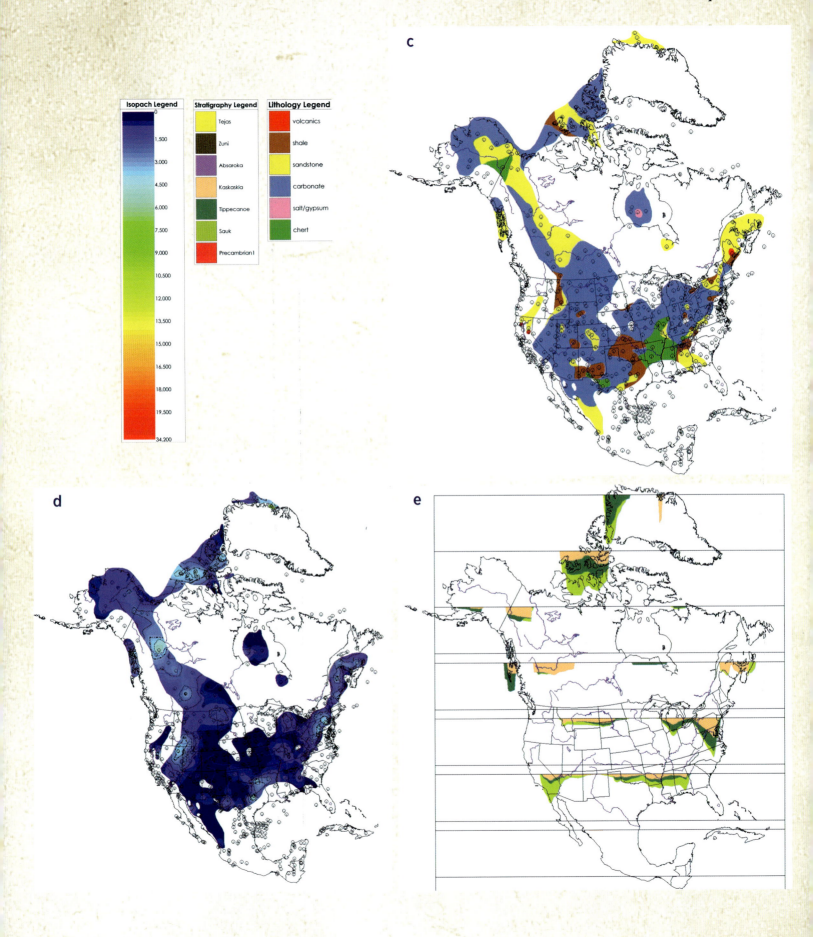

South America

The base of this megasequence also consists of a sandstone layer that mirrors the locations of the Tippecanoe Megasequence below it. Both the Tippecanoe and the Kaskaskia Megasequences cover slightly greater than 4 million square kilometers in total surface area. And like the Sauk and Tippecanoe, much of the remainder of South America shows little evidence of Kaskaskia deposition. Most of the Kaskaskia is also spread relatively thin, only averaging a thickness of 720 meters. Siliciclastics continue to dominate deposition during the Kaskaskia Megasequence. Similar to the Sauk and Tippecanoe Megasequences, shale-rich rocks compose about 61% of the total sedimentary volume. Sandstones are again the second-most common rock type at 36% by volume. Carbonate layers are still extremely limited, composing about 2% of the total megasequence volume. Overall, the Kaskaskia Megasequence approximately doubled the volume of sediment deposited during the earlier Tippecanoe Megasequence and tripled the volume deposited during the Sauk Megasequence across South America. The Kaskaskia Megasequence only shows a slight increase in surface area coverage compared to the preceding Tippecanoe, at 4.4 million square kilometers. The Kaskaskia also has a total volume across South America of nearly 3.2 million cubic kilometers, an amount that is almost twice that of the Tippecanoe.

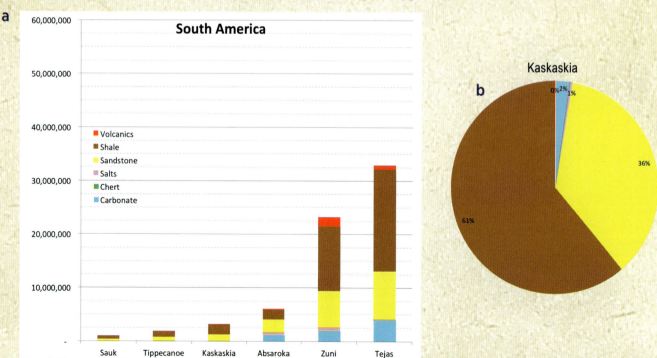

Compilation of Kaskaskia Megasequence data for South America: (a) Histogram of the type and volume of rocks in place across South America for each megasequence; (b) pie chart of the total Kaskaskia Megasequence by rock type; (c) map of the bottommost (basal) Kaskaskia rock type; (d) thickness map of the total Kaskaskia Megasequence; (e) stratigraphic cross-sections showing the relative thickness of the Kaskaskia Megasequence in tan on top of the earlier megasequences. Histogram measurement is in cubic kilometers, isopach map measurement is in meters.

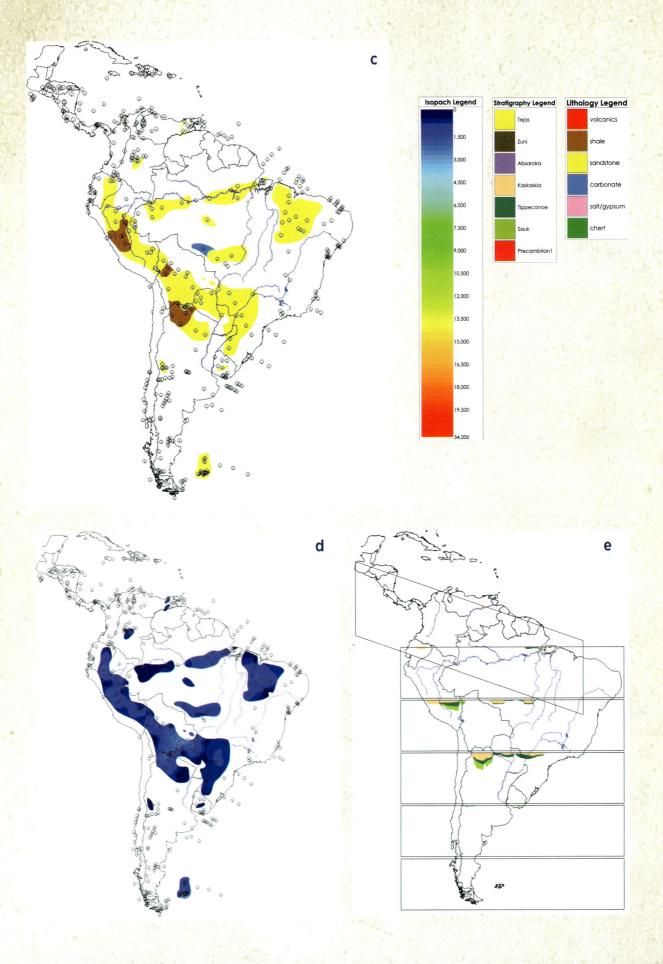

Africa

The base of this megasequence consists of an extensive blanket sandstone that covers much of North Africa and the Saudi Arabian Peninsula, mirroring the locations of the Sauk and Tippecanoe Megasequences below it. And like the Sauk and Tippecanoe, much of the rest of Africa shows no evidence of Kaskaskia deposition other than the Cape Fold Belt in South Africa. Most of the Kaskaskia is also spread relatively thin, only exceeding a thickness of 1 kilometer in isolated locations. The thickest depocenter is centered near Algeria in North Africa, where the Kaskaskia exceeds 2 kilometers thick. Shales dominate the strata of the Kaskaskia Megasequence with 56.9% of the total rock volume, very similar to the Tippecanoe. Sandstone is the second-most common rock type at 35.2% by volume. Carbonate layers are still limited, composing just 7.8% of the total megasequence volume. The Kaskaskia Megasequence exhibits the least volume of any of the six sequences across Africa.

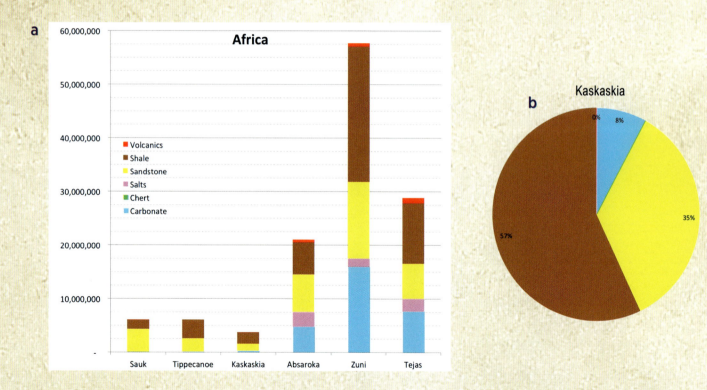

Compilation of Kaskaskia Megasequence data for Africa: (a) Histogram of the type and volume of rocks in place across Africa for each megasequence; (b) pie chart of the total Kaskaskia Megasequence by rock type; (c) map of the bottommost (basal) Kaskaskia rock type; (d) thickness map of the total Kaskaskia Megasequence; (e) stratigraphic cross-sections showing the relative thickness of the Kaskaskia Megasequence in tan on top of the earlier megasequences. Histogram measurement is in cubic kilometers, isopach map measurement is in meters.

Chapter 11 • 241

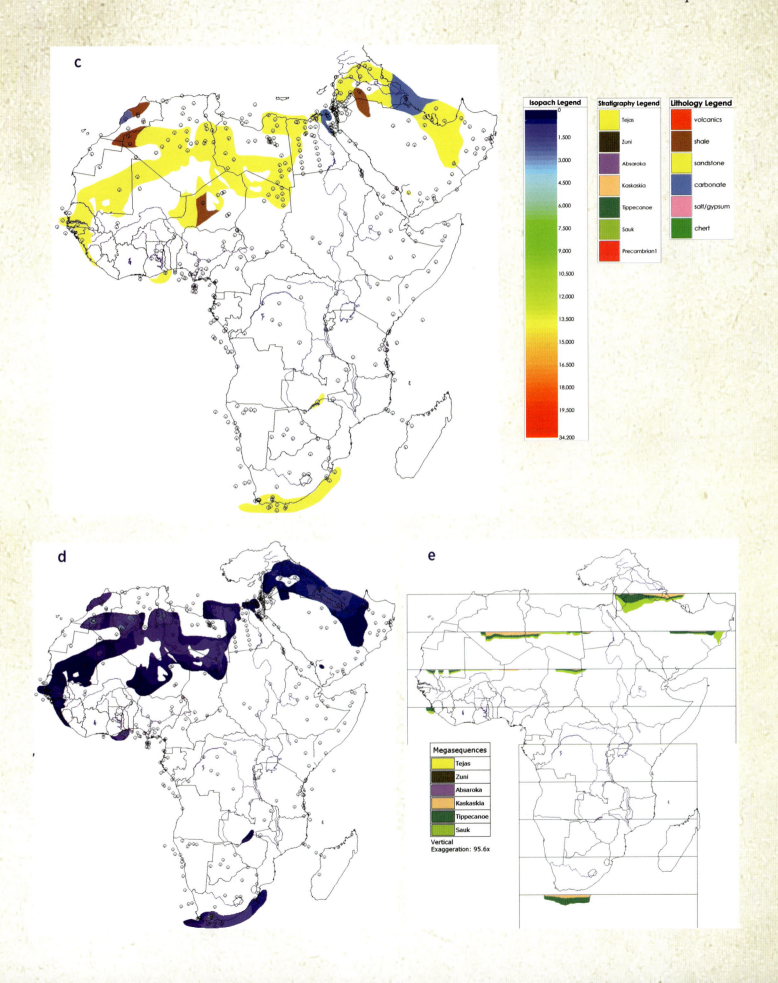

Current Data for the Kaskaskia Megasequence

Figure 11 shows the paleocurrent data collected by Dr. Art Chadwick and his students across North America, South America, and Africa within Devonian system rocks, the lower part of the Kaskaskia Megasequence.[2]

The Kaskaskia, like the Sauk and Tippecanoe Megasequences before it, shows similar current directions for each of the three continents in this study. The Devonian rocks of the lower Kaskaskia Megasequence continue to demonstrate a dominant southwesterly pattern across North America, although they appear to be less developed in the West. The Devonian rocks across South America show a similar but more strongly developed westerly pattern compared to the preceding Tippecanoe. And the lower Kaskaskia rocks across North Africa indicate a shift from mostly northerly in the Sauk and Tippecanoe to a dominant northwesterly pattern in the Kaskaskia. Overall, the patterns are much the same as the preceding megasequences.

Summary of the Rock Data

Across North America, the Sauk Megasequence shows a complete, ideal sequence of rocks, beginning with sandstone across most of North

Figure 11.1. Current directions compiled across (a) North America, (b) South America, and (c) Africa for the Devonian system (Lower Kaskaskia). Courtesy of Art Chadwick.

America, followed by shale and then carbonates. The Tippecanoe begins with some sandstone in the Midwest (St. Peter Sandstone), but in the East and West it begins with carbonate, not sandstone at all. Recall that the Tippecanoe had the highest percentage of carbonate rock of any megasequence across North America at 67%. The Kaskaskia Megasequence, however, still contains a high percentage of carbonate rock (36%, the Redwall Limestone and equivalent) with slightly less sandstone (24%) and shale (33%). Sloss noted similar observations about the Kaskaskia.[3] But the numbers show that the percentage of carbonate rock is dropping at this point in the Flood, at least across North America.

It's possible that the basal carbonate rocks were deposited while the floodwaters were still covering much of North America. These rocks in the Tippecanoe and Kaskaskia Megasequences may indicate the waters never fully drained off the continent after each megasequence cycle (i.e., after the Sauk and Tippecanoe Megasequences). In other words, the water may have continued to rise as each new megasequence advanced before the previous megasequence had fully withdrawn, leaving part of the continent submerged even between the megasequences.

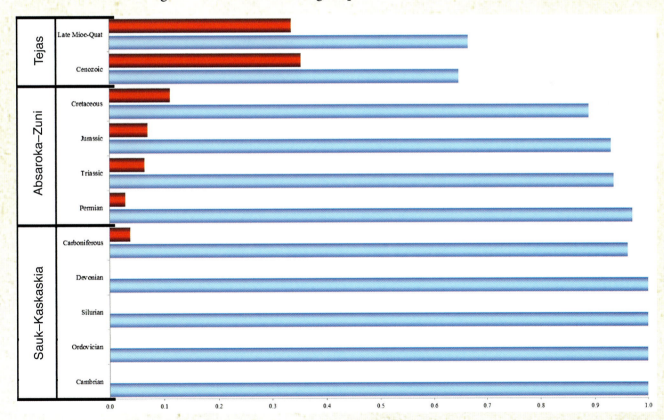

Proportion of marine (blue) vs. terrestrial (red) fossil occurrences from the Paleobiology Database. Note nearly 100% of fossil occurrences are marine in the Sauk to Kaskaskia Megasequences. Courtesy of Nathaniel Jeanson.

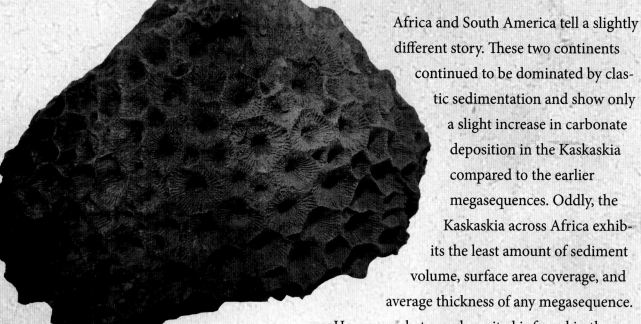

Michigan's Petoskey stone, a coral found in Devonian system rocks

Africa and South America tell a slightly different story. These two continents continued to be dominated by clastic sedimentation and show only a slight increase in carbonate deposition in the Kaskaskia compared to the earlier megasequences. Oddly, the Kaskaskia across Africa exhibits the least amount of sediment volume, surface area coverage, and average thickness of any megasequence. However, what was deposited is found in the same locations as the two previous megasequences, just a little thinner and slightly less extensive.

Overall, the patterns of the sediments for the first three megasequences are very similar across each of the three continents in this study. This implies that the same general processes and environments were being inundated during this stage of the Flood. All three of these megasequences exhibit fossils that indicate nearly exclusively shallow marine environments. Few land plant fossils and land animals are found in any of these megasequences.

Continental Configuration for the Kaskaskia Megasequence

The plate configuration during the Kaskaskia was likely still close to the original pre-Flood arrangement except for limited subduction along the North American East Coast, Europe, and a few other locations globally. The remainder of the northern Iapetus Ocean was consumed during the Kaskaskia Megasequence, forming the more traditional Pangaea. Secular geologists refer to this activity in North America as the Acadian Orogeny.[1] A new clastic wedge of sediment filled the trough created by this runaway subduction process, called the Catskill Delta.[1] This deposit, although up to three times thicker, overlaps the clastic wedge deposited during the preceding Tippecanoe Megasequence, termed the Queenston Delta.[1] As noted previously, this wedge was formed and infilled as a consequence of the drawdown of continental crust

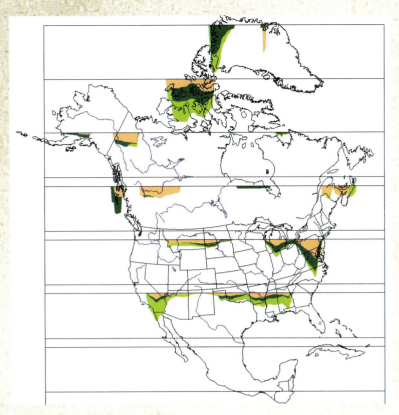

Figure 11.2. East-west stratigraphic cross-sections across North America showing the relative thickness of the Kaskaskia Megasequence in tan on top of the earlier megasequences. Note the thick Catskill Delta clastic wedge in Pennsylvania.

due to the high rate of subduction, as predicted by Dr. John Baumgardner.[4] Figure 11.2 shows this Kaskaskia wedge on top of the preceding Tippecanoe wedge in Pennsylvania and New York.

The southern part of the Iapetus Ocean was also nearly completely consumed during the Kaskaskia. South America and Africa (still part of one landmass called Gondwana) began to close up the narrow ocean between North America, South America, and Africa.[1] This collision resulted in the Ouachita Mountains of Oklahoma and Arkansas that reached their maximum activity level a bit later in the Absaroka Megasequence. Figure 11.3 shows the plate configuration and Kaskaskia extent and thickness across the three continents in this study.

Similar to the Tippecanoe, the global sea level seemed to rise slightly during the Kaskaskia compared to previous megasequence cycles, resulting in slightly higher volumes of Kaskaskia sediment globally (although slightly less in Africa). Many of the same types of marine invertebrates were deposited during the Kaskaskia as in previous megasequences, such as trilobites and brachiopods. However, many new types of marine fossils were also deposited in the

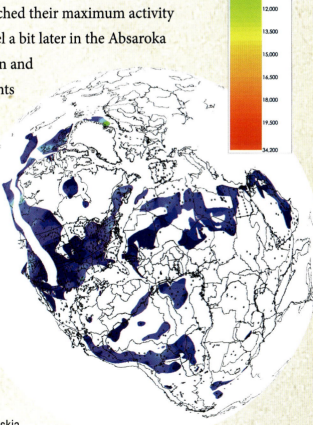

Figure 11.3. Continental configuration during the deposition of the Kaskaskia Megasequence and its sedimentary thickness. Measurement is in meters.

Devonian and Mississippian system rocks, including many more vertebrates like fish. These differences indicate the tsunami waves were reaching new and higher levels and/or affecting areas in the deeper parts of the offshore regions surrounding the continents. The lack of significant numbers of terrestrial fossils indicates the floodwaters had yet not begun to cover much of the pre-Flood land areas.

The 40 Days of Rain

Genesis 7:12 records the initial 40 days of rain, and verse 17 indicates the Ark began to float on, or immediately after, Day 40 of the Flood. Why are the initial 40 days described as predominantly rainfall? And why 40 days? For one thing, God used the number 40 for some significant events in biblical history. The Israelites were 40 years in the wilderness, and Jesus fasted 40 days before He began His earthly ministry. And Jesus resided on the earth for 40 days after His resurrection before ascending to heaven. Forty is one of those special numbers God chooses to use.

> "The Bible suggests in Genesis 7:17 that it wasn't until after these first 40 days that the Ark began to float, thereby verifying that the flooding of the land had commenced."

It is possible that until Day 40 the floodwaters still had not affected the land. The Sauk, Tippecanoe, and Kaskaskia Megasequences contain nearly 100% marine fossils. Very few land animals or plants were trapped by these three megasequence cycles. Apparently, the intense rain was the major factor affecting the dry land portions of the continents up to this point in the Flood. Humans would have known the first 40 days as a time of intense rainfall without significant flooding. The Bible suggests in Genesis 7:17 that it wasn't until after these first 40 days that the Ark began to float, thereby verifying that the flooding of the land had commenced.

Why Dinosaur Fossils Are Found Only in Later Flood Rocks

Evolutionary scientists view Earth's rock layers as a chronological record of millions of years of successive sedimentary deposits. Creation scientists, on the other hand, see them as a record of the geological work accomplished during the great Flood's year-long destruction of the earth's surface. But if that is the case, then why don't we find dinosaur fossils in the earliest North American Flood sediment layers? Why do we find them only in later Flood rocks? The answers to these questions are being provided by our data-based research.

Deposition of the earliest Flood sediments (the Sauk, Tippecanoe, and Kaskaskia Megasequences) was thickest in the eastern half of the U.S.—often deeper than two miles! In contrast, the early Flood deposits across much of the West are commonly less than a few hundred yards deep, and in many places there was no deposition at all (Figure 11.4).

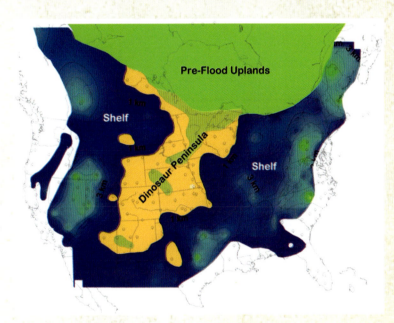

Figure 11.4. Dinosaur Peninsula

The dinosaurs evidently could survive through the early Flood in the West because they were able to move to the elevated remnants of land, places where the related sedimentary deposits aren't as deep. I call this high ground Dinosaur Peninsula. These dinosaurs were able to escape burial, at least in the early stages of the Flood.

However, later in the Flood (during deposition of the Absaroka and Zuni Megasequences), things changed dramatically. Pangaea, the former supercontinent made up of most of today's continents, began to break up. This change in tectonics, combined with increasing water levels, caused great changes in the ways the rock layers were deposited. Violent, tsunami-like waves washed across western North America, while virtually no sedimentation was occurring in the East. This is a complete reversal of the pattern observed earlier in the Flood.

Rock sequence data show that more than three miles of sediment rapidly accumulated across the American West during the Absaroka and Zuni Megasequences. This apparently overwhelmed and buried the Triassic, Jurassic, and Cretaceous dinosaurs that couldn't escape the Flood. As the waters rose, Dinosaur Peninsula began flood-

ing from south to north. We also find the largest herds of dinosaurs, in the form of dinosaur fossil graveyards, in the Upper Cretaceous system sediments in northern Wyoming, Montana, and Alberta, Canada. It's as if the dinosaurs were fleeing northward up the peninsula as the waters advanced from the south. By Day 150 of the Flood (Genesis 7:24), even the uplands area to the north, in present Canada, was covered by the floodwaters (Figure 11.4).

In his book *Digging Dinosaurs*, American paleontologist Jack Horner reported the discovery of a huge dinosaur graveyard—over 10,000 adult *Maiasaura* in a small area, and yet no young were mixed in with them.[5] What could have caused this odd sorting? In a Flood model, this is easily explained: The adult dinosaurs were likely stampeding away from the imminent danger of raging floodwaters. Their young could not keep up and became engulfed in some lower part of the peninsula.

More research is being done on the stages of the Flood and the order in which the continents were submerged. But each answer provides new insight into the great catastrophe that forever altered the topography of our world.

Why Aren't There Dinosaurs in Grand Canyon?

Most Grand Canyon rocks are from the Sauk through Kaskaskia Megasequences (Lower Paleozoic rocks). The uppermost rocks from the Absaroka Megasequence are, in fact, a bit younger, but the Tapeats Sandstone through Redwall Limestone are from the earliest megasequences. Many geologists draw long cross-sections from Grand Canyon, Arizona, north to Bryce Canyon, Utah, and simply and naively place each rock sequence atop the preceding sequence, resulting in a pile of sedimentary rocks over 10,000 feet thick with the bottom 5,000 feet composed of Grand Canyon rocks (Figure 11.5).[6]

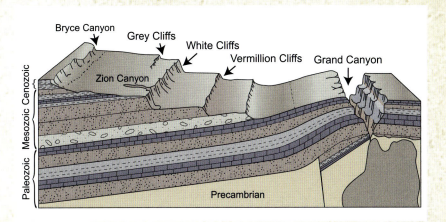

Figure 11.5. Traditional (but wrong) cross-section showing the rock layers of Grand Canyon maintaining their same thickness and extending under Bryce Canyon

Because dinosaurs are found only in the upper layers of this cross-section (Mesozoic rocks), some old-earth geologists have wondered how dinosaurs were able to stay afloat in a global flood while 5,000 feet of Grand Canyon sediment were deposited before becoming buried later in Mesozoic rocks. These geologists scoff at the notion of a global flood, but they

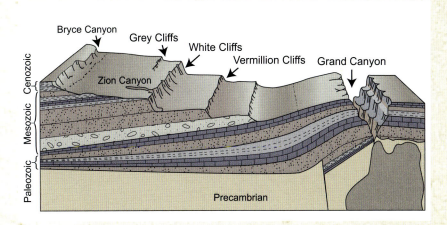

Figure 11.6. Correctly drawn cross-section showing the rock layers of Grand Canyon thinning dramatically (sometimes even to zero) under Bryce Canyon

have not looked closely at the actual rock data. The oil wells north of Grand Canyon tell a completely different story from theirs.

North of Grand Canyon, and especially northeast, the rocks of the Lower Paleozoic section (the rocks of Grand Canyon) thin dramatically and in many places are absent altogether. This is how we identified Dinosaur Peninsula discussed above. A true cross-section showing the proper thicknesses of rocks in place from Grand Canyon north to Bryce Canyon should show dramatically less Grand Canyon rocks at the base. We drew such a section, shown in Figure 11.6. According to this interpretation, the dinosaurs didn't even live in the area of Grand Canyon before the Flood. The pre-Flood Grand Canyon area would have been south of the Dinosaur Peninsula and likely a shallow sea. This is why the fossils in Grand Canyon are marine fossils. Dinosaurs aren't found until later as the Flood sediments of the upper Absaroka and Zuni Megasequences were being deposited. Dinosaurs didn't have to swim while Grand Canyon rocks were being flooded and deposited. They were still on dry land until the floodwaters rose higher in the Absaroka and Zuni, inundating Dinosaur Peninsula. At that point, they became buried in Flood sediments (Mesozoic rocks).

This is the same reason that no dinosaurs are in the American Midwest. This area was east of the Dinosaur Peninsula that ran from Minnesota to New Mexico, which secular geologists call the Transcontinental Arch and later the Ancestral Rockies. Most old-earth and secular geologists recognize these features but ignore them when they draw their cross-sections. We just have to use the actual rocks when we make our cross-sections. Rock data don't lie.

Figure 11.7 illustrates the thinning of the Sauk, Tippecanoe, and Kaskaskia Megasequences from the southwest to the northeast of Grand Canyon. This is drawn directly from our stratigraphic column data and reflects exactly what the oil wells show.

The Mystery of Devonian Black Shales

Black shale deposits are ubiquitous in the Middle and Upper Devonian rocks across the U.S. and Canada and even globally. The Woodford Shale of the Permian Basin, the Marcellus Shale of the Appalachian Basin, the New Albany Shale of the Illinois Basin, and the Antrim Shale of the Michigan Basin are all Middle to Upper

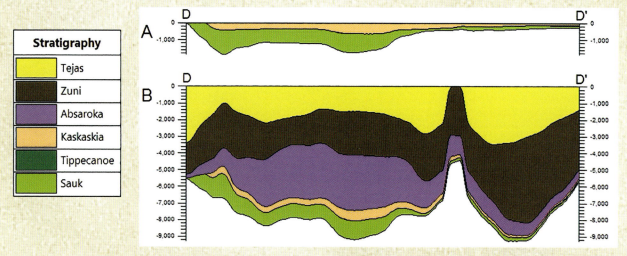

Figure 11.7. Stratigraphic cross-section D–D' (A) showing Grand Canyon megasequence rocks (Sauk and Kaskaskia) thinning to the northeast, and (B) all megasequence rocks along the same profile

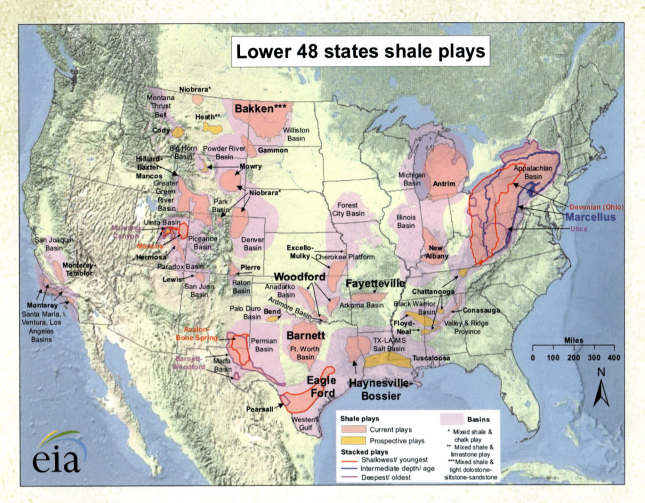

Map of oil and gas industry shale plays across the United States

Devonian black shales that formed nearly simultaneously across North America.[7]

The reason for the black color of these shales is usually their high concentration of organic material. Approximately 95% of all organics found in sedimentary rocks are found in shales.[8] Shales can contain up to 40% organic matter by weight but average much lower values.[8] Oil and gas geologists refer to organic-rich black shales as source rocks because they are assumed to be the source of kerogen and hydrocarbons. Many of these black shales are actively being exploited with modern fracking techniques, literally going right to the source.

"Rock data don't lie."

Black shales are thought to form in stagnant water bodies like the Black Sea where the water column becomes sufficiently stratified and oxygen levels are low at depth, allowing organic debris to accumulate faster than its consumption by bacteria. How-

Devonian black shale

ever, as modern research is demonstrating with flume tanks, laminated shales (the majority of the Devonian black shales) are deposited by moving water, creating difficulty in meeting the aforementioned stagnant and low-oxygen conditions.[9]

The massive extent of these black shales at all the same time in the geological record remains a mystery to secular geologists. They realize that you cannot have anoxic conditions everywhere at the same time or most ocean life would be destroyed. And the laminated nature of these shales shows the water column was not stagnant, eliminating the assumed uniformitarian conditions for their formation.

Wicander and Monroe conclude:

> Although the origin of these extensive black shales is still being debated, the essential features required to produce them include undisturbed anaerobic bottom water, a reduced supply of coarser detrital sediment [sand], and high organic productivity in the overlying oxygenated waters.[10]

So, the uniformitarian model requires stagnant water and extremely low oxygen to preserve the organic material, whereas the laminations observed in the rocks require moving water to form. No wonder there is a mystery here. You cannot simultaneously satisfy both of these conditions with the current secular explanation for black shales. Nor is there a satisfactory explanation for their massive extent globally.

In contrast, the Flood model provides a better explanation for the formation of laminated black shales. Moving floodwaters transported and buried organic debris faster than it could be consumed, resulting in highly organic black laminated shales. All it takes is rapid burial. These are exactly the conditions we have in the Flood.

Human Perspective

Humans living on the earth at this time were experiencing continued intense rainfall, massive earthquakes, and tsunami waves reaching higher and higher elevations. Although most tsunamis were still not impacting the highest elevations, we do see a bit more coal and even a few land animal fossils appearing in the Kaskaskia Megasequence, indicating that some tsunamis were reaching the edges of the formerly dry land. Humans were most likely beginning to move to higher elevations in preparation for the inevitable rising water level. Animals would also migrate to higher elevations, including the dinosaurs and animals that lived at lower elevations. The humans not on the Ark were likely full of regret and remorse for not heeding the warnings of Noah.

Conclusion: The Water Continued to Rise in the Kaskaskia

The Kaskaskia is similar in many ways to the Sauk and Tippecanoe before it. The stratigraphic data show that a significant rise in sea level advanced across all three continents simultaneously, similar to the previous two megasequences but even higher. This megasequence contains many of the same shallow marine types of fossils, with few land animals or plants. In addition, the Kaskaskia follows the same general patterns as the previous megasequences. This similarity is reflected in the consistency in total surface area covered for each of the first three megasequences, found to vary from only 22.6 to 23.8 million square kilometers. If this is true, then most of the sedimentary rocks in the Kaskaskia were merely piled on the rocks deposited earlier, essentially only covering the same areas of the continents. In contrast, the surface area coverage of the Absaroka Megasequence goes up dramatically and covers many new locations of the continents.

The Kaskaskia is also similar to the Sauk and Tippecanoe in the total volume of rock and in the total thickness deposited. The average thickness of each of the first three megasequences falls within the range of 462 to 541 meters, and the total rock volume varies from 10.4 to 12.4 million cubic kilometers. We will see that in the next megasequence, the Absaroka, this thickness goes up dramatically, nearly doubling, and the total volume increases to over 33 million cubic kilometers.

The plate configuration is only slightly different in the Kaskaskia as well. The Iapetus Ocean (pre-Atlantic) was essentially gone by the time the Kaskaskia is finished. The northern and southern portions were subducted away, placing Europe, North

America, Africa, and South America together in a traditional Pangaea configuration. Note that we originally placed Florida and the southeastern U.S. as part of West Africa.

The lack of surface coverage of most of the continents, particularly across Africa and South America, supports the notion of only limited subduction through this point in the Flood. The more subduction, the more new seafloor created. And more new seafloor rises and displaces more water, pushing sea level higher and higher (see chapter 6). Through the first three megasequence cycles, we do not see physical evidence of massive continental flooding. It all seemed to be confined to the lowest pre-Flood locations, affecting mostly shallow seas, since these megasequences contain about 99% marine fossils. North America (the contiguous U.S. in particular) has a lot more coverage across the continent during these first three megasequences, probably because it was mostly a shallow sea in the pre-Flood world (see chapter 17).

Finally, I have lumped the first three megasequence cycles into the first 40 days of the Flood primarily because the fossils show only marine influence and the fact that the Ark was not afloat until Day 40 (Genesis 7:17). These two factors tell me that the

Surface Area (km²)	North America	South America	Africa	Total
Sauk	12,157,200	1,448,100	8,989,300	22,594,600
Tippecanoe	10,250,400	4,270,600	9,167,200	23,688,200
Kaskaskia	11,035,000	4,392,600	7,417,500	22,845,100
Absaroka	11,540,300	6,169,000	17,859,900	35,569,200
Zuni	16,012,900	14,221,900	26,626,900	56,861,700
Tejas	14,827,400	15,815,200	24,375,100	55,017,700
Volume (km³)	North America	South America	Africa	Total
Sauk	3,347,690	1,017,910	6,070,490	10,436,090
Tippecanoe	4,273,080	1,834,940	6,114,910	12,222,930
Kaskaskia	5,482,040	3,154,390	3,725,900	12,362,330
Absaroka	6,312,620	6,073,710	21,075,040	33,461,370
Zuni	16,446,210	23,198,970	57,729,600	97,374,780
Tejas	17,758,530	32,908,080	28,855,530	79,522,140
Average Thickness (km)	North America	South America	Africa	Total
Sauk	0.275	0.703	0.675	0.462
Tippecanoe	0.417	0.430	0.667	0.516
Kaskaskia	0.497	0.718	0.502	0.541
Absaroka	0.547	0.985	1.180	0.941
Zuni	1.027	1.631	2.168	1.712
Tejas	1.198	2.081	1.184	1.445

Surface area, sediment volume, and average thicknesses for North America, South America, and Africa for each of the six megasequences

Flood hadn't affected much of the dry land parts of the continents up to this point and that subduction was limited to only selected areas with limited amounts of new oceanic lithosphere created. These conclusions explain the limited amounts of coverage across the continents that are observed for the Sauk through Kaskaskia Megasequences.

In the next chapter, we will turn our attention to the Absaroka Megasequence. This is the point in the Flood when things went from really bad to much, much worse. During the Absaroka, the land portions of the continents begin to be flooded, and runaway subduction starts to operate on a grander scale. The net result is more new seafloor, which pushes up the ocean from below, creating a higher overall sea level. Tsunami waves were then able to race across the continents, reaching higher elevations and burying land and marine flora and fauna in the process. And this is all supported by the actual rock data. It's all beginning to make sense.

> "The Ark was not afloat until Day 40."

References
1. Wicander, R. and J. S. Monroe. 2013. *Historical Geology,* 7th ed. Belmont, CA: Brooks/Cole.
2. Chadwick, A. V. 1993. Megatrends in North American Paleocurrents. *Society of Economic Paleontologists and Mineralogists Abstracts.* 8: 58.
3. Sloss, L. L. 1963. Sequences in the Cratonic Interior of North America. *Geological Society of America Bulletin.* 74 (2): 93-114.
4. Baumgardner, J. 2005. Recent Rapid Uplift of Today's Mountains. *Acts & Facts.* 34 (3).
5. Horner, J. R. and J. Gorman. 1988. *Digging Dinosaurs.* New York: Workman Publishing, 128.
6. Stearley, R. 2016. Fossils of the Grand Canyon and Grand Staircase. In *The Grand Canyon, Monument to an Ancient Earth: Can Noah's Flood Explain the Grand Canyon?* C. Hill et al, eds. Grand Rapids, MI: Kregel Publications, 131-143.
7. Dong, T. et al. 2018. Relative sea-level cycles and organic matter accumulation in shales of the Middle and Upper Devonian Horn River Group, northeastern British Columbia, Canada: Insights into sediment flux, redox conditions, and bioproductivity. *Geological Society of America Bulletin.* 130 (5/6): 859-880.
8. Blatt, H., G. Middleton, and R. Murray. 1980. *Origin of Sedimentary Rocks,* 2nd ed. Englewood Cliffs, NJ: Prentice-Hall, Inc.
9. Schieber, J. et al. 2007. Accretion of mudstone beds from migrating floccule ripples. *Science.* 318 (5857): 1760-1763.
10. Wicander and Monroe, *Historical Geology,* 237.

12 Things Go from Bad to Worse: Absaroka Megasequence

> **Summary:** The Sauk, Tippecanoe, and Kaskaskia Megasequences were deposited during the first 40 days of the Flood. The Absaroka Megasequence marks a critical turning point. Things went from bad to worse. Land was now being flooded. Land animals and plants begin to appear as fossils in these rock layers. The Absaroka increases dramatically in thickness, more than double any of the earlier megasequences. This shows that floodwaters washed much higher on land than during the Sauk, Tippecanoe, or Kaskaskia. The Ark began lifting off the ground. Increasing global subduction, creation of new seafloor, and increasing thickness of the Absaroka all remarkably correspond, logging a critical chapter in the story of the Flood.

Flooding of the Land Begins

The Absaroka Megasequence marks a critical juncture in the Flood account when things went from bad to worse. After 40 days of intense rain and tsunamis crashing across the shallow seas on the edges of the continents, the seas begin to rise higher and flood some of the land, including lifting up the Ark. Early in the Absaroka, the collision of North America with Europe and South America (Gondwana) was complete. These resulted in the deformation and folding and thrusting of many of the earliest Flood sediments known as the Hercynian (Variscan)-Alleghenian Orogeny.[1] This completed the formation of the traditional supercontinent of Pangaea as most people know it. Before this, the continents were in a slightly different pre-Flood supercontinent configuration I refer to as modified Pangaea. Creation geophysicist Dr.

John Baumgardner calls this supercontinent Pannotia, but they are really about the same thing.[2]

Later in the Absaroka Megasequence cycle, subduction along the U.S. West Coast commenced, and the various plates of the Pacific Ocean began their rapid development. The supercontinent of Pangaea was wrenched apart once again, beginning with rifting that separated North America from West Africa, creating the new seafloor of the North Atlantic Ocean.

> Now the flood was on the earth forty days. The waters increased and lifted up the ark, and it rose high above the earth. (Genesis 7:17)

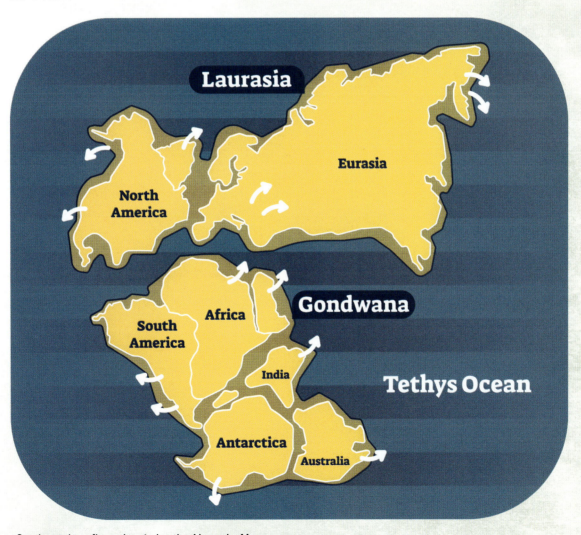

Continental configuration during the Absaroka Megasequence

What Do the Rocks Show?

North America

The Absaroka Megasequence extends from the Pennsylvanian to the Lower Jurassic system. This megasequence marks a major shift in depositional pattern and initiates the dominance of siliciclastic (sand and shale) deposition across North America. The basal layer is predominantly a mixture of sandstone and shale, but significant deposits of volcanic rocks also mark some locations along the West Coast and Alaska. These volcanic rocks are part of the subduction and accretion process that initiated along the western coast during the Absaroka sequence. This is the time in the Flood when massive amounts of the original ocean lithosphere began to be rapidly destroyed through runaway subduction and a new seafloor was formed at the ocean ridges, including the formation of the seafloor for the Pacific Ocean. This megasequence also recorded the development of the new seafloor forming the Atlantic Ocean off the East Coast and the formation of a new passive margin and ridge system. The thickest sedimentary layers were deposited across the American Southwest, where many areas received over 3 kilometers of mostly siliciclastic rocks, including the Coconino Sandstone. Uplift of the Appalachians may have prevented widespread deposition across much of the eastern U.S. Sandstones make up about 32% of this megasequence by volume. Shale rocks were a little more common at 35%, and carbonate rocks diminished to about 23% of the total volume deposited. Over 6.3 million cubic kilometers of Absaroka sediment was spread over a surface area of about 11.5 million square kilometers, averaging about 547 meters thick.

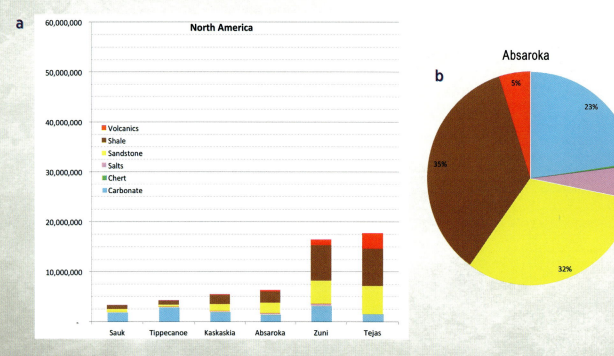

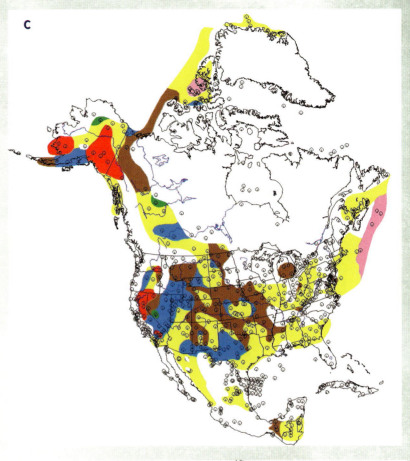

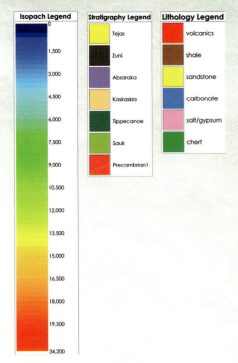

Compilation of Absaroka Megasequence data for North America: (a) Histogram of the type and volume of rocks in place across North America for each megasequence; (b) pie chart of the total Absaroka Megasequence by rock type; (c) map of the bottommost (basal) Absaroka rock type; (d) thickness map of the total Absaroka Megasequence; (e) stratigraphic cross-sections showing the relative thickness of the Absaroka Megasequence in purple on top of the earlier megasequences. Histogram measurement is in cubic kilometers, isopach map measurement is in meters.

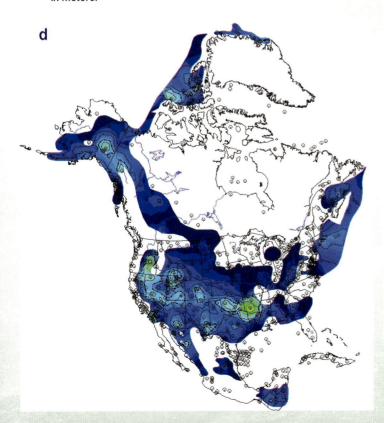

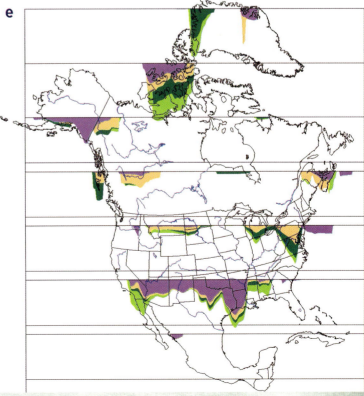

South America

This megasequence also marks a major shift in depositional pattern across South America, with the volume of sedimentary rock doubling the amount deposited during the Kaskaskia sequence. The amount of surface area covered by this sequence also increased to over 6 million square kilometers. The Absaroka Megasequence also shows a transition to more carbonate, salt, and rock gypsum deposition, lowering the proportion of siliciclastics. The basal layer is again predominantly sandstone, but some basal carbonate is present in the Acre and Madre de Dios Basins of Peru. The average thickness of the Absaroka Megasequence is nearly 1,000 meters. Limited amounts of sediment were also deposited offshore in a couple of locations along the eastern edge of the continent. These may be an early indication of the eventual breakup with Africa. Sandstones make up about 39% of the entire Absaroka Megasequence. Shale is the second-most common rock type by volume at 30%, followed by carbonate at 19%. The Absaroka also has the highest percentage of salt and gypsum deposits (8%) of any of the sequences.

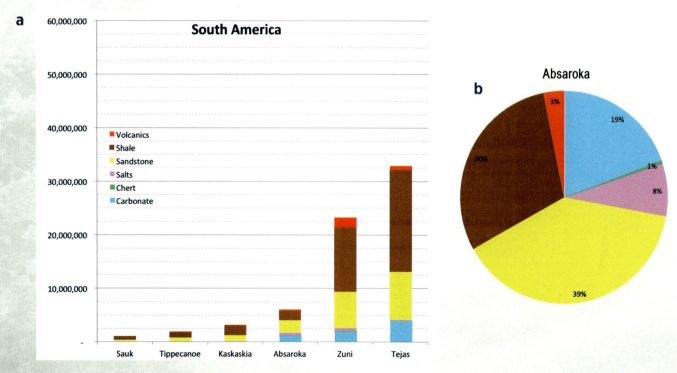

Compilation of Absaroka Megasequence data for South America: (a) Histogram of the type and volume of rocks in place across South America for each megasequence; (b) pie chart of the total Absaroka Megasequence by rock type; (c) map of the bottommost (basal) Absaroka rock type; (d) thickness map of the total Absaroka Megasequence; (e) stratigraphic cross-sections showing the relative thickness of the Absaroka Megasequence in purple on top of the earlier megasequences. Histogram measurement is in cubic kilometers, isopach map measurement is in meters.

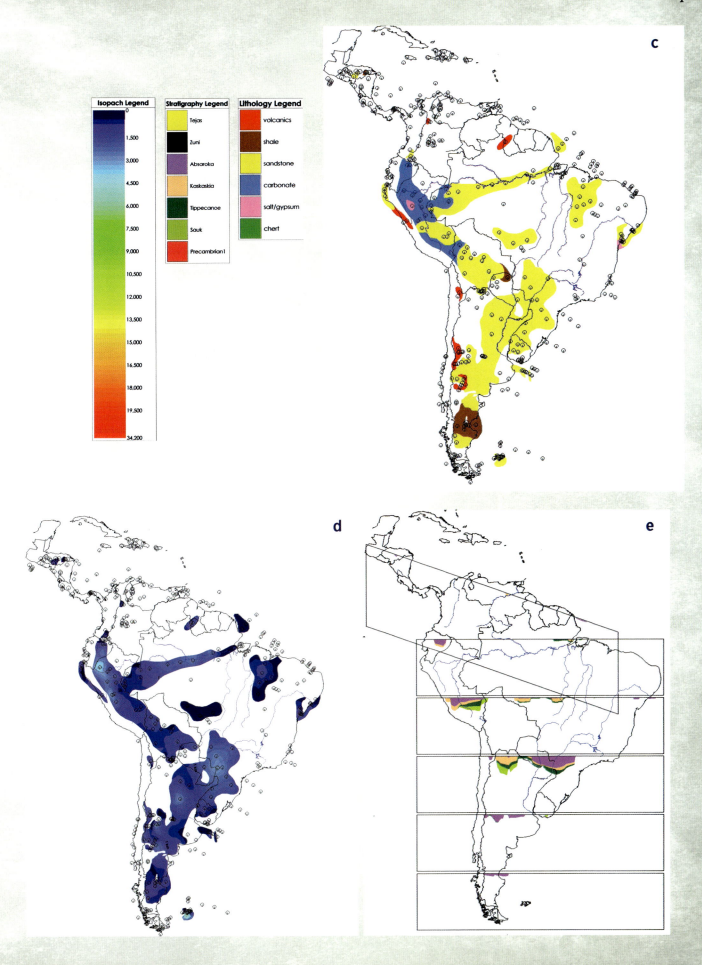

Africa

This megasequence also marks a major shift in depositional pattern across Africa, with the volume of sedimentary rock more than tripling the amount deposited in any single earlier sequence. The Absaroka Megasequence also shows a transition to more carbonate, salt, and rock gypsum deposition, and less siliciclastics. The basal layer is predominantly sandstone (Karoo Supergroup and equivalent) and is spread more uniformly across the continent, except for the immediate sub-Saharan region, which is largely devoid of Absaroka deposition. Depo-centers are centered in the Congo, Kenya, Madagascar, and across the North African coast and the Middle East. Deposition off the northwest coast of Africa marks the breakup of Pangaea into Gondwana and Laurasia, and the beginning of the Atlantic Ocean. Sandstones make up about 33.3% of the entire Absaroka Megasequence. Shale is the second-most common rock type by volume at 28.4%, followed by carbonate at 22.8%. The Absaroka also has the highest percentage of salt and gypsum deposits (12.9%) of any of the megasequences in Africa.

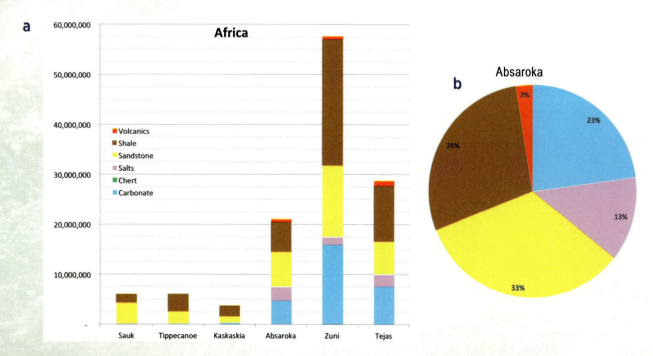

Compilation of Absaroka Megasequence data for Africa: (a) Histogram of the type and volume of rocks in place across Africa for each megasequence; (b) pie chart of the total Absaroka Megasequence by rock type; (c) map of the bottommost (basal) Absaroka rock type; (d) thickness map of the total Absaroka Megasequence; (e) stratigraphic cross-sections showing the relative thickness of the Absaroka Megasequence in purple on top of the earlier megasequences. Histogram measurement is in cubic kilometers, isopach map measurement is in meters.

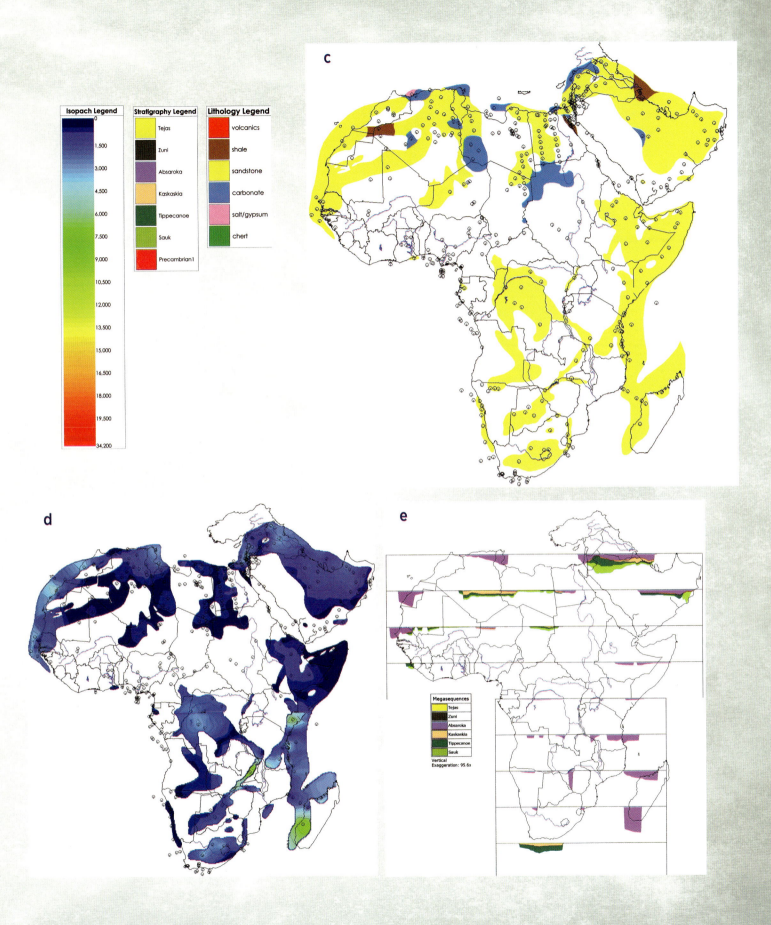

Current Data for the Absaroka Megasequence

Figure 12.1 shows the paleocurrent data collected by Dr. Art Chadwick and his students across North America, South America, and Africa for Permian system rocks within the lower part of the Absaroka Megasequence.[3]

Across North America, the Absaroka shows a much stronger easterly directed current pattern than in previous megasequences, especially in the Southwest. South America also shows a westerly directed flow across eastern and southern Brazil, but a more pronounced easterly current pattern was observed across Argentina. Africa had very little current data in the Permian section except for a strong northeasterly flow direction mapped across the Saudi Arabian Peninsula.

Figure 12.1. Current directions compiled across (a) North America, (b) South America, and (c) Africa for the Permian system (Lower Absaroka). Courtesy of Art Chadwick.

Current data within the three previous megasequences were much more consistent than that observed in the Absaroka. The Sauk, Tippecanoe, and Kaskaskia all showed only slight shifts in current direction from megasequence to megasequence. However, the Absaroka was a time of sudden change. New patterns in flow direction indicate something different was occurring. Changes in plate motion and the onset of subduction along the Pacific margin were most likely the driving factors in changing the Absaroka current directions.

Summary of the Rock Data

The Absaroka data show a rather abrupt global increase in the volume of sediment deposited and in the amount of surface area that it covered. In fact, the global volume of sediment deposited in the Absaroka Megasequence more than doubled the amount deposited by any of the earlier individual megasequences. This increase is also reflected in the thickness of the Absaroka, showing an increase to an average of 941 meters. Africa showed the most dramatic increase in sedimentary volume during the Absaroka, with a nearly sixfold increase in volume compared to the previous Kaskaskia Megasequence.

Glen Canyon, Utah

The percent by volume of salt and gypsum-rich rocks reached their highest levels across all three continents during the Absaroka Megasequence. Exactly why this occurred is unclear. It may be because of the increase in runaway subduction globally and the corresponding sudden increase in seafloor spreading. Salt and gypsum-rich rocks seem to result from ocean rifting (seafloor spreading). Whether these are hydrothermally derived salts or some type of supercritical flows needs further investigation by geochemists. However, there is without a doubt a relationship between an increase in ocean rifting and an increase in salt and gypsum deposition.

The Absaroka also shows a major shift in the locations of deposition compared to the previous three megasequences. This is also reflected in the paleocurrent directions. The three previous megasequence cycles all seemed to align and stack one on top of the other across each of the three continents. In contrast, the Absaroka was deposited across many new locations of the continents. For example, we see much more extensive coverage across southern Africa. And we see a completely different depositional picture developed across North America, with most of the deposition shifting to the southwestern U.S. South America again shows more substantial areal coverage, as it did for each of the previous megasequences, including new massive deposits across Argentina.

Continental Configuration for the Absaroka Megasequence

By the earliest part of the Absaroka Megasequence cycle, the major continents of the world had completed the formation of the supercontinent Pangaea. This resulted in renewed deformation along the Appalachian Mountains (including many overthrusts) and the intense folding within the Hercynian mobile belt across western Europe.[1] These deformational events folded and faulted many of the earliest deposits of the Flood. We can determine the relative timing of these events based on the rocks they deform and the overlying rocks that are less affected or completely undeformed.

Erosion and renewed deposition as each megasequence cycle progressed do not require vast amounts of time. It was this claim of requiring vast amounts of time between one deformational cycle and another depositional cycle that put secular geologists on the wrong track in the late 18th century. James Hutton was wrong in his analysis of the angular unconformity at Siccar Point, Scotland. Folding does not require a long period of time if the sediments are still unlithified and pliable. Nor does erosion take vast amounts of time to plane off the tops of folded sediments, especially with tsunami wave energy. In fact, in most places of the world we see each megasequence of rocks deposited on another megasequence of rocks in parallel, exhibiting no difference in their angles of deposition. In other words, the norm is flat-lying rocks deposited on flat-lying rocks like bricks in a wall. There is no deep time required between each megasequence, as Hutton declared. These early geologists forgot about the effects of the global Flood, or maybe they never quite understood the geology as well as they thought.

The rock record is not the result of slow, uniformitarian rates of deposition. Most of the sedimentary rocks were deposited rapidly, as discussed in chapter 2.

Later in the Absaroka (the Triassic system), rifting was renewed along the U.S. East Coast as North America began to split with West Africa. North America was pulled apart from Africa as subduction began along the West Coast. We see evidence of this

Unconformity at Siccar Point, Scotland

in the offshore deposits along the East Coast of North America and the West Coast of Africa. In this megasequence cycle, things changed dramatically across the globe. The U.S. West Coast was now actively subducting, and new seafloor was being created in the Pacific Ocean at a high rate of speed. Rapid movement of North America to the west was caused by runaway subduction along the West Coast, forcing the continent to overrun the ocean ridge (East Pacific Rise). This is one of the only locations in the world where this has occurred on this vast of a scale. A hot, buoyant ridge system is normally difficult to subduct. However, the rapid runaway subduction that occurred in the Flood makes this possible.

The subduction process also began to cause accretion of pre-Flood and early Flood ocean sediments and/or volcanic islands along the West Coast of North America. This caused the continent to expand westward as the subduction process left the deposits from the seafloor smeared onto the continent as the ocean crust was pulled away and consumed beneath, adding California, Oregon, and Washington, as well as parts of Alaska, sliver by sliver. The more massive scale of subduction that was occurring during the Absaroka and its associated volcanism caused a noticeable spike in global volcanic activity. Volcanoes were also adding new pieces of granitic, continental crust over the active subduction zones around the Pacific Rim.

Continental configuration during the deposition of the Absaroka Megasequence and its sedimentary thickness

World-Changing Tectonic Activity

I think that at this point in the Flood the tectonics of the world dramatically changed. The cause of this shift can be tied directly back to a sudden increase in the amount of runaway subduction occurring globally. Prior to the Absaroka, it seems that runaway subduction was limited to only selected locations, like the destruction of the narrow Iapetus Ocean between North America, Europe, and possibly a few other

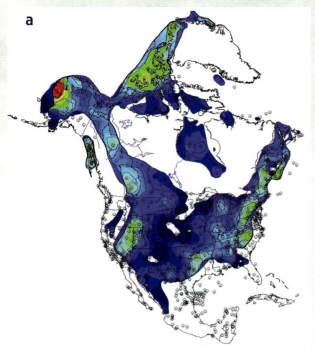
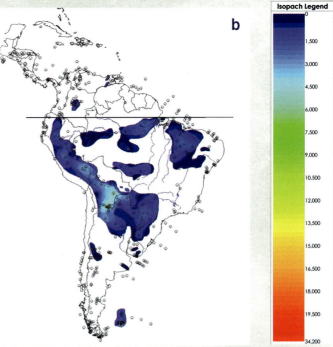
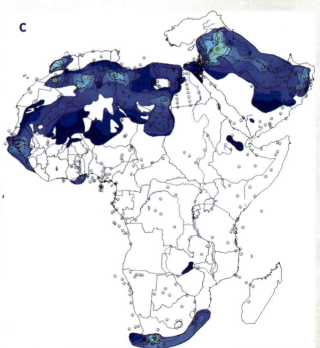

locations globally. And this is reflected in the limited flooding that occurred in the Sauk through Kaskaskia Megasequences (Figure 12.2). During the first three megasequence cycles, only areas that were primarily pre-Flood shallow seas were inundated and little dry land was affected, as exhibited by the fossils found in these megasequences. This was likely because there was not enough new seafloor created from the limited amount of subduction to push the global sea level high enough to flood much of the land.

Dr. John Baumgardner mentioned that one of the consequences of runaway subduction that we should be able to observe is a pulling down of the continental crust below sea level by as much as

Figure 12.2. Cumulative thickness and extent of the Sauk through Kaskaskia Megasequences across (a) North America, (b) South America, and (c) Africa. Note many of the continents had thin to limited coverage up to this point in the Flood. Measurement is in meters.

3 kilometers as subduction begins (and subsequent infilling with sediments into the wedge-shaped drawdown).[4] During the Absaroka, subduction of the pre-Flood oceanic lithosphere was initiated along the rim of the Pacific on a massive scale. This caused

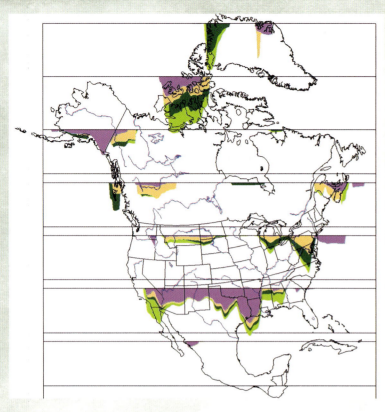

Stratigraphic cross-sections across North America showing the thick Absaroka deposition (purple) in the American Southwest

tremendous amounts of new seafloor to be created, pushing up the global sea level dramatically. The earthquake-generated tsunami waves could now strike across the formerly dry land portions of the continents. Keep in mind that the continental crust was also being pulled down about 3 kilometers by the runaway subduction process along the West Coast of North America. This caused the dramatic shift in the location of deposition that is observed during the Absaroka and a massive amount of sediment spread across the American Southwest. Note that several kilometers were deposited there at this time, again confirming Baumgardner's prediction of 3 kilometers of drawdown.[4]

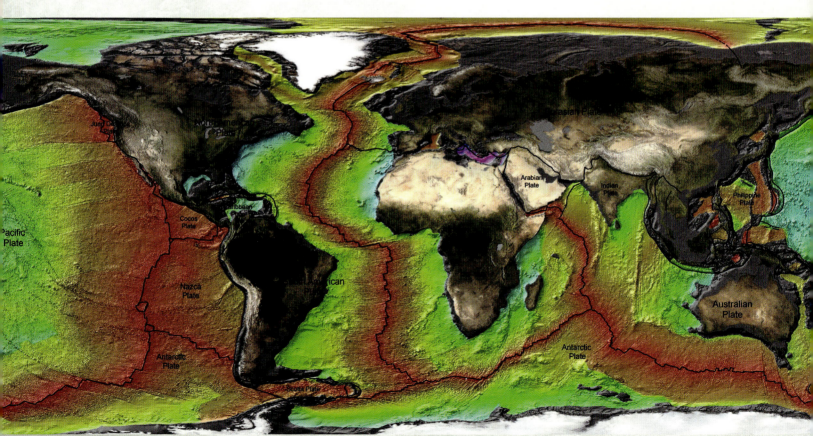

Figure 12.3. Map of the age of the ocean crust. The oldest crustal rocks (light blue) are part of the Absaroka Megasequence.

Cathedral Rock in Sedona, Arizona

Not Just a Coincidence

The Absaroka introduces a lot of firsts to the geologic record. It is not just a coincidence that so much occurs at the same time at this point in the Flood. These events had a common cause. Sea level was pushed upward dramatically in the Absaroka, resulting in the first areas of dry land being inundated across the globe. This began to change everything in the rock record.

During the Absaroka, the entire ocean floor began to be created anew (Figure 12.3). Runaway subduction was now happening all over the globe on a massive scale. As Pangaea began to break apart, the Pacific Ocean plates began to subduct along the edges and simultaneously create the beginnings of an entirely new global seafloor at the ocean ridges. The oldest ocean crust in the world only goes back to the Absaroka Megasequence. I think the creation of this new seafloor, led primarily by the activity in

Pennsylvanian system coal bed in Arkansas

Cross-bedding in Coconino Sandstone

the Pacific region, ultimately drove the water high enough to Flood the entire globe. But that wouldn't happen until a bit later in the Zuni Megasequence.

As the newly created seafloor pushed up more of the water from below, sea level rose dramatically in step with the volume that was created. We see evidence of this in the increased surface coverage of the continents during the Absaroka and in the volume of sediment that nearly tripled from the previous megasequence.

The Absaroka exhibits the first appearance of the so-called aeolian (desert) sandstones across the globe. These massive, cross-bedded sand wave deposits have been interpreted as aeolian by secularists because they cannot imagine water waves big enough to create them in their uniformitarian models. However, Dr. John Whitmore and Stephen Chueng have shown quite clearly that the minerals found in these sand deposits and the angles of the cross-beds fit better as water deposition.[5,6] Some of these deposits include the Pennsylvanian-Permian Tensleep Sandstone of Wyoming, the Permian Coconino Sandstone of Arizona, the Permian-Triassic New Red Sandstone of England, the Permian Lyons Sandstone of Colorado, the Triassic-Jurassic Navajo Sandstone of Utah and Colorado, and the Pennsylvanian-Permian Casper Sandstone of Wyoming.[7] All of these units were deposited in the Absaroka Megasequence, with many of them in the lower part of the megasequence. These formations are evidence of the massive high-energy sand waves that migrated across the continents at this point in the Flood.

Next, there is the first appearance in the Absaroka Megasequence of prolific coal beds across the globe. These first coal beds share a common type of plant flora, primarily lycopods and ferns. These plants are typical of coastal wetlands and swampy areas in lowland environments. Although there are some earlier coals in the rock record, they do not become prolific and extensive until the Pennsylvanian system, the lowermost part of the Absaroka (Upper Carboniferous).

I do not think there were "floating forests" growing across the pre-Flood oceans as some creationists have suggested. The geology doesn't support this hypothesis.[8] Neither were these trees as hollow as they have been portrayed.[9] Dr. Jeffrey Tomkins and I described a "rooted" lycopod tree site in Glasgow, Scotland, that shows clearly that these trees were, in fact, rooted in soil.[9] There will be more on this in chapter 18.

The Absaroka also saw the first and sudden appearance of massive amounts of terrestrial animal fossils. Amphibians show up in great numbers in the base of the Absaroka, followed by reptiles in the layers above. Even dinosaurs and mammals make sudden appearances before the Absaroka is over (Triassic system). And as described below, most of these terrestrial fossils are mixed with marine fossils.

Large marine reptiles also make their first appearance in the Triassic system of the Absaroka Megasequence. Ichthyosaurs were common fossils beginning suddenly in the Lower Triassic, and are found in rocks as high in the Flood strata as the Cretaceous system.

Finally, the so-called Permian extinction occurs in the early portion of the Absaroka. This has been hailed by secular scientists as the largest extinction of all geologic time, or at least exhibiting the most abrupt changes in fossil species. Many of the fossils found above and below this horizon are, in fact, vastly different. However, creation geologists explain extinction events as the last occurrence of organisms in the Flood record. We explain them as a result of rapid changes in water level that buried completely new types of organisms from new biozones. In this view, the so-called extinctions are merely a record of abrupt disappearances of many organisms at the same level in the fossil record. The Permian-Triassic event may correlate with the high-water mark of the Absaroka, or possibly one of the highest water levels.

The coinciding of these seemingly unrelated events is not by chance. As runaway

Swimming reptiles like the plesiosaur (near right), aquatic reptile (far right), and ichthyosaur (below) were quite specialized, having been equipped to thrive in hostile waters.

Science learned from the ichthyosaur fossil that this reptile gave birth to live young rather than laying eggs. The ichthyosaur must have been buried extremely quickly since the entire process of giving birth would not have taken a long period of time. This is not a record of the slow accumulation of sediments around a dead carcass but a record of death by rapid burial as sediments cascaded around the struggling animal.

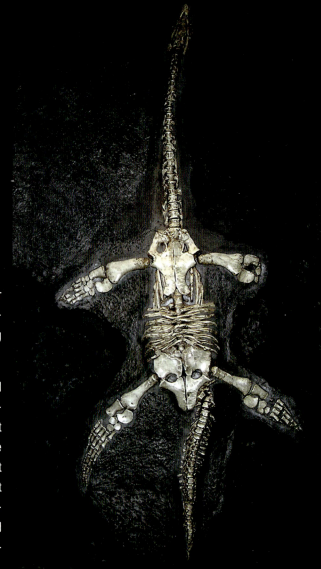
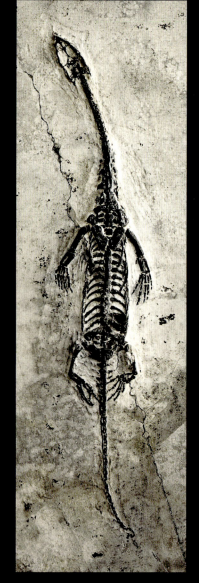

subduction began to operate on a massive scale, the newly created seafloor pushed the water levels higher. This allowed the tsunami waves to rapidly move higher and further inland, inundating land animals and plants and leaving their remains behind in the rock record. At the same time, the tsunami waves also left behind vast deposits of huge, high-energy cross-bedded sandstones across the continents.

Cyclothems: What Are They and How Do They Form?

Cyclothems were first named by the European geologists working with coal seams.[10] The term was introduced to the U.S. through the work of Wanless and Weller and their study of the Illinois Basin.[11] Cyclothems are defined as alternating layers of sandstone, shale, and limestone exhibiting frequent alternations between marine and nonmarine sediments. They are most commonly observed in the early Absaroka Megasequence, in particular in Pennsylvanian strata. Secular geologists attempt to explain these occurrences as the result of fluctuations in sea level due to the expansion and melting of continental glaciations.[12] However, Haq and Schutter concluded that cyclothems cannot be solely the result of glaciation cycles since many appear in the rock record without evidence of any glacial episodes.[13] This leaves secular geologists without a satisfactory explanation for cyclothems.

In their comprehensive literature review of Paleozoic sea levels, Rygel and his colleagues found that maximum sea level fluctuations occurred during the deposition of Pennsylvanian system rocks, when many of the cyclothems formed.[12] They estimated that sea level may have risen by as much as 250 meters during the accumulation of these sediments globally. The generation of tsunami-like waves from increased runaway subduction during the Absaroka Megasequence seems to offer a better explanation for the origin of cyclothems.

As sea level rose from the production of much new seafloor, tsunami-like waves began reaching the higher parts of the U.S. midcontinent, repetitively depositing alternating layers of limestone and shale as each wave came and went. The activity of these individual waves, with each tsunami consisting of many wave pulses, provides a logical rationale for the cyclothems in the middle of the continental U.S. The results are similar to the repetitive nature of turbidites, deep-sea landslide deposits that leave behind a common sedimentological pattern with each occurrence. In contrast, cyclothems formed in the shallower water sections of the flooding continent where the waves broke and then retreated prior to the next wave advance. Cyclothems can then be thought of as the swash zone of the tsunami waves that were advancing across the North American continent.

Mixing of Land and Marine Fossils Common in the Same Rocks

As mentioned above, the Absaroka is the first megasequence to contain a widespread mixing of land and marine fossils. Every megasequence thereafter (Zuni and Tejas) also exhibits this same phenomenon. A global flood washing in huge waves from the sea onto land is the best explanation for these mixed environments.

On a recent visit to the Royal Tyrell Museum in Alberta, Canada, I headed straight to the famous Dinosaur Hall that houses over 40 mounted specimens, including the *Tyrannosaurus rex* known as Black Beauty. However, a seemingly insignificant pair of fossil fish caught my eye. They illustrate the fallacy inherent in uniformitarian thought and interpretation.

The first display was a beautifully preserved fossil herring with signage stating "Modern herring live in saltwater, but close Eocene relatives were abundant in the freshwater lakes of western North America." A second display fea-

"Black Beauty" *T. rex* fossil

tured a spectacularly preserved fossil ray and claimed, "Rays are rarely preserved as fossils, in part because their skeletons are made of cartilage rather than bone. Most rays prefer saltwater, making this freshwater form an even more remarkable fossil."

Why are these fossil fish, which look nearly identical to modern herring and rays that live exclusively in the marine realm, claimed to be ancient freshwater fish in this museum? Uniformitarian scientists make this claim because these fish were found in the Green River Formation of Wyoming, and this rock unit also contains a lot of fish—like gar, paddlefish, and sand fish—that are found only in freshwater. Therefore, they have to conclude that the entire rock unit represented an ancient freshwater lake deposit.

Stingray fossil

In an *Acts & Facts* article, I discussed several similar examples of marine and terrestrial environments mixing within the same rock layer. Six species of sharks have been found in the same strata as *T. rex* fossils, and the deep-ocean-dwelling coelacanth fish has been found in rock layers with the dinosaur *Spinosaurus*.[14]

Contrary to the claims of uniformitarian scientists, there is no evidence that these fish lived in fresh water and somehow evolved to live in saltwater—the claims are entirely speculative. These fossils are nearly identical to modern fish found only in the ocean. The spectacular preservation of these specimens serves as a stunning testament to rapid burial and the globe-sweeping catastrophic nature of the floodwaters. Huge tsunami-like waves must have transported these marine fish onto the continents, mixing them in the same sedimentary deposits as the dinosaurs and other land animals.

Coelacanth

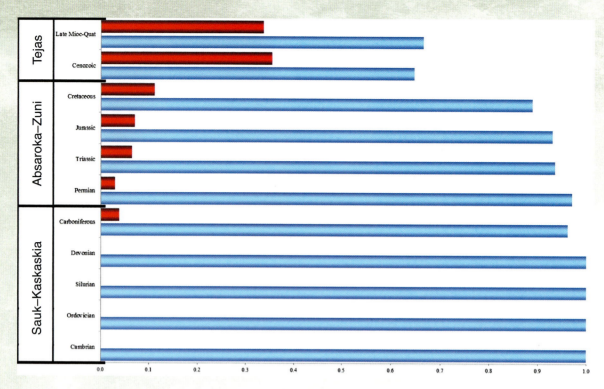

Proportion of marine (blue) vs. terrestrial (red) fossil occurrences from the Paleobiology Database. Note the sudden increase in terrestrial (land) animal fossils in the Absaroka and upward (Zuni and Tejas). Also note that the land animals are always mixed with marine fossils. Courtesy of Nathaniel Jeanson.

There are many other examples of land animals transported by floodwaters into the sea, many miles from shore. The deepest dinosaur bone discovered to date was found in an oil well core taken from the North Sea between Greenland and Norway.[15] Coal fragments from land plants have been found in marine sediments thousands of feet below the surface in an oil well off the coast of Labrador, Canada.[16] Finally, terrestrial plant debris and lignite have been found hundreds of miles east of the Falkland Islands in nearly 10,000 feet of water[17] and also in deep water off the coast of California.[18]

Uniformitarian scientists often ignore or downplay these discoveries because they are not readily explainable in their worldview. By refusing to accept the Word of God, these scientists have no recourse but to explain away the evidence as insignificant anomalies or use a rescuing device such as claiming these marine fish lived in fresh water. They forget that only the great Flood can explain the turbulence necessary to transport marine animals hundreds of miles onto the continents and sweep terrestrial organisms into the deep ocean.

How do secular scientists explain mixed environments? They usually interpret these rocks as delta deposits on the edge of land/sea. They have a rescuing device for everything. But there cannot be deltas everywhere.

Human Perspective

Conditions were rapidly eroding for humans outside the Ark at this point in the Flood. Waves were now striking higher and flooding vast portions of the lowlands. During the Absaroka, humans would likely have seen the land take its first major hit from the rising water as tsunami waves began crashing across the coastal areas. These waves would have shoved debris, mud, and sand inland, destroying, transporting, and burying everything in their path. Whole forests of lycopod trees disappeared in the advancing waves to become the first major coal seams of the Pennsylvanian system. Land animals residing in coastal and swampy terrains were swept away by the waves coming in from the oceans, mixing land and sea creatures in many of the same rock layers. Each day would have brought new waves, new earthquakes, new volcanic eruptions, and water levels climbing higher and higher across the land. Humans would probably have begun swarming together and heading toward the highest ground of the continents. Fighting and competition for the remaining dry land would have been commonplace. Many might have considered building some type of raft or small boat. It was a slow, agonizing judgment that they could sense drawing closer each day. They undoubtedly knew the end was near.

Conclusion: God Had a Plan Through the Chaos

The stratigraphic data show that a significant rise in sea level advanced across all three continents simultaneously, similar to the previous megasequences but even more extensive and higher. The Absaroka Megasequence was a pivotal moment in the Flood. At this point, things went from bad to worse, and from a rather ordered chaos to complete chaos. I suspect that the Absaroka might have lasted from Day 40 to Day 100 or so of the Flood. It is unclear exactly when the Absaroka ended. I do not think this was the highest point in the Flood. Instead, it was likely still many days away from Day 150 (Genesis 7:24; 8:3).

In our ongoing study, the geology was beginning to make sense. The megasequence data were revealing a pattern. The volume of sediment and the surface coverage across the continents were beginning to match what was happening in the seafloor. As more subduction occurred globally and new seafloor was created, the water levels were pushed higher. As the water levels rose, land plants and animals made their first appearances in great numbers. Massive cross-bedded sandstones suddenly showed up globally. A whole new seafloor began to be created. God had a plan. He also had a plan for human salvation too, if people had heeded Noah's warning. There was room for more people on the Ark, but only eight believed and walked through the door before God shut it and exercised His planned judgment.

In the next chapter we will review the rock record of the Zuni Megasequence. I am convinced that this is the high point of the Flood that was reached on Day 150. No land-dwelling, air-breathing animal outside the Ark survived beyond that point.

References
1. Wicander, R. and J. S. Monroe. 2013. *Historical Geology,* 7th ed. Belmont, CA: Brooks/Cole.
2. Baumgardner, J. 2018. Understanding how the Flood sediment record was formed: The role of large tsunamis. In *Proceedings of the Eighth International Conference on Creationism*. J. H. Whitmore, ed. Pittsburgh, PA: Creation Science Fellowship, 287-305.
3. Chadwick, A. V. 1993. Megatrends in North American Paleocurrents. *Society of Economic Paleontologists and Mineralogists Abstracts.* 8: 58.
4. Baumgardner, J. 2005. Recent Rapid Uplift of Today's Mountains. *Acts & Facts.* 34 (3).
5. Whitmore, J. H. 2004. An alternative to the mud crack origin for sand-filled cracks at the base of the Coconino Sandstone, Grand Canyon, Arizona. *Geological Society of America Abstracts with Programs.* 36 (5): 55.
6. Cheung, S. et al. 2009. Occurrence of dolomite beds, clasts, ooids and unidentified microfossils in the Coconino Sandstone, Northern Arizona. *Geological Society of America Abstracts with Programs.* 41 (7): 119.
7. Blatt, H., G. Middleton, and R. Murray. 1980. *Origin of Sedimentary Rocks*, 2nd ed. Englewood Cliffs, NJ: Prentice-Hall, Inc.
8. Clarey, T. L. 2015. Examining the floating forest hypothesis: a geological perspective, *Journal of Creation*. 29 (3): 50-55.
9. Clarey, T. L. and J. P. Tomkins. 2016. An investigation into an *in situ* lycopod forest site and structural anatomy invalidates the floating forest hypothesis. *Creation Research Society Quarterly*. 53 (2): 110-122.
10. Hampson, G., H. Stollhofen, and S. Flint. 1999. A sequence stratigraphic model for the Lower Coal Measures (Upper Carboniferous) of the Ruhr district, north-west Germany. *Sedimentology.* 46 (6): 1199-1231.
11. Wanless, H. R. and J. M. Weller. 1932. Correlation and extent of Pennsylvanian cyclothems. *Geological Society of America Bulletin.* 43: 1003-1016.
12. Rygel, M. C. et al. 2008. The magnitude of late Paleozoic glacioeustatic fluctuations; a synthesis. *Journal of Sedimentary Research.* 78 (8): 500-511.
13. Haq, B. U. and S. R. Schutter. 2008. A chronology of Paleozoic sea-level changes. *Science.* 322 (5898): 64-68.
14. Clarey, T. 2015. Dinosaurs in Marine Sediments: A Worldwide Phenomenon. *Acts & Facts.* 44 (6): 16.
15. More information on this discovery is discussed in Clarey, T. 2015. *Dinosaurs: Marvels of God's Design.* Green Forest, AR: Master Books.
16. These coal fragments were found in the Chevron Skolp E-07 well.
17. Harris, W. K. 1977. Palynology of cores from Deep Sea Drilling project Sites 327, 328, and 330, South Atlantic Ocean. In Barker, P. et al. *Initial Reports DSDP, 36.* Washington, DC: U.S. Government Printing Office, 761-815.
18. Rullkötter, J. et al. 1981. Organic petrography and extractable hydrocarbons of sediments from the eastern North Pacific Ocean, Deep Sea Drilling Project leg 63. In Yeats, R. S. et al. *Initial Reports DSDP, 63.* Washington, DC: U.S. Government Printing Office, 819-836.

Sea caves on the Algarve coast of Portugal

13 Covering the Highest Hills: Zuni Megasequence

> **Summary:** During the Zuni, the fifth megasequence, the Flood reached its high point, probably marking Day 150. The continents were completely submerged, and no land-dwelling creatures outside the Ark remained alive. The Zuni began at around Day 100 of the Flood and nearly tripled in thickness compared to the Absaroka. Fossils such as dinosaurs and mosasaurs are found in the Zuni on every continent, supporting the global nature of these deposits. The mixing of massive terrestrial and marine creatures also supports their catastrophic deposition. At this point in the Flood, 10,000 duck-billed dinosaurs are found buried together catastrophically. The rock data tell the story of an ever-increasing global flood event.

> And the waters prevailed exceedingly on the earth, and all the high hills under the whole heaven were covered. The waters prevailed fifteen cubits upward, and the mountains were covered. And all flesh died that moved on the earth: birds and cattle and beasts and every creeping thing that creeps on the earth, and every man. All in whose nostrils was the breath of the spirit of life, all that was on the dry land, died.… And the waters prevailed on the earth one hundred and fifty days. (Genesis 7:19–22, 24)

The High-Water Mark Is Reached

The Zuni Megasequence appears to be the point in the Flood when the waters reached their maximum, at least by the end of the megasequence (end Cretaceous/earliest Paleogene). I suggest that this megasequence may have started about Day 100 of the Flood, depending on when the Absaroka ended. Either way, the Zuni likely began within days of the end of the Absaroka. I picked Day 100 to divide the two megasequences because there are a lot of diverse rock types and rapid shifts in fossil content within each megasequence, justifying the 50 to 60 days for each.

Pangaea continued to split up in the Zuni, with South America splitting from Africa. Also, the massive-scale runaway subduction we saw in the Absaroka continued through the Zuni, with no signs of slowing down. The plates of the Pacific Ocean were likely still driving global activity, both through their subduction and their seafloor spreading processes.

Most secular geologists do not think the entire world ever completely flooded, at least not during the Phanerozoic Eon (Paleozoic, Mesozoic, and Cenozoic). But the rocks tell a different story. Thus far, the data have revealed the story of an ever-increasing global Flood event. Each megasequence seemed to have risen a bit higher than the last. As we examine the data for the Zuni Megasequence, we see that this trend continues. In fact, it is here in the Zuni that we see the flooding of the continents reached a maximum level.

Did the Zuni Megasequence completely flood all of the continents? I believe the evidence answers yes! Do we see Zuni deposits everywhere today? The answer to that question is no. However, the Bible tells us that the highest water level rose only 15 cubits over the highest mountains. Fifteen cubits is about 22.5 to 30 feet, depending on the length of a cubit. So, with only 25 feet of water column, we shouldn't expect to find a lot of sediment covering the pre-Flood uplands. I think post-Flood erosion removed a lot of these thinner deposits and left vast areas with little or no Zuni accumulation. However, there are still some erosional remnants, indicating there was more extensive, likely global coverage of all continents as described in the Bible. Let's look at the data.

Grand Staircase, Utah

What Do the Rocks Show?

North America

The Zuni Megasequence extends from the Middle Jurassic to the lowermost Paleogene system (post-Cretaceous). This sequence continued the dominance of siliciclastic deposition across western North America, with a slight shift in pattern to the northern Rocky Mountains and Canada. The Zuni deposits also buried the last of the dinosaurs across the West. The basal Zuni layer is predominantly sandstone and shale but shifted to massive salt deposition in the northern Gulf of Mexico (Louann Salt). Siliciclastic deposition continued to spread across the passive Atlantic margin, recording the split of Greenland and Canada. Although the Appalachian uplift prevented thick deposition across the eastern states, there are limited Zuni deposits preserved in the Illinois and Michigan Basins and erosional remnants near Hudson Bay. I refer to these remnants as the "bathtub ring." These remnants indicate that there was a thin layer deposited across much of this area that has since been eroded. The thickest Zuni deposits are found across the western portion of the continent and in the Gulf of Mexico, where many of these locations received sediments in excess of 3 kilometers. Sandstones make up about 28% of the Zuni Megasequence by volume. Carbonate rocks have further diminished to only 19%, and shale is by far the dominant lithology, comprising 43% of the total sequence volume.

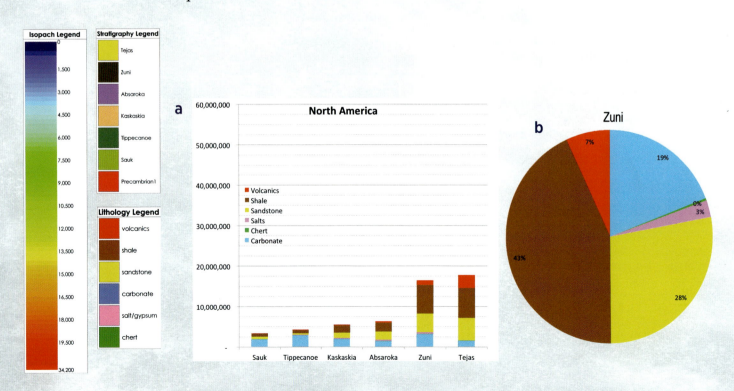

The Zuni shows a major increase in deposition over all previous megasequences in North America. The total volume of sediment in the Zuni (excluding the volcanic rocks) is the highest of any megasequence, depositing over 15 million cubic kilometers of sedimentary material. When volcanic rocks are included, this megasequence has a total volume of over 16.4 million cubic kilometers, second only to the Tejas Megasequence. The volume deposited by the Zuni Megasequence is over 10 million cubic kilometers greater than any previous megasequence totals. In fact, the Zuni nearly triples the amount deposited in the preceding Absaroka Megasequence. The Zuni Megasequence also records the highest coverage in terms of surface area of any megasequence across North America, at 16 million square kilometers. The Zuni Megasequence averages just over 1,000 meters thick, nearly double the thickness of any previous megasequence.

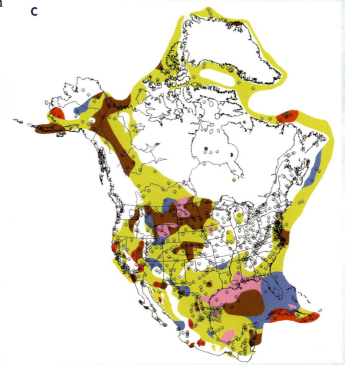

Compilation of Zuni Megasequence data for North America: (a) Histogram of the type and volume of rocks in place across North America for each megasequence; (b) pie chart of the total Zuni Megasequence by rock type; (c) map of the bottommost (basal) Zuni rock type; (d) thickness map of the total Zuni Megasequence; (e) stratigraphic cross-sections showing the relative thickness of the Zuni Megasequence in dark brown on top of the earlier megasequences. Histogram measurement is in cubic kilometers, isopach map measurement is in meters.

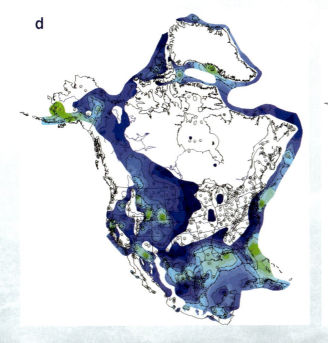
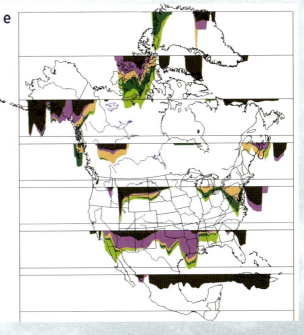

South America

By volume, the Zuni records a massive increase in deposition across South America and Central America (23.2 million cubic kilometers), more than doubling the combined total of the first four megasequences. The Zuni deposits also engulfed the last of the dinosaurs across the continent, particularly in Argentina. The basal rock layer is predominantly sandstone across the center of the continent and mixed sandstone and volcanic rocks along the continental margins. Siliciclastic deposition continued to infill the passive Atlantic margin, recording the split of South America from Africa. Zuni deposits are thickest along the flanks of the continent and are relatively thin across the center, particularly Brazil. The Zuni Megasequence averages about 1.6 kilometers thick, about 650 meters thicker than the preceding Absaroka. The total volume of sediment (23.2 million cubic kilometers) deposited during the Zuni across South America was about four times greater than the Absaroka Megasequence. And the surface area covered by the Zuni is more than twice that of any previous megasequence at 14.2 million square kilometers. Sandstones make up about 29% of the Zuni Megasequence. Carbonate rocks make up about 8%, whereas shale is again the dominant lithology, comprising 52% of the total megasequence volume.

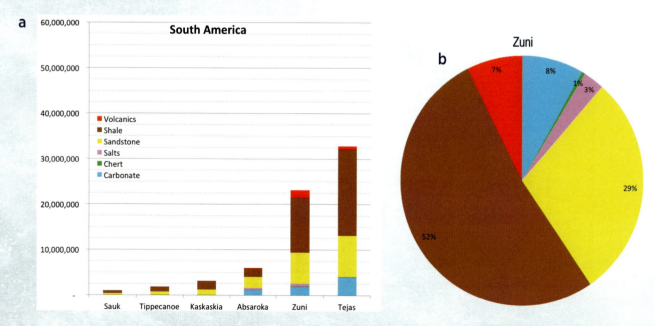

Compilation of Zuni Megasequence data for South America: (a) Histogram of the type and volume of rocks in place across South America for each megasequence; (b) pie chart of the total Zuni Megasequence by rock type; (c) map of the bottom-most (basal) Zuni rock type; (d) thickness map of the total Zuni Megasequence; (e) stratigraphic cross-sections showing the relative thickness of the Zuni Megasequence in dark brown on top of the earlier megasequences. Histogram measurement is in cubic kilometers, isopach map measurement is in meters.

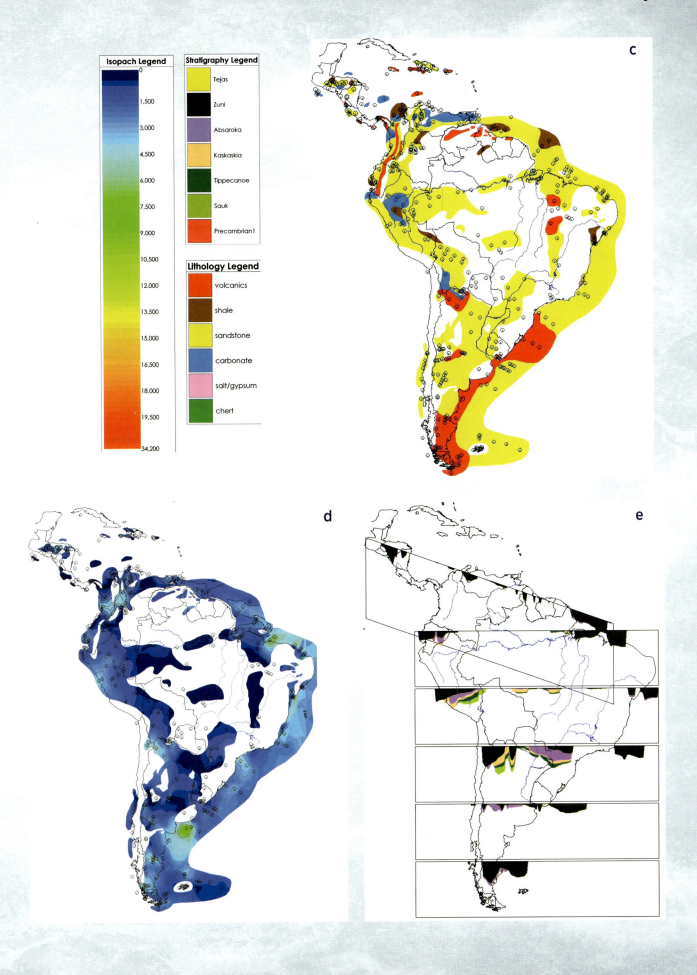

Africa

Across Africa, this megasequence continued to show an increase by percentage and volume of carbonate deposition. By volume, the Zuni records the most deposition of any sequence across Africa. The Zuni deposits also engulfed the last of the dinosaurs across the continent. The basal layer is predominantly sandstone in the center of the continent and carbonate around the edges. Siliciclastic deposition continued to infill the passive Atlantic margin, recording the split of South America from Africa. Zuni deposits are thickest in the center of the continent in Niger, Chad, and southern Sudan. Deposits are also thick on the flanks of the continent and along the offshore shelves. The average thickness of the Zuni across Africa is nearly 2.2 kilometers thick, close to double the preceding Absaroka Megasequence. The total volume deposited during the Zuni across Africa was more than double the Absaroka and was by far the greatest volume of any megasequence across any of the continents at 57.7 million cubic kilometers. And the Zuni also exhibited the maximum surface area coverage of any of the megesequences, at 26.6 million square kilometers. Sandstones make up about 30% of the Zuni Megasequence. Carbonate rocks have diminished to only 20%, and shale is again the dominant lithology, comprising 43% of the total megasequence volume.

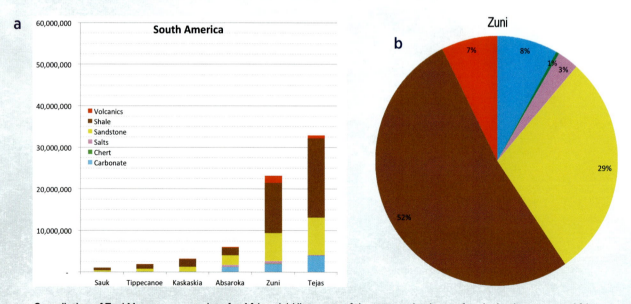

Compilation of Zuni Megasequence data for Africa: (a) Histogram of the type and volume of rocks in place across Africa for each megasequence; (b) pie chart of the total Zuni Megasequence by rock type; (c) map of the bottommost (basal) Zuni rock type; (d) thickness map of the total Zuni Megasequence; (e) stratigraphic cross-sections showing the relative thickness of the Zuni Megasequence in dark brown on top of the earlier megasequences. Histogram measurement is in cubic kilometers, isopach map measurement is in meters.

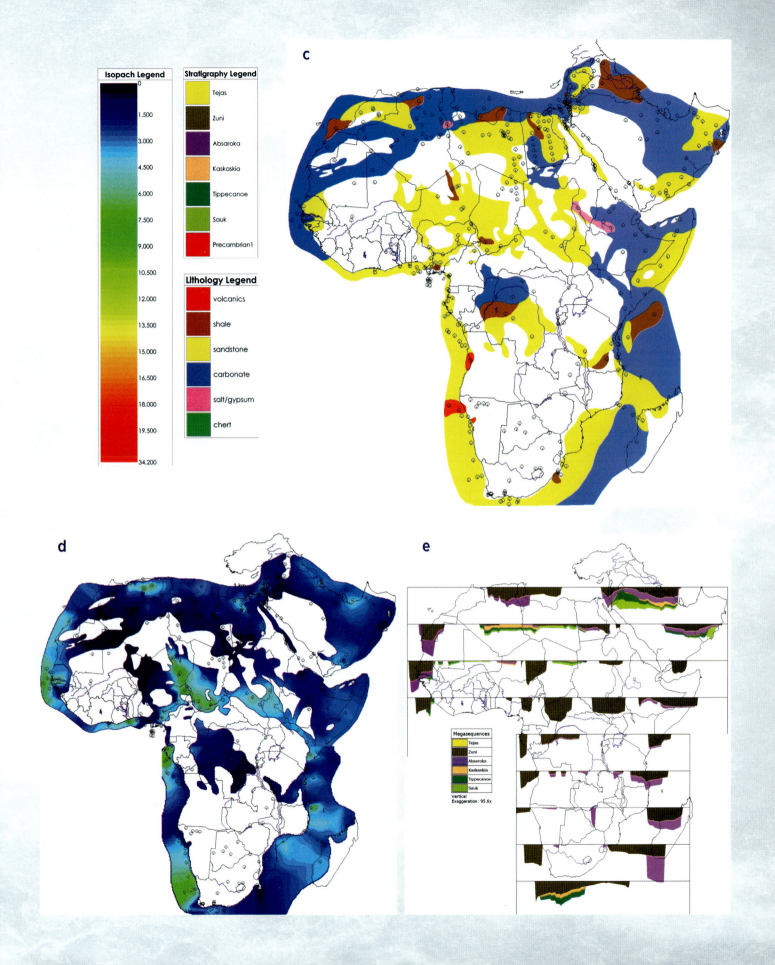

Current Data for the Zuni Megasequence

Figure 13.1 shows the paleocurrent data collected by Dr. Art Chadwick and his students across North America, South America, and Africa for Lower Cretaceous system rocks within the middle part of the Zuni Megasequence.[1]

We observe an easterly current direction across the southeastern North American continent, but a westerly direction seems to dominate the central Midwest and West Coast regions. These directions are similar to that observed in the Absaroka. Neither megasequence shows a dominant unidirectional flow like observed in the earliest three megasequences, where the currents were mostly directed to the west and southwest.

South America also exhibits a current pattern similar to the Absaroka, with a westerly directed current in eastern Brazil and northern South America and an easterly directed pattern in southern South America. Africa shows a strong northeasterly current in the Middle East.

Overall, the Zuni current data are more consistent with the preceding Absaroka than with the earliest three megasequences. This is likely due to the dramatic change in tectonics that began in the Absaroka, including subduction along the West Coast of North America and South America.

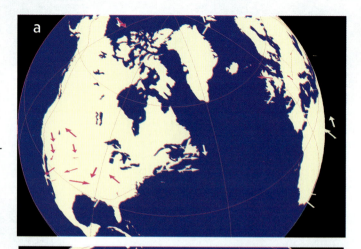

Figure 13.1. Current directions compiled across (a) North America, (b) South America, and (c) Africa for the Lower Cretaceous system (Middle Zuni). Courtesy of Art Chadwick.

Summary of the Rock Data

Across all three continents, we observe a massive spike in the amount of sediment deposited in the Zuni Megasequence. Fossils like dinosaurs and mosasaurs are also found in the Zuni on every continent, supporting the universal nature of the deposits. The Zuni has the highest total surface area covered, the highest total volume deposited, and highest average total thickness of any of the six megasequences. In fact, the total volume in the Zuni increased by nearly three times that of any of the earlier megasequences, and the average thickness nearly doubled.

By far, North America and Africa exhibit the highest amount of surface coverage in the Zuni, including a couple of small remnants near Hudson Bay and the Michigan and Illinois Basins. In contrast, South America has the second-most surface area in the Zuni, with only the Tejas Megasequence showing a slightly greater amount. This is likely due to the new land and offshore regions that formed from subduction activity in Central America during the Tejas. And it is likely influenced by the erosion and redistribution of earlier Flood sediment as the Andes Mountains became uplifted in the early Tejas. Both of these factors contributed vast amounts of Tejas sediment load.

Africa exhibits the greatest increase in the volume deposited in the Zuni, with over 36 million cubic kilometers of additional sediment compared to the preceding Absaroka. Africa's 57.7 million cubic kilometers of Zuni sediment is the greatest volume of any continent at any time in the Flood.

Most of the continents also show a near doubling of the average thickness in the Zuni compared to the Absaroka.

The major locations of the Zuni deposits also changed from earlier megasequences, with each of the continents showing much more sediment deposited offshore. This was caused by the creation of

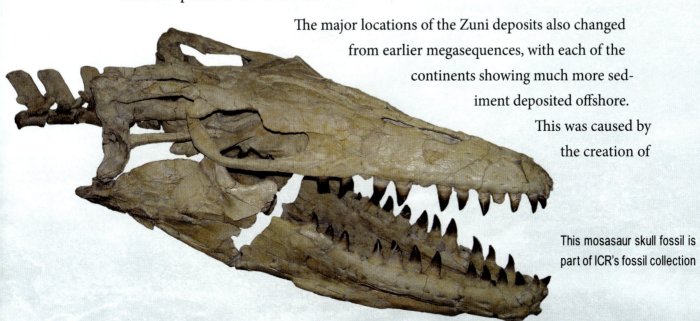

This mosasaur skull fossil is part of ICR's fossil collection

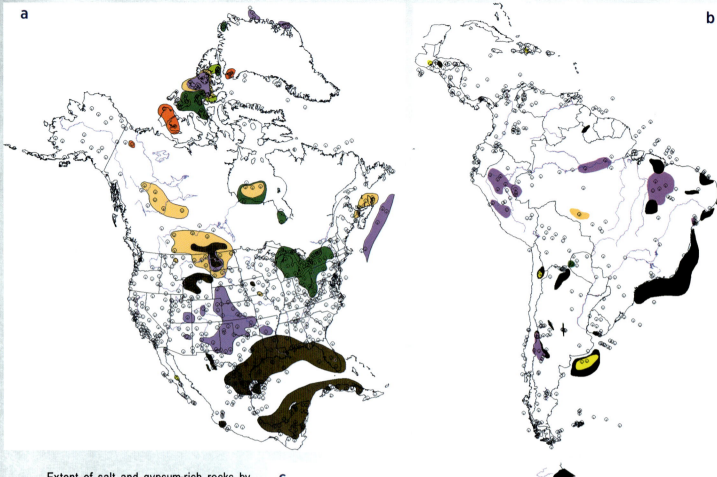
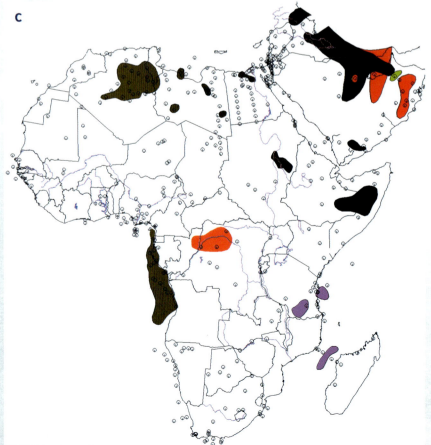

Extent of salt and gypsum-rich rocks by megasequence for (a) North America, (b) South America, and (c) Africa. Note the massive Zuni salt deposits in dark brown near the margins of the continents.

new shelf areas (passive margins) as the former supercontinent continued to separate.

These data collectively support that the Zuni was the high-water point of the Flood, which is why I place the end of the Zuni at Day 150. The thin remnants of Zuni sediment found spread across each of the continents tell me that the Zuni was once more continuous and has since been eroded (the bathtub ring). Recall that the Bible says that the water only rose 15 cubits over even the highest hills (Genesis 7:20). That doesn't allow a lot of room for thick Zuni sediment to be deposited in these highest locations. For that reason, we presently do not find all of the continents covered completely by the Zuni Megasequence.

Massive salt deposits also developed in the Zuni in the northern (Louann Salt) and southern Gulf of Mexico and offshore Brazil and West Africa. These deposits were thousands of feet thick and make many of the traps for oil deposits in these offshore regions. As suggested in chapter 8, these massive salt deposits seem tied to the rifting events that formed these regions.

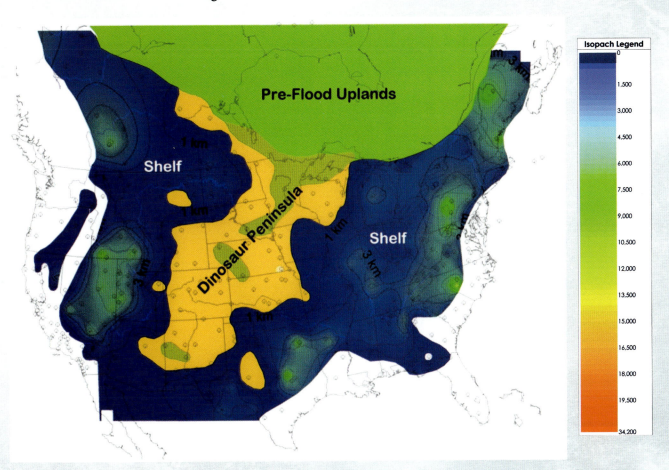

Pre-Flood interpretation for the continental United States showing Dinosaur Peninsula, a lowland extending diagonally across the American West. Measurement is in meters.

Delicate Arch within Arches National Park, Utah. The arch is composed of Middle Jurassic Entrada Sandstone, a Zuni deposit.

The record of the dinosaurs in the Zuni and Absaroka in the American West roughly follows the path of Dinosaur Peninsula described in chapter 11. Because this lowland environment was fairly narrow, the dinosaurs had a limited range of mobility, with shallow seas on the east and west. As subduction along the West Coast began in the Absaroka, the continent was pulled downward and the southwestern U.S. was progressively flooded mile by mile. Initial flooding began at the southern tip of this peninsula, systematically flooding from south to north, depositing sediments of the Upper Triassic (Absaroka), and burying dinosaurs like *Coelophysis*. Next, dinosaurs farther up the peninsula were overcome as floodwaters advanced across Utah, Colorado, and southern Wyoming. These were the Jurassic dinosaurs like the sauropods and stegosaurs. Finally, the last massive herds of stampeding dinosaurs are found in the Cretaceous rocks of northern Wyoming, Montana, and Alberta, Canada, as they were inundated by the tsunami waves. These were the hadrosaurs and ceratopsians, and even the tyrannosaurs. Many of these dinosaurs were swept off the peninsula by the wave energy prior to burial. This general pattern of burial may explain the reason why dinosaurs are found in progressively younger and younger rocks from south to north in North America. These fossil observations are best explained by the Flood, especially since dinosaurs are commonly found mixed with marine fossils and in marine rocks, as described below.

Continental Configuration for the Zuni Megasequence

The Zuni saw the continuation of the massive runaway subduction process that began globally in the Absaroka Megasequence, particularly in the Pacific region. The creation of tremendous amounts of new seafloor at the ocean ridges as a consequence of this subduction was pushing sea level higher and higher until the water was able to

Continental configuration during the deposition of the Zuni Megasequence and its sedimentary thickness

completely flood all of the landmasses in the Zuni. This was the culmination of a 150-day rise in global sea level. The peak in surface area coverage, average thickness, and total volume shown by the Zuni Megasequence data verify this interpretation.

Many of the remaining pieces of Pangaea began to separate during the Zuni as the runaway subduction process continued to wrench apart the plates of the earth. Early in the Zuni, South America and Africa began to separate, splitting the last big piece of Pangaea. In North and South America, the subduction along the West Coast was creating large reservoirs of magma and fueling volcanic eruptions from Alaska to Chile, including the beginnings of many parts of Central America. In North America, this early Zuni activity is known as the Nevadan Orogeny.[2]

Accretion of new terranes, like the Franciscan Complex, added to the U.S. West Coast as the Farallon Plate continued to subduct beneath the North American continent. Most of the Franciscan Complex is composed of sediments and volcanic rocks suggestive of offshore regions. These terranes were added to the continent as the ocean crust beneath was subducted.[2]

Later in the Zuni, Greenland began to separate from both Europe and Canada, depositing Zuni sediments on the eastern and western flanks. Renewed uplift along the West Coast of North America caused many of the earlier Flood sediments to slide eastward as thin-skinned overthrust faults. These overthrust faults are claimed by secular geologists to result from compression caused by subduction.

However, the mechanical difficulty of moving large, coherent sheets of rock (overthrusts) great distances down fairly flat slopes has never been fully explained in the secular geologic literature.[3] Lithified sedimentary rock will not fold and behave plastically at surface conditions,[4] yet we see the clear geometric results in overthrust belts around the globe. Creationists in the past—like John Whitcomb and Henry Morris,[5] Walter Lammerts,[6,7] and Clifford Burdick[8-10]—have been right to criticize secular explanations for overthrusts. Today, however, creationists must accept the results of hundreds of drill hole penetrations and thousands of kilometers of seismic reflection data, collected since the 1970s, proving the existence of overthrust faults.

Before the development of the plate tectonics theory, overthrusts were generally thought to be gravity slides.[11-14] More recently, and after the advent of plate tectonics, overthrusts became thought of as compressional features that are pushed. Davis et al

OVERTURNED SYNCLINE OF CAMBRIAN SCHIST, 50 FEET EAST OF OVERTHRUST; TOP OF BOARDMAN HILL, CLARENDON, VT.
Looking N. 35° E.   Hammer, 21 inches long.

and Chappel have pointed out that thrust belts are commonly wedge-shaped and move only when the wedge reaches a critical taper angle.[15,16] Davis and Reynolds explain that "the critical shear stress required for sliding to occur is equal to the product of the coefficient of sliding friction and the effective stress."[17] However, published experiments were performed with unconsolidated sediments where the basal detachment was pulled out from beneath the sediments.[15] Actual pushing of rocks from the rear, as commonly believed, results in crushing of the rocks at the point of compression with no detachments and no thrust development.[18] Gravity seems to remain the only viable force to move overthrusts.[4] Uniformitarian geologists are coming back to gravity tectonics to explain some overthrusts. Walter Alvarez, in his discussion of the development of the Alps, believes gravity spreading of uplifted areas drive collapse, stating, "Gravity carries the rising mountains away, thrust sheet by thrust sheet."[19]

High fluid pressures, developing during dewatering reactions, have the ability to create overpressured zones and float large thrust sheets downslope.[11,20,21] The formation of supercritical carbon dioxide seems to be an additional method to move carbonate-rich sediments rapidly.[22]

All data suggest overthrust faults moved rapidly. Some fault breccia and/or fault gouge is observed along the surface contact of many overthrusts. In addition, clastic dikes are commonly observed in the hanging wall of these fault systems, further supporting a catastrophic interpretation. Unlithified sediments are essential to the development of overthrust faults in order to explain the tightly folded geometries that are observed in the toe areas. These conditions must have occurred *after* many of the Flood sediments were deposited but *while* they were still uncemented. Rapid deposition during the Flood, combined with compaction and dewatering of clay-rich sediments and gypsum layers, created overpressured zones along impermeable boundaries.

> "Gravity carries the rising mountains away, thrust sheet by thrust sheet."

Secular explanations of overthrusts, using slow movement and maintenance of overpressured horizons over great distances, still cannot resolve the glaring mechanical paradox.[23] However, the Flood model of overthrusting, involving rapid downhill movement of unlithified sediments, provides both a cause and a mechanism for the development of large thrust sheets and the resulting tightly folded geometries in the toe areas. These overthrusts developed late in the Zuni Megasequence across North America, from Alaska to Mexico, as part of the Sevier Orogeny.[2] These overthrusts also formed east of the Andes Mountains in South America.

During the latter part of the Zuni, the final episode of uplift began across the American West and western South America. This is known as the Laramide Orogeny in North America.[2] The culmination of this uplift event occurred in the next megasequence cycle, the Tejas, so I will discuss it more in the next chapter. Suffice it to say that the Laramide was mostly a basement-cored orogenic event, with the crust moving up and down many thousands of feet across the American West.

The K-T or K-Pg Extinction? The Chicxulub Crater

Was there really a major extinction near the end of the Zuni as suggested by secular geologists? Did this really cause the demise of the dinosaurs?

Flood geologists merely view these so-called extinctions as the last appearance of particular organisms in the rock record. In other words, these surfaces represent merely the highest occurrence of a fossil organism as a particular ecological zone was inundated by the floodwaters. As I have noted before, these so-called extinction events seem to have a relation to the maximum water level and/or to rapid changes in sediment type within each megasequence.

Nonetheless, in secular literature and movies, the most popular explanation for the dinosaurs' extinction is an asteroid impact. The Chicxulub crater in Mexico is often referred to as the "smoking gun" for this idea. But do the data support an asteroid impact at Chicxulub? I reviewed the evidence and found some surprising results.[24]

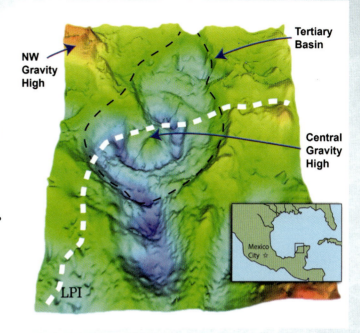

Figure 13.2. Gravity map showing density differences across the claimed Chicxulub crater, Yucatán Peninsula, Mexico

The Chicxulub crater isn't visible on the surface because it is covered by younger, relatively undeformed sediments. It was identified from a nearly circular gravity

Morrison Formation at Dinosaur National Park

anomaly along the northwestern edge of the Yucatán Peninsula (Figure 13.2).[25,26] There's disagreement on the crater's exact size, but its diameter is approximately 110 miles—large enough for a six-mile-wide meteorite to have caused it.

Although some of the expected criteria for identifying a meteorite impact are present at the Chicxulub site—such as high-pressure and deformed minerals—not enough of these materials have been found to justify a large impact.[24] And even these minerals can be caused by other circumstances, including rapid crystallization[27] and volcanic activity.[28]

The biggest problem is what is *missing*. Iridium, a chemical element more abundant in meteorites than on Earth, is a primary marker of an impact event. A few traces were identified in the cores of two drilled wells, but no significant amounts have been found in any of the ejecta material across the Chicxulub site.[29] The presence of an iridium-rich layer is often used to identify the K-Pg (Cretaceous-Paleogene) boundary, yet ironically there is virtually no iridium in the ejecta material at the very site claimed to be the smoking gun!

In addition, secular models suggest melt-rich layers resulting from the impact should have exceeded a mile or two in thickness beneath the central portion of the Chicxulub crater.[30] However, the oil wells and cores drilled at the site don't support this. The thickest melt-rich layers encountered in the wells were between 330 and 990 feet—nowhere near the expected thicknesses of 5,000 to 10,000 feet—and several of the melt-rich layers were much thinner than 300 feet or were nonexistent.

Finally, the latest research even indicates that the tsunami waves claimed to have been generated by the impact across the Gulf of Mexico seem unlikely.[31]

The thinner-than-expected melt-rich layers, the lack of any substantial iridium anomaly, and the alternative explanations for the high-pressure and deformed minerals and the gravity anomaly all raise concerns about the Chicxulub crater and ultimately the asteroid extinction theory itself. An impact may have occurred at Chicxulub during the Flood, but if so, then it seems to have been much smaller than commonly claimed, creating a mere fraction of the postulated effects. And it's entirely possible there was never an impact at Chicxulub in the first place. All of the data can have non-impact explanations.[24]

The Chicxulub impact has become the iconic tale most secular scientists use for the so-called major extinction event that wiped out the dinosaurs.[32] They need such a story because they categorically reject Earth history as described in the Bible, including the global Flood. In order to explain the disappearance of the dinosaurs and other creatures, they uphold the Chicxulub impact as one of the principle factors in their tale—even though the evidence for the impact at the Chicxulub site is not nearly as strong as they claim.

After careful examination of the data, the smoking gun appears to be mostly smoke. There was no true extinction event at the end of the Cretaceous, just a rapid shift in the fossils as the Zuni Megasequence was reaching its maximum water level.

Dinosaurs in Marine Sediments: A Worldwide Phenomenon

Although we discussed this earlier in the Absaroka Megasequence (chapter 12), this issue justifies repetition. Throughout the Absaroka we find land animals mixed with marine fossils. This has become the norm, not the exception. The rocks of the Zuni Megasequence again show that not only are land fossils found mixed with marine fossils, but they are often found mixed together in marine rocks.

For many years, paleontologists have known of marine fossils within various dinosaur-bearing rock units in the American West. These occurrences are largely ignored by mainstream scientists, who deny that dinosaurs were buried in the global and recent Flood as described in Genesis.

The Hell Creek Formation in eastern Montana has yielded many *T. rex* specimens, including well-documented dinosaur soft

Sinosauropteryx and a fish fossil at ICR's headquarters in Dallas, Texas

tissue fossils. Surprisingly, in two volumes of papers published specifically on the Hell Creek discoveries, little is mentioned of the six species of shark and 14 species of fish fossils that are indicative of marine influence.[33,34] Secular scientists either ignore these findings or dismiss them all as freshwater sharks and freshwater fish, in spite of the more likely conclusion that they represent marine organisms.

Other authors have studied the fauna of the Hell Creek Formation since the 1950s and found ample evidence of a mixture of marine and nonmarine fossils.[35,36] As Joseph Hartman and James Kirkland stated, "Although previously reported, knowledge of the continuation of marine conditions above the Fox Hills Formation [in the Hell Creek Formation] is not well or widely known."[33]

It is now becoming obvious that the mixing of terrestrial and marine environments is not a rare occurrence in the rock record. Recent discoveries in Morocco and Europe have shown that most dinosaurs are found with marine fossils or buried in marine sediments.

Nizar Ibrahim et al reported that sharks, sawfish, ray-finned fishes, and coelocanths were found in the same rock layers as a *Spinosaurus* dinosaur in Morocco.[37] How can this be? Today's coelocanths live about 500 feet below the ocean surface and not in freshwater rivers, as many paleontologists have proposed. They dismiss the blatant physiological evidence from living specimens and insist that ancient coelocanths must have lived in fresh water simply because they are found in strata with dinosaurs. Where is the logic in this conclusion?

Zoltan Csiki-Sava and his colleagues surveyed all the recent research on dinosaur occurrences in Europe within the six accepted stages of the Late Cretaceous system. The team reported that "although isolated occurrences of continental [terrestrial] vertebrate fossils were occasionally reported from the Cenomanian to lower Santonian [lower four Upper Cretaceous stages] of Europe, these were mainly from marginal marine deposits."[38] And the vast majority of these dinosaur occurrences were even found in open marine chalk and limestone deposits mixed with marine invertebrates.

Their survey of the upper two stages of the Cretaceous also showed nearly all dinosaur fossils were located in marine rocks. Here, too, the paleontologists reported numerous discoveries of dinosaur remains in open marine chalk beds that are difficult to explain in a uniformitarian context. "Although these are isolated skeletal elements [individual bones] that washed out to sea, they are remarkably common and have been reported in surprisingly large numbers since the early discoveries."[38]

Dinosaur fossils found in rock strata with marine fossils are commonplace, not the exception. The mounting empirical evidence cannot be ignored or simply explained away as a rare occurrence. The fossil evidence supports a catastrophic and global flood that mixed the marine realm with the terrestrial realm as tsunami-like waves spread ocean fauna and sediments across the continents. Genesis 7 and 8 describe this process better than any secular scientist could imagine.

Another Whopper Sand?

I will discuss the Whopper Sand in the next chapter because it is a Tejas Megasequence deposit. But there are similar massive sandstone deposits in earlier megasequences as well. One of these was deposited in the Gulf of Mexico region during the early part of the Zuni Megasequence.

In the U.S. Gulf Coast region, the Upper Jurassic Norphlet Sandstone rests right on top of thousands of feet of Middle Jurassic salt known as the Louann Salt. Secular geologists believe this sand layer was deposited by the windblown accumulation of sand in an arid environment, commonly labeled as an aeolian deposit.[39]

Exactly how this claimed desert deposit was supposed to have formed across much of the offshore Gulf of Mexico region and directly on

"Eddie," a juvenile *Edmontosaurus* at ICR's headquarters in Dallas, Texas

top of thick salt beds is never addressed or explained. And recent discoveries in deep water make this interpretation even more improbable because the sand layer is far more extensive than originally thought.[39]

The Upper Jurassic Norphlet Sandstone was first described onshore at Norphlet, Arkansas. It has been drilled for natural gas and oil for many years in places like Mobile Bay, Alabama, and in the shallower water regions of the Gulf of Mexico. The Norphlet averages less than 300 feet thick near the Gulf Coast but thickens to closer to 1,000 feet in the deepwater Gulf of Mexico.[39]

The first well targeting the Norphlet Sandstone in the deepwater Gulf of Mexico was located in the De Soto Canyon protraction in 2003.[40] Since then, numerous wells have continued the search in deeper and deeper water.

In 2018, several large discoveries in the Norphlet in deeper-water portions of the Gulf of Mexico were reported by Shell and Chevron. Chevron announced a discovery in the Norphlet at their Ballymore prospect in 6,536 feet of water in the area of the Gulf of Mexico known as Mississippi Canyon, located south of Louisiana. This well drilled to 29,194 feet and found 670 feet of net pay in the Norphlet.[39] Net pay is the oil-producing interval and is less than the full thickness of a stratigraphic unit.

Also in 2018, Shell announced a Norphlet discovery at their Dover prospect about 170 miles south of the Louisiana coastline. This well was drilled in 7,500 feet of water in the Mississippi Canyon protraction, finding 800 feet of net pay.[39]

Louise Durham reported, "This is the company's sixth discovery in the Jurassic-age Norphlet geologic play in the Gulf. It joins such high-profile finds as Appomattox, approximately 13 miles away." She continued, "The Norphlet ranks as the first extensive post-salt deposit in the central Gulf Coast."[41]

Not only has the Norphlet been extended into the deepwater portion of the Gulf of Mexico, but it may also continue right across the Gulf of Mexico as a blanket sandstone bed to the offshore region of the Yucatán Peninsula, Mexico.[40] This sand bed covers an area that is "several hundred thousand square kilometers across Louisiana, Mississippi and Alabama and outward into the deepwater Gulf."[42]

Essentially, this is another Whopper Sand,[43] but it was deposited earlier in the rock record. It is comparable in size to the Ogallala Sandstone bed that extends for 174,000 square miles across the Great Plains states.[44]

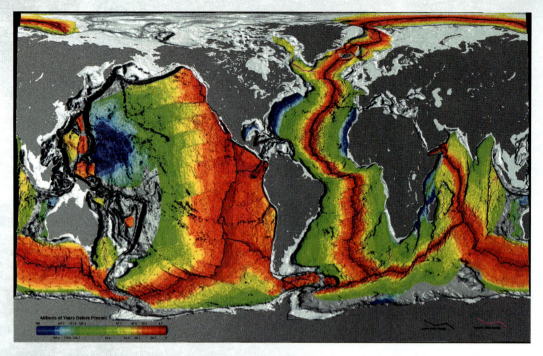

Map of the age of the ocean crust. The light blue, green, and yellow areas formed during the deposition of the Zuni Megasequence.

The Norphlet is the newest deepwater oil and gas target. And yet its massive extent cannot be adequately explained by conventional models, let alone with an aeolian interpretation. How did a thick desert sandstone just happen to become spread across nearly the entire deepwater Gulf of Mexico and on top of thousands of feet of pure salt? And how did the thick sandstone bed get nearly 200 miles offshore? Uniformitarian scientists have provided no answers, yet they continue to find more sand and more oil in the Norphlet.

Salt beds are only deposited in an underwater marine setting. So, how did the ocean mysteriously and suddenly disappear, exposing a massive pure salt layer, and then become a terrestrial desert all in the blink of an eye?

Sadly, secular geologists merely apply the aeolian label to any sandstone that has large cross-beds, like they did with the Coconino Sandstone in Grand Canyon. And yet the Coconino contains minerals like carbonate ooids and flakes of mica minerals that cannot form in an aeolian environment.[45,46] The ooids and minerals are best explained by a marine deposit.

The Norphlet is another case where uniformitarian geology fails. Blanket sands are best explained as massive Flood deposits as the waters were advancing and/or retreating. The Norphlet can be better explained as a mid-Flood deposit near the base of the

Zuni Megasequence. Rapid shifts in rocks types (i.e., from salt to sandstone) are best explained by sudden shifts in direction of tsunami-like waves, bringing in sediment from different locations and sources during the Flood.

Human Perspective

Most humans were likely still alive for much of the Zuni Megasequence. But by the end, they had all perished. Between these moments, people must have experienced horrendous conditions. Massive earthquakes would have been ongoing. Rain was still falling. The advancing tsunami waves would have been pushing tremendous amounts of debris and destroying everything in their path. Trees, sediment, and dead animals would have been mixed together in the advancing waves. Adding to the geologic chaos were massive volcanic eruptions, spewing tremendous volumes of ash and debris, choking the air and making it difficult to breathe. These ash-rich volcanic eruptions were caused by the massive amounts of runaway subduction that was occurring during the deposition of the Absaroka and Zuni Megasequences.

As the Zuni progressed and the time drew closer to Day 150, the tsunami waves would reach progressively higher and higher as more seafloor was created, pushing up water levels even farther. This would have forced the remaining animals and humans to rapidly migrate toward the highest elevations. The last vestiges of dry land would have been getting smaller each day and with each tsunami wave. This is the point in the rock record where the fossil remains of 10,000 *Maiasaura* (duck-billed dinosaurs) are all found buried together. These animals, and many just like them, were stampeding to avoid the advancing floodwaters. Humans were also fiercely competing for the last remnants of dry land. It was a life-or-death situation for every air-breathing animal and human. And each day the waves just kept rising higher, with no relief.

As Day 150 approached, the humans would probably have been the last remaining life form swarmed atop the final tiny pieces of remaining dry land. Thousands of people were undoubtedly clustered together on these little peaks. Each wave would have washed away more humans, never to be seen again. The chaos, violence, and the hopelessness of the situation would have been unimaginable. Finally, the last people were washed off the final remnant of land by the rising waves. Water had completely covered the earth. The high velocity of the waves and the battering by moving material would have drowned, smashed, or capsized any humans attempting to survive on floating debris. By Day 150 of the Flood, all air-breathing life forms that walked the earth were dead. The tsunami waves rose 15 cubits above the highest peaks, wiping humanity from the earth and eroding the land surface right down to the underlying crust.

Conclusion: The Zuni Is the High-Water Point of the Flood

I have suggested that the Zuni may have been deposited from about Day 100 to Day 150 of the Flood. The exact timing of when the Absaroka ended and the Zuni began is rather subjective. The Bible gives no clues of changes between Day 40 and Day 150 other than the water was prevailing higher and higher. However, the rocks do indicate that the Zuni ended at or near Day 150 of the Flood. It is quite evident that the Zuni records the highest sea level of all the megasequences. It is not a coincidence that the Zuni exhibits the maximum surface area coverage, the highest volume, and the greatest average thickness across three continents. The Zuni was the culmination of a fairly continuous rise in global sea level that began in the Sauk. By this point in the Flood, Pangaea had completely separated. The massive runaway subduction that began in the Absaroka continued unimpeded.

> "The Zuni brought on the end of the dinosaurs and many other animals both on and off the land. It was literally the last gasp for these creatures."

The subduction process caused tremendous amounts of seafloor to be created during the Zuni. This pushed the ocean water level high enough to completely cover the earth. The tsunami waves could go right over the tops of the highest hills unimpeded. The aforementioned erosional remnants found across each of the continents verify this interpretation. The "bathtub ring" is still there to observe.

The Zuni brought on the end of the dinosaurs and many other animals both on and off the land. It was literally the last gasp for these creatures. Next, we turn to the Tejas Megasequence, where we see the floodwaters drained off the continents and back into the oceans. This was a slow process that took a total of about 164 days. It also

included the last rapid movements of the tectonic plates and the end of runaway subduction. The earth was settling back into a new post-Flood equilibrium.

View of Cretaceous Liscomb bone bed on Colville River, Alaska

References

1. Chadwick, A. V. 1993. Megatrends in North American Paleocurrents. *Society of Economic Paleontologists and Mineralogists Abstracts.* 8:58.
2. Wicander, R. and J. S. Monroe. 2013. *Historical Geology,* 7th ed. Belmont, CA: Brooks/Cole.
3. Briegel, U. 2001. Rock mechanics and the paradox of overthrusting tectonics. In *Paradoxes in Geology.* U. Briegel and W. Xiao, eds. Amsterdam, Netherlands: Elsevier, B.V., 231-244.
4. Snelling, A. A. 2009. *Earth's Catastrophic Past: Geology, Creation & the Flood.* Dallas, TX: Institute for Creation Research.
5. Whitcomb, J. C. and H. M. Morris. 1961. *The Genesis Flood: The Biblical Record and Its Scientific Implications.* Grand Rapids, MI: Baker Book House.
6. Lammerts, W. E. 1966. Overthrust faults of Glacier National Park. *Creation Research Society Quarterly.* 3 (1): 61-62.
7. Lammerts, W. E. 1972. The Glarus overthrust. *Creation Research Society Quarterly.* 8 (4): 251-255.
8. Burdick, C. L. 1969. The Empire Mountains–a thrust fault? *Creation Research Society Quarterly.* 6 (1): 49-54.
9. Burdick, C. L. 1974. Additional notes concerning the Lewis thrust-fault. *Creation Research Society Quarterly.* 11 (1): 56-60.
10. Burdick, C. L. 1977. Heart Mountain revisited. *Creation Research Society Quarterly.* 13 (4): 207-210.
11. Hubbert, M. K. and W. W. Rubey. 1959. Role of fluid pressure in mechanics of overthrust faulting. I. Mechanics of fluid-filled porous solids and its application to overthrust faulting. *Geological Society of America Bulletin.* 70: 115-166.
12. Eardley, A. J. 1963. Relation of uplifts to thrusts in Rocky Mountains. In *Backbone of the Americas.* American Association of Petroleum Geologists Memoir 2, 209-219.
13. Roberts, R. J. 1968. Tectonic Framework of the Great Basin. *University of Missouri Research Journal.* 1: 101-119.
14. Mudge, M. R. 1970. Origin of the Disturbed belt in northwestern Montana. *Geological Society of America Bulletin.* 81: 377-392.
15. Davis, D. M., J. Suppe, and F. A. Dahlen. 1983. Mechanics of fold-and-thrust belts and accretionary wedges. *Journal of Geophysical Research.* 88: 1153-1172.
16. Chapple, W. M. 1978. Mechanics of thin-skinned fold-and-thrust belts. *Geological Society of America Bulletin.* 89: 1189-1198.
17. Davis, G. H. and S. J. Reynolds. 1996. *Structural Geology of Rocks and Regions*, 2nd ed. New York: John Wiley and Sons, Inc., 338.
18. Personal communication with John Baumgardner, 2009.
19. Alvarez, W. 2009. *The Mountains of Saint Francis: Discovering the Geologic Events That Shaped Our Earth.* New York: Norton, 166.
20. Guth, P. L., K. V. Hodges, and J. H. Willemin. 1982. Limitations on the role of pore pressure in gravity gliding. *Geological Society of America Bulletin.* 93 (7): 606-612.
21. Clarey, T. L. 2012. South Fork fault as a gravity slide: its break-away, timing, and emplacement, northwestern Wyoming, U.S.A. *Rocky Mountain Geology.* 47 (1): 55-79.
22. Beutner, E. C. and G. P. Gerbi. 2005. Catastrophic emplacement of the Heart Mountain block slide, Wyoming and Montana, USA. *Geological Society of America Bulletin.* 117 (5/6): 724-735.
23. Price, R. A. 1988. The mechanical paradox of large overthrusts. *Geological Society of America Bulletin.* 100: 1898-1908.
24. Clarey, T. L. 2017. Do the Data Support a Large Meteorite Impact at Chicxulub? *Answers Research Journal.* 10: 71-88.

25. Penfield, G. T. and Z. A. Camargo. 1981. Definition of a major igneous zone in the central Yucatán platform with aeromagnetics and gravity. In *Technical Program, Abstracts and Bibliographies*. Tulsa, OK: Society of Exploration Geophysicists, 51st Annual Meeting, 37.
26. Hildebrand, A. R. et al. 1991. Chicxulub Crater: A possible Cretaceous/Tertiary boundary impact crater on the Yucatán Peninsula, Mexico. *Geology*. 19 (9): 867-871.
27. Huffman, A. R. and W. U. Reimold. 1996. Experimental constraints on shock-induced microstructures in naturally deformed silicates. *Tectonophysics*. 256 (1-4): 165-217.
28. Alexopoulos, J. S., R. A. F. Grieve, and P. B. Robertson. 1988. Microscopic lamellar deformation features in quartz: Discriminative characteristics of shock-generated varieties. *Geology*. 16 (9): 796-799.
29. Keller, G. et al. 2004. More evidence that the Chicxulub impact predates the K/T mass extinction. *Meteoritics & Planetary Science*. 39 (7): 1127-1144.
30. Morgan, J. V. et al. 2000. Peak-ring formation in large impact craters: geophysical constraints from Chicxulub. *Earth and Planetary Science Letters*. 183 (3-4): 347-354.
31. Boslough, M. et al. Asteroid-Generated Tsunami and Impact Risk. Abstract NH13A-1763. 2016 American Geophysical Union Fall Meeting, San Francisco.
32. A limited number of dinosaur bones have been found above the K-Pg boundary in several locations across the earth. However, most secular scientists think these bones were merely "reworked" by erosion from earlier bone deposits.
33. Hartman, J. H. and J. I. Kirkland. 2002. Brackish and marine mollusks of the Hell Creek Formation of North Dakota: Evidence for a persisting Cretaceous seaway. In *The Hell Creek Formation and the Cretaceous-Tertiary Boundary in the Northern Great Plains: An Integrated Continental Record of the End of the Cretaceous*. J. H. Hartman, K. R. Johnson, and D. J. Nichols, eds. Geological Society of America Special Paper 361, 271-296.
34. Clemens, W. A. and J. H. Hartman. 2014. From *Tyrannosaurus rex* to asteroid impact: Early studies (1901-1980) of the Hell Creek Formation in its type area. In *Through the End of the Cretaceous in the Type Locality of the Hell Creek Formation in Montana and Adjacent Areas*. G. P. Wilson et al, eds. Geological Society of America Special Paper 503, 1-87.
35. Archibald, J. D. 1996. *Dinosaur Extinction and the End of an Era: What the Fossils Say*. New York: Columbia University Press.
36. Lucas, S. G. 2007. *Dinosaurs: the Textbook*, 5th ed. Boston, MA: McGraw-Hill Higher Education.
37. Ibrahim, N. et al. 2014. Semiaquatic adaptations in a giant predatory dinosaur. *Science*. 345 (6204): 1613-1616.
38. Csiki-Sava, Z. et al. 2015. Island life in the Cretaceous-faunal composition, biogeography, evolution, and extinction of land-living vertebrates on the Late Cretaceous European archipelago. *ZooKeys*. 469: 1-161.
39. Durham, L. S. 2018. Shell makes large discovery on Deepwater GoM. *AAPG Explorer*. 39 (7): 24-25.
40. Saunders, M., L. Geiger, and A. Steier. Mapping the Jurassic Norphlet Sandstone along the Northern Margin of the Yucatan Peninsula. AAPG Datapages/Search and Discovery Article #90291. AAPG Annual Convention and Exhibition, Houston, Texas, April 2-5, 2017.
41. Durham, Shell makes large discovery on Deepwater GoM, 24.
42. Ibid.
43. Clarey, T. 2015. The Whopper Sand. *Acts & Facts*. 44 (3): 14.
44. Clarey, T. 2018. Palo Duro Canyon Rocks Showcase Genesis Flood. *Acts & Facts*. 47 (7): 10.
45. Cheung, S. et al. 2009. Occurrence of dolomite beds, clasts, ooids and unidentified microfossils in the Coconino Sandstone, Northern Arizona. *Geological Society of America Abstracts with Programs*. 41 (7): 119.
46. Whitmore, J. H. and R. Strom. 2010. Petrographic analysis of the Coconino Sandstone, Northern and Central Arizona. *Geological Society of America Abstracts with Programs*. 41 (7): 122.

14 The Receding Phase: Tejas Megasequence

Summary: The floodwaters receded during the sixth and final megasequence, the Tejas. It seems clear from the biblical text that the receding began on Day 150, so this is probably when the Tejas began to be deposited. Even though the water was receding, there was still significant plate tectonic activity—about one-third to one-half of the ocean crust formed during the deposition of the Tejas. This megasequence also shows a reversal in water-flow direction. Secular researchers found that significant amounts of water were draining off the continents during the deposition of the Tejas. The massive drainage probably carved Grand Canyon. The sea level also suddenly dropped, corresponding with the recession of the floodwaters. Geologic rock data give deep insights into the final recession chapter of the Flood story.

Then God remembered Noah, and every living thing, and all the animals that were with him in the ark. And God made a wind to pass over the earth, and the waters subsided. The fountains of the deep and the windows of heaven were also stopped, and the rain from heaven was restrained. And the waters receded continually from the earth. At the end of the hundred and fifty days the waters decreased. Then the ark rested in the seventh month, the seventeenth day of the month, on the mountains of Ararat. And the waters decreased continually until the tenth month. In the tenth month, on the first day of the month, the tops of the mountains were seen....And it came to pass in the six hundred and first year, in the first month, the first day of the month, that the waters were dried up from the earth. (Genesis 8:1-5, 13)

The Waters Recede

The Tejas Megasequence includes most of the Tertiary system, now split into the Paleogene and Neogene systems. The Tejas Megasequence most likely represents the time when the floodwaters were receding. It seems quite clear in the biblical text that the recession of the water began on Day 150, after reaching a maximum level that same day. Genesis 8:13 suggests that the floodwaters had completely dried up across the entire earth by the first day of the month of that year. Depending on the length of the calendar year used by the ancients, this equates to about 314 days after the Flood began. (Day 1 was the 17th day of the second month of the previous year.)

What may not be so clear is exactly which day the fountains, the windows of heaven, and the rain were stopped. All we are told is that the water began to recede on Day 150. It doesn't say with exact certainty when these other events occurred, just that they did at some point.

How exactly did the water recede? The Bible tells us that God sent a wind, the fountains were stopped, the windows of heaven were stopped, and the rain was restrained. Geologically, the Tejas shows that significant plate activity was occurring

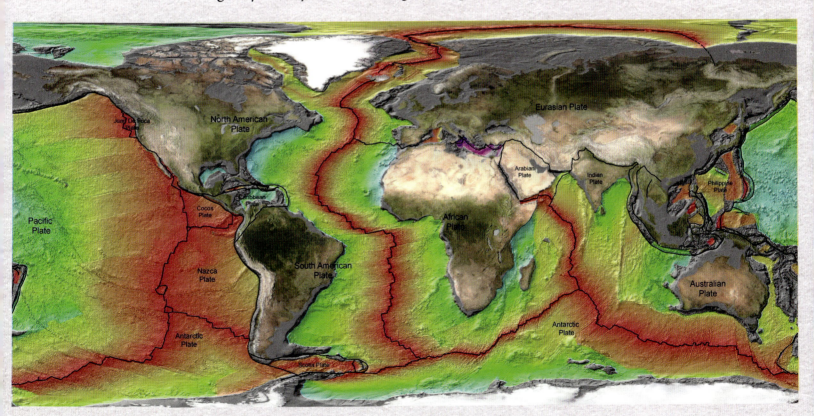

Map of the age of the ocean crust. The orange areas formed during the Tejas Megasequence.

right up until the end of the megasequence. In fact, about one-third to one-half of the ocean crust (depending on the ocean) formed during the deposition of the Tejas Megasequence. This means that seafloor spreading and runway subduction were still traveling at meters per second with no signs of stoppage until the end of the Tejas, or close to the level of the Pliocene. So, if the fountains of the deep are referring to rifting and plate activity, then they were not stopped until closer to Day 314 of the Flood year. This may also be true of the windows of heaven and the rain. All of these activities may have continued until near the end of the Tejas.

The recession of the water also seems tied to the cooling of the newly created ocean seafloor/lithosphere. Recall that as ocean lithosphere cools it becomes more dense, contracts, and sinks a bit deeper, pulling the water depth in the ocean down with it. This likely was the primary mechanism for drawing the waters off the continents and back into the ocean basins. As noted above, ocean seafloor was still being created at an astounding rate from the Absaroka right through much of the Tejas. However, the older ocean lithosphere that was created in the Absaroka and early Zuni was apparently cooling fast enough to subside significantly, lowering the seafloor in those areas. The result of this seafloor subsidence surpassed the rate of production of new buoyant seafloor, causing a net lowering of seal level. This process continued throughout the

Grand Teton peak, Wyoming

Tejas Megasequence and contributed greatly to the withdrawal of the floodwaters off the continents.

Obviously, the wind provided by God pushed the waters off the continents too. Eventually, the Bible tells us the fountains and rainfall stopped for good. This seems to have been late in the Tejas, at about the time of Pliocene deposition. Today, the plates are barely moving in comparison to the velocities at which they moved in the Flood because the cold, dense pre-Flood ocean lithosphere has been completely consumed. There is no longer a great enough density difference to drive runaway subduction. This essentially stopped plate motion and the rapid rifting associated with the fountains of the great deep.

Sixty years ago, few geologists believed the earth's crust ever moved at all. Now, nearly all secular scientists accept plate tectonics and plate movement. And yet today, most secular geologists refuse to accept the concept of runaway subduction and plate movement rates that were much faster in the past. They cannot imagine the plates moving fast because they have never witnessed it in historic times, just like they could not image plates moving in the first place. They are stuck in their uniformitarian mindset. However, Dr. John Baumgardner has shown that runaway subduction would have proceeded at speeds of meters per second during the Flood,[1] explaining:

Dr. John Baumgardner

> The energy to drive this event was readily available in the form of gravitational potential energy of the cold, pre-Flood ocean floor rocks. The stress-weakening tendency of silicate minerals comprising mantle rocks allows the process to unfold in a runaway manner. Laboratory experiments document that these minerals can weaken by as much as 8-10 orders of magnitude for shear stress levels that can occur in the mantles of planets the size of the earth.[2]

The cold, original ocean crust (lithosphere) would have dropped into the mantle like a weight in water, rapidly destroying all of the older ocean crust while simultaneously making a whole new ocean crust in less than a year. Dr. Baumgardner's computer models and math and physics calculations have never been shown to be wrong. The ocean seafloor even bears witness to the runaway subduction process. All of the current ocean seafloor is young compared to the continents, only going back to the Absaroka Megasequence at its oldest. It's as if the entire ocean floor was consumed and recreated anew in the not-too-distant past. Recall from chapter 6 that numerous cold lithospheric slabs are visible with seismic tomography deep in the mantle, supporting a recent subduction. Although the seafloor data and the math support that runaway subduction did, in fact, occur during the Flood, Dr. Baumgardner's ideas are still ignored by nearly all secular scientists. On that note, let's examine the rock data for the Tejas.

What Do the Rocks Show?

North America

The Tejas Megasequence extends from near the base of the Paleogene system to the top of the Neogene (top of the Tertiary). This megasequence documents another change in the type of deposition that occurred across North America, recording the lowest percentage of carbonate sediment of any megasequence. The Tejas also shows the highest percentage of volcanic rocks of any megasequence. The uplift of the Rocky Mountains during the Laramide Orogeny shed millions of cubic kilometers of shale and sandstone across the western states. And the Yellowstone Supervolcano and subduction zone volcanoes along the West Coast reached their peak in activity.

A notable shift in the continental drainage direction from north to south occurred at the onset of the Tejas, pouring tremendous quantities of siliciclastics into the Gulf of Mexico, including the basal Tejas Whopper Sand (Lower Wilcox). Siliciclastic deposition continued to spread across the continental shelf area along much of the Atlantic seaboard, offshore northern Canada, and Greenland. Few deposits were preserved in the eastern U.S. and across Canada, other than offshore. The thickest Tejas deposits are found in the Gulf of Mexico, where the entire area received siliciclastic sediment in excess of 3 kilometers. Sandstones make up about 31% of the Tejas sequence. Carbonate rocks diminished to only 9%, and shale is again the dominant lithology, comprising 42% of the total sequence volume.

The total surface area covered by the Tejas across North America is over 14.8 million square kilometers, second only to the Zuni Megasequences. The total volume deposited was over 17.7 million cubic kilometers. This is the greatest volume of all megasequences across North America. However, if volcanic rocks are excluded, the Zuni Megasequence edges out the Tejas in the total volume of sediment deposited. The average thickness of the Tejas is about 1.4 kilometers, second only to the Zuni.

Chapter 14 ★ **317**

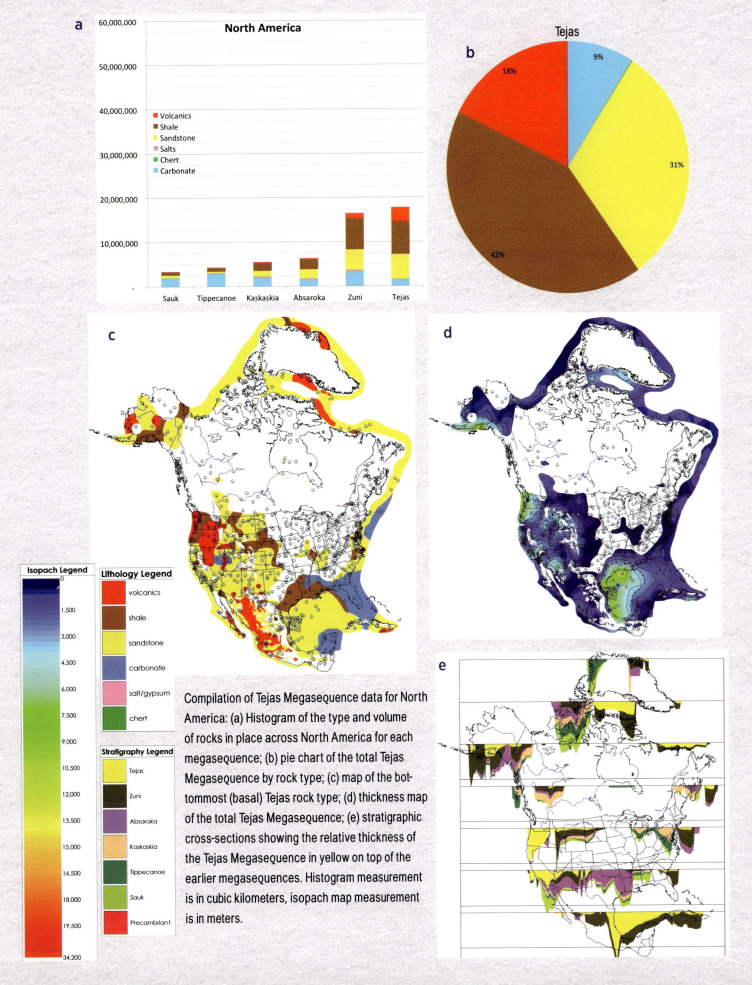

Compilation of Tejas Megasequence data for North America: (a) Histogram of the type and volume of rocks in place across North America for each megasequence; (b) pie chart of the total Tejas Megasequence by rock type; (c) map of the bottommost (basal) Tejas rock type; (d) thickness map of the total Tejas Megasequence; (e) stratigraphic cross-sections showing the relative thickness of the Tejas Megasequence in yellow on top of the earlier megasequences. Histogram measurement is in cubic kilometers, isopach map measurement is in meters.

South America

This megasequence documents another sizable increase in the volume of sediment deposited across South America, even more than the Zuni. Similar to North America and the Rocky Mountains, millions of cubic kilometers of shale and sandstone were shed during the Tejas from erosion of the Andes Mountains across western South America. In addition, a major influx of sediments and volcanic rocks was generated by the Cenozoic development of portions of southern Central America. Siliciclastic deposition continued to be the dominant sediment type across South America. Sandstones made up approximately 27% of the total Tejas by volume, and shales and mudrocks were about 58%. Carbonate deposition was only about 12% of the sediment by volume. Siliciclastic deposition was again thin across the center of the continent (Brazil), with thicker deposits along the flanks and also east of the Andes Mountains. The surface area covered by the Tejas was greater than any of the earlier megasequences, with over 15.8 million square kilometers. The average thickness of the Tejas was slightly greater than 2 kilometers, even more than the thickness of the Zuni Megasequence. The total volume deposited across South America was nearly 33 million cubic kilometers, the most of any megasequence.

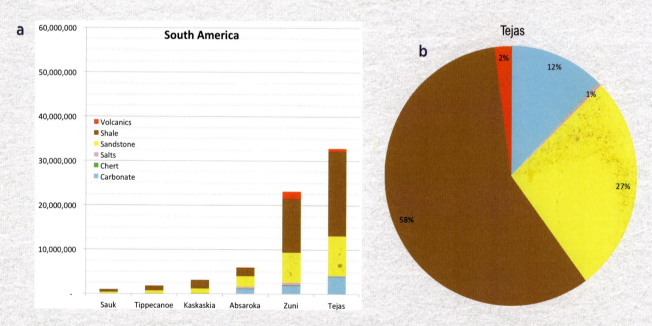

Compilation of Tejas Megasequence data for South America: (a) Histogram of the type and volume of rocks in place across South America for each megasequence; (b) pie chart of the total Tejas Megasequence by rock type; (c) map of the bottom-most (basal) Tejas rock type; (d) thickness map of the total Tejas Megasequence; (e) stratigraphic cross-sections showing the relative thickness of the Tejas Megasequence in yellow on top of the earlier megasequences. Histogram measurement is in cubic kilometers, isopach map measurement is in meters.

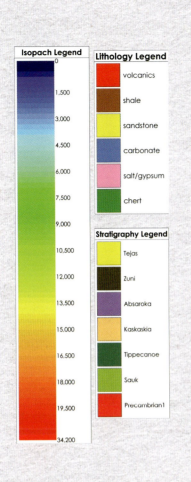
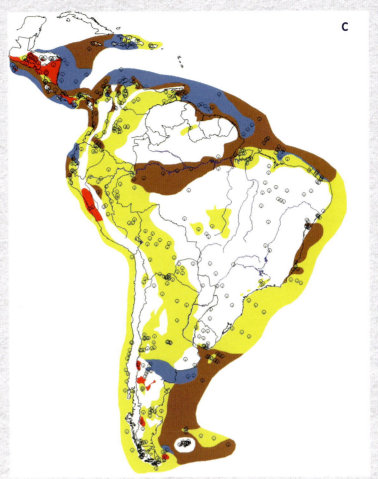

c

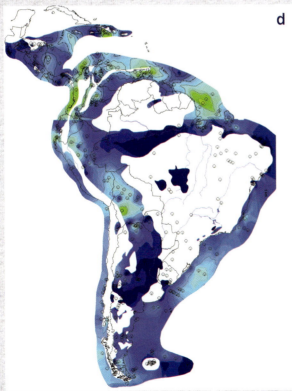

d

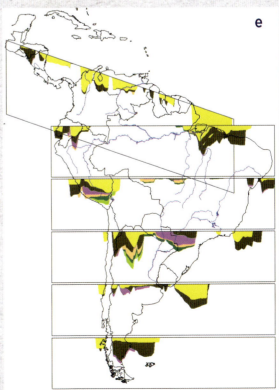

e

Africa

Across Africa, the Tejas Megasequence documents a rather large drop in the volume of deposition compared to the Zuni. Unlike North America, there was no major mountainous uplift similar to the Rocky Mountains to shed millions of cubic kilometers of shale and sandstone across the continent. Carbonate deposition continued to be important across portions of Africa, particularly North Africa, making up 27% of the sediment by volume. Tejas sediments also included thick Miocene salt deposits along the southern boundary of the Mediterranean Sea. Salt, exceeding 3 kilometers, was deposited in the Red Sea, marking the separation of the Saudi Arabian Peninsula from the main continent. Siliciclastic deposition continued to spread across the center of the continent, with depo-centers, presumably from subsidence, again in Niger and Sudan. New, thick siliciclastic wedges formed offshore as deltas for the Niger and Nile Rivers. Shales make up about 39% of the Tejas sequence, and sandstone about 23% by volume. The total surface area, total volume deposited, and average thickness deposited were all second to the preceding Zuni Megasequence. The Tejas across Africa covered an area of over 24.4 million square kilometers. The total volume was nearly 28.9 million cubic kilometers, and the average thickness was close to 1.2 kilometers.

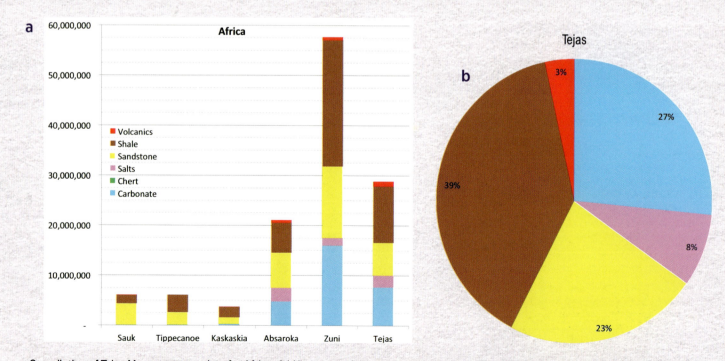

Compilation of Tejas Megasequence data for Africa: (a) Histogram of the type and volume of rocks in place across Africa for each megasequence; (b) pie chart of the total Tejas Megasequence by rock type; (c) map of the bottommost (basal) Tejas rock type; (d) thickness map of the total Tejas Megasequence; (e) stratigraphic cross-sections showing the relative thickness of the Tejas Megasequence in yellow on top of the earlier megasequences. Histogram measurement is in cubic kilometers, isopach map measurement is in meters.

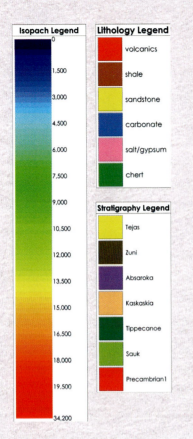
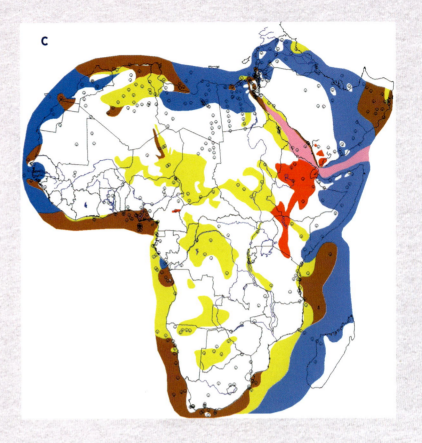
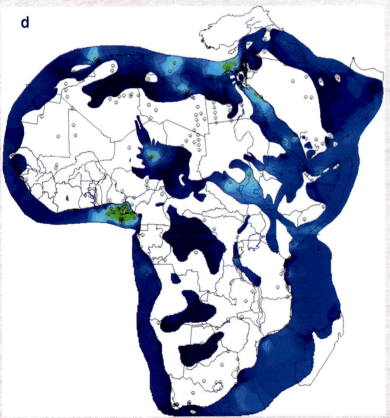
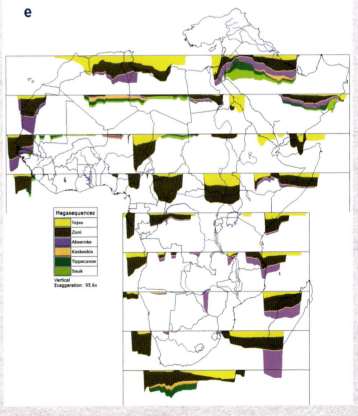

Current Data for the Tejas Megasequence

Figure 14 shows the paleocurrent data collected by Dr. Art Chadwick and his students across North America, South America, and Africa for Tertiary system (Paleogene and Neogene) rocks, which make up the Tejas Megasequence.[3]

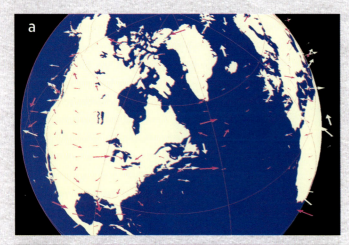
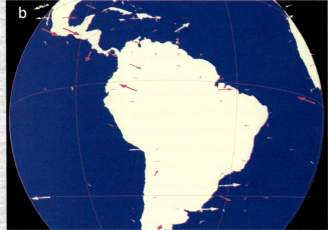

Figure 14.1. Current directions compiled across (a) North America, (b) South America, and (c) Africa for the Tejas Megasequence. Courtesy of Art Chadwick.

The currents across North America show a more dominant easterly flow developed across the middle of the continent compared to the earlier Zuni. Flow directions along the edges of the continent also exhibit a much more pronounced offshore flow direction.

Around the edges of South America, the current pattern exhibits a dominant direction of transport toward the offshore. There is very little current data available in the center of South America. Africa has a few strong current directions (longer arrows) indicating flow toward the offshore regions.

Overall, the Tejas current data consistently indicate flow directions off the continents and toward the offshore regions. This fits well with the Tejas representing the receding phase of the Flood.

Summary of the Rock Data

Totaling the rock data across the three continents, we find the Tejas is second only to the Zuni in total surface area coverage, total volume deposited, and total average thickness. This makes sense if the end of the Zuni was the high-water mark of the Flood (Day 150). The subsequent Tejas should be close to the same area covered if it started on Day 150-151, imme diately after the Zuni ended. And in fact, the 55 million square kilometers covered by the Tejas is very close to the 56.9 million square kilometers of the preceding Zuni Megasequence. The numerical difference can be easily explained by the likelihood of a bit more erosion of the more exposed rocks of the Tejas. No other megasequence is even close to this amount of surface area coverage. The megasequence with the third-most total surface coverage is the Absaroka, with 35.6 million square kilometers. The nearly identical amounts of surface coverage for both the Zuni and the Tejas strengthen the interpretation that the end of the Zuni was the high-water mark of the Flood (Day 150) and that the Tejas was the receding phase.

Much of the Tejas Megasequence likely represents material washed off the highest pre-Flood hills that became backwashed onto the Zuni strata as the floodwaters began to recede (Day 150+). Fossils in the Tejas Megasequence also contain increasingly more angiosperms (flowering plants) and mammal fossils compared to the Zuni deposits, indicative of higher terrains. These areas were apparently wiped free of all life, removing even the pre-Flood soil and any rock layers that might have existed there.

> "I will *wipe off* man whom I have created *from* the face of the land, from man to animals to creeping thing and to birds of the sky; for I am sorry that I have made them."

Dr. Russ Humphreys, in his translation of Genesis 6:7 and Genesis 7:23, suggests the term "wiped off" to explain this stripping of the land surface right down to the crust:

> And the Lord said, "I will *wipe off* man whom I have created *from* the face of the land, from man to animals to creeping thing and to birds of the sky; for I am sorry that I have made them." Thus He *wiped off* every living thing that was upon the face of the land, from man to animals to creeping things and to birds of the sky, and they were *wiped off from* the earth.[4]

Humphreys goes on to suggest this "wiping off" meant no earth (or soil) was left behind, as in the way one wipes a dish clean (2 Kings 21:13). "Taking these verses straightforwardly means the waters swept mud, plants, the animals completely off the formerly dry land, the pre-Flood continental surface."[4] And this is exactly what we see across large portions of the continents. The pre-Flood high hills include the major shield areas of Canada, Greenland, Brazil, and Central and Western Africa. When placed back together in a Pangaea-like configuration, the upland areas become quite substantial.

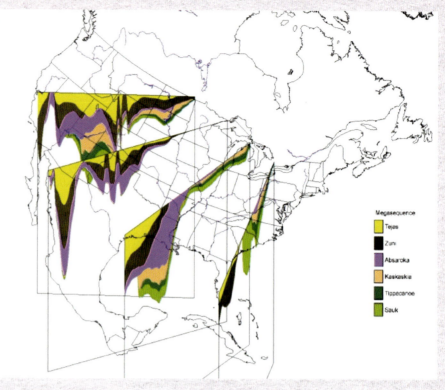

Cross-sections across North America showing the thinning to the north toward Canada

Deposits in the Tejas also include the thickest and most extensive coal seams in the world.[5] These huge mats of transported trees (non-lycopods) likely represented plants swept off the upland areas and buried within Tejas sediments, becoming vast coal seams. Some of these seams are over 200 feet thick in western North America.[5]

God wiped off these areas of highest elevation where most of the large mammals, flowering plants, and humans likely existed in the pre-Flood world, spreading their remains in sedimentary layers on top of the earlier buried dinosaurs, creating Tejas (Cenozoic) strata.

Animals may have been buried closer to their place of origin as the floodwaters were rising (Sauk through Zuni Megasequences) until Day 150 was reached. The water and sediment likely engulfed the animals nearly *in situ* as the water level increased. But the Tejas depositional pattern appears to have been different. It was apparently the result of a reversal in flow direction as God began to remove the waters off the continents after Day 150. This not only transported the flora and fauna from off of the highest hills, it spread those deposits outward toward the continental margins. Animals and plants that lived in areas that are now exposed crystalline rock (Precambrian shields) were transported great distances and deposited on top of the Zuni strata and sometimes older exposed strata too.

And there is evidence of a reversal of water flow direction at the Zuni/Tejas boundary. Recent research using detrital zircons shows a rapid shift in the direction of drainage from the Cretaceous (uppermost Zuni) to the Paleocene (lowermost Tejas) across North America.[6] The researchers found that during deposition of the Cretaceous, the drainage pattern was to the north/northwest across much of North America. They also determined that very little area was draining to the Gulf of Mexico during this time.

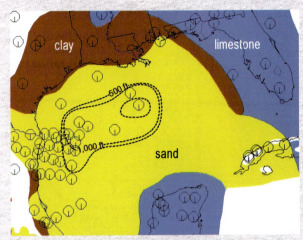

The Whopper Sand in the Gulf of Mexico

In contrast, they found that the Paleocene (Tejas) drainage shifted dramatically southward to the Gulf of Mexico.[6] This was not a single river like the modern Mississippi but a series of rivers, effectively behaving like sheet wash, draining into the Gulf of Mexico across a broad area. This sudden shift in drainage coincides with the end of the Zuni Megasequence and the onset of the Tejas Megasequence.

My colleague and I used this documented change in drainage direction at the Zuni/Tejas boundary to explain the Whopper Sand in the deep water of the Gulf of Mexico.[7] The Whopper Sand is unusual because its location is in ultra-deep water, nearly 200 miles from the Lower Wilcox shelf margin, and far from any conventional sand source.[8]

We believe the Whopper Sand is a consequence of this rapid shift in drainage at the Zuni/Tejas boundary, when water suddenly began draining off the North American continent into the Gulf of Mexico, reversing the earlier direction of flow.[7] This drainage shift is marked by a sudden change in deposition from the uppermost Zuni layer (Lower Paleocene Midway Shale) to the lowermost Tejas (Paleocene-Eocene Whopper Sand). In a Flood model, this coincides with the change in water direction described

Surface Area (km²)	North America	South America	Africa	Total
Sauk	12,157,200	1,448,100	8,989,300	22,594,600
Tippecanoe	10,250,400	4,270,600	9,167,200	23,688,200
Kaskaskia	11,035,000	4,392,600	7,417,500	22,845,100
Absaroka	11,540,300	6,169,000	17,859,900	35,569,200
Zuni	16,012,900	14,221,900	26,626,900	56,861,700
Tejas	14,827,400	15,815,200	24,375,100	55,017,700
Volume (km³)	**North America**	**South America**	**Africa**	**Total**
Sauk	3,347,690	1,017,910	6,070,490	10,436,090
Tippecanoe	4,273,080	1,834,940	6,114,910	12,222,930
Kaskaskia	5,482,040	3,154,390	3,725,900	12,362,330
Absaroka	6,312,620	6,073,710	21,075,040	33,461,370
Zuni	16,446,210	23,198,970	57,729,600	97,374,780
Tejas	17,758,530	32,908,080	28,855,530	79,522,140
Average Thickness (km)	**North America**	**South America**	**Africa**	**Total**
Sauk	0.275	0.703	0.675	0.462
Tippecanoe	0.417	0.430	0.667	0.516
Kaskaskia	0.497	0.718	0.502	0.541
Absaroka	0.547	0.985	1.180	0.941
Zuni	1.027	1.631	2.168	1.712
Tejas	1.198	2.081	1.184	1.445

Table 14.1. Surface area, sediment volume, and average thicknesses for North America, South America, and Africa for each of the six megasequences

for Day 150 of the Flood. Initial drainage rates in the Paleocene, coinciding with a sudden drop in sea level at the onset of the Tejas, were likely high volume and highly energetic, providing a mechanism to transport the Whopper Sand into deep water. As the flow volume diminished with time, the drainage volume lessened, lowering the energy available for transport until the present-day pattern developed. We now observe relatively small flow volumes compared to what was probably occurring during the initial draining of North America at the start of the Tejas. There are more details on the Whopper Sand later in this chapter.

Note that South America has a greater volume of Tejas than Zuni. This may be caused by erosion from the newly formed landmass of southern Central America, which developed mostly in the Cenozoic (Tejas). And the uplift of vast mountain ranges running the length of both North and South America played a major role in the volume deposited during the Tejas. These major mountain ranges shed tremendous amounts of sediment during their uplift, creating great volumes of Tejas sedimentary rock east of the mountain ranges. And combining that with the increased amount of sediment generated by the newly formed subcontinent of Central America, we get a greater volume of Tejas deposition for South America. In contrast, Africa has no significant Tejas-age (Cenozoic) mountain ranges running the length of the continent to provide vast volumes of additional Tejas sediment.

Finally, the Tejas isopach maps of North and South America show areas where no Tejas exists in the regions of the Rocky Mountains and the Andes Mountains. Erosion of these early Cenozoic uplifts has exposed the underlying basement rocks in these locations. The rapid uplift of the mountains and the adjacent subsidence separated the various sedimentary basins from one another, particularly in North America.

Continental Configuration for the Tejas Megasequence

The rapid separation of the continents that began in earnest in the Absaroka continued throughout much of the Tejas, forming new seafloor right up to the latest part of the megasequence cycle in the Pliocene. Runaway subduction showed no signs of slowing until all of the original pre-Flood ocean crust was subducted. By the end of the Tejas, the continents of the world were pretty close to their present locations.

The Saudi Arabian Peninsula also split with Africa during the Tejas. One of the first rocks deposited in the newly formed Red Sea was salt, over 10,000 feet of salt in places. Again, we see a close relationship of salt and rifting. Secular geologists still have no adequate explanation for the formation of this much pure salt, deposited so quickly. The Tejas also saw the beginning of the East African Rift Valley that completed the so-called triangle zone of rifting that connects the Red Sea Rift and the Indian Ocean Rift.

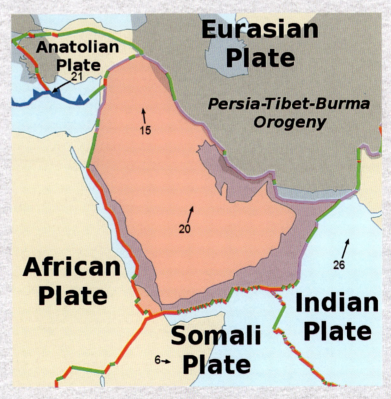

Continental configuration during the deposition of the Tejas Megasequence and its sedimentary thickness

Tectonics map of the Red Sea Rift, the East African Rift, and the Indian Ocean Rift

Subduction-related volcanoes of Central America (above) and South America (below)

In South America, the subducting crust beneath the west coast caused the Andes Mountains to reach their zenith in development in the Tejas. Magmas were intruded and volcanic activity abounded. Central America showed similar volcanic activity from the subducting ocean crust along the western boundary. The coalescence of many of the volcanoes down the length of Central America completed the connection between North and South America through the Costa Rican/Panamanian Isthmus.

In North America, most of the uplift of the Rocky Mountains occurred in the Tejas. The Laramide Orogeny that began at the end of the Zuni caused great basement-cored mountains to be pushed upward and many of the adjacent areas to be down-dropped. These uplifts became massive mountains like the Front Range in Colorado and the Wind River Mountains in Wyoming. The areas between the mountains subsided nearly equally, forming great basins that accumulated upward of 10,000 feet of sediment in the Tejas. Some of these areas became the Bighorn Basin and the Green River Basin of Wyoming.

Nearly 80% of the mountain ranges of the world developed in the Tejas, including the Circum-Pacific Orogenic Belt and the Alpine-Himalayan Belt. These include the major mountains like the Andes, the Rockies, the Alps, and the Himalayas.[9] In fact, most of these mountain ranges have undergone several kilometers of uplift even since the Pliocene (uppermost Tejas).[1]

The reason for these simultaneous uplifts may be tied to the thickening of the continental crust above subduction zones and the eventual cessation of runaway subduction at the end of the Tejas. For example, the continental crust along the west coast of South America has reached a thickness close to 70 kilometers due to the subduction process, whereas most continental crust is only about 35 kilometers thick.[1] Crust is thickened above subduction zones because of the melt generated, the intrusion of these magmas into the crust, and the physical dragging of portions of the warm and ductile lower crust underneath the continental margin. Thicker crust causes more isostatic uplift such that a 60-kilometer-thick crust will tend to have an elevation 14,500 feet higher than a 35-kilometer-thick crust.[1]

John Baumgardner explained it this way:

> In a nutshell, the catastrophic processes unleashed in the Flood not only deposited thousands of feet of fossil-bearing sediments on all the

Himalayas

continents and moved North and South America some 3000 miles westward relative to Europe and Africa, but also increased the thickness of the buoyant crustal rock in the belts where high mountains now exist….

But when the process of rapid subduction shuts down, these dynamical forces disappear, and the buoyancy forces take over to elevate the zone of thickened crust toward a state of isostatic balance. The uplift of high mountains at the close of this episode of rapid subduction is therefore a logical after effect of this runaway process. Within the Flood framework, the timing of the uplift, unfolding in the centuries following the cataclysm, is just what one should expect based on simple mechanics considerations. On the other hand, no mechanical response in terms of uplift during tens of millions of years of tectonic forcing followed by a sudden pulse of uplift poses a serious problem for the uniformitarian framework.[2]

In addition to the formation of most of the world's mountains, the Tejas saw the formation of the San Andreas Fault, the eruptions of the Yellowstone Supervolcano, and the outpouring of massive amounts of basaltic lava to make the Columbia River Plateau.

Why Are the Rocky Mountains So Wide Compared to Other Ranges?

Secular scientists today cannot adequately explain why the Rocky Mountains of North America are so far inland from the coast, extending all the way to the Black Hills of South Dakota. Note that north of Wyoming, these basement-cored mountain ranges disappear and become thin-skinned thrusts (only involving sedimentary deformation), like at Glacier National Park and in the Canadian Rockies. In contrast, most mountain ranges that develop from

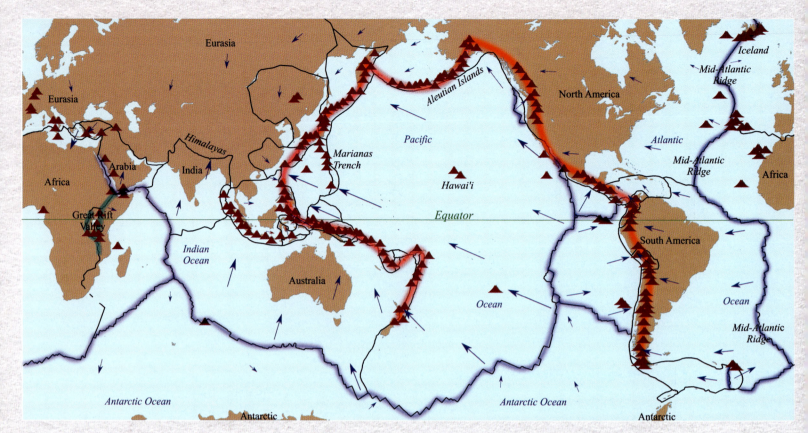

Ring of Fire in red. Ridges in blue.

subduction are relatively narrow, like the Andes Mountains, and form right along the coast. There must have been something about the subduction along the west coast of North America that was different.

Today, we see a ridge system coming north up the Gulf of California (along the Baja Peninsula) and a second ridge system off the west coast of Washington and Oregon, creating the subduction zone there and the Cascade Mountains. These two ridge systems are now connected through the San Andreas Fault system. The former section of the ridge system in between has apparently been subducted beneath North America. It's possible that this subducted ridge has something to do with the location of the Rocky Mountains now spread across New Mexico, Colorado, and Wyoming.

For many years, secular geologists have speculated that the Rocky Mountains were caused by low-angle or near-flat subduction.[10-12] One of my former professors expanded on this idea by suggesting that a subducted ridge system may have caused the transmission of stresses great distances inland. He could offer no other explanation for the vertical uplift and subsidence observed in the Wyoming Foreland (area of basement-cored mountains).[13]

A subducted ridge system tends to be more buoyant and would exert pressure upward on the base of the continent, possibly providing a cause for the basement-cored mountain ranges so far inland from the subduction zone that existed along the West Coast. Kelleher and McCann first noticed this off the west coast of South America where ridges have been subducted nearly horizontally beneath the continent for hundreds of kilometers.[14]

If North and South America were moving westward at the velocities suggested by runaway subduction, it is likely they possessed sufficient momentum to override ridge systems that are normally thought to be too buoyant to be subducted. One of the objections I have with hydroplate theory involves the claimed sliding of North and South America westward away from the Mid-Atlantic Ridge system late in the Flood.[15] The East Pacific Rise (ridge system in the Pacific that connects to the Baja Ridge) is much higher and wider than the Mid-Atlantic Ridge. How could a continent slide away from a smaller ridge to the east and travel west, overrunning a larger, higher, and wider ridge system in the process? This makes no sense. In fact, it is impossible. However, rapid plate motion caused by runaway subduction and catastrophic plate tectonics solve the apparent dilemma. North America could fairly easily overrun a ridge system moving at velocities of meters per second.

A subducted ridge system so far inland beneath North America could also explain the trail of the Yellowstone Hot Spot from Oregon to northwest Wyoming in the Cenozoic. This line of volcanism parallels the northern edge of the subducted ridge system. The highest point of the ridge system may have funneled magma along the crest of the subducted ridge, like the peak of a tent, releasing vast quantities of magma to burn a hot spot trail through the overlying crust. In similar fashion, the Yellowstone Supervolcano eruptions could also be a consequence of the pooling of magma beneath the subducted ridge.

Grand Prismatic Spring, Yellowstone National Park, Wyoming

Finally, the eventual sinking of this subducted ridge system beneath North America may have released and funneled the massive quantities of magma needed to create the Columbia River Plateau and associated basaltic volcanism late in the Cenozoic. Both the Yellowstone region and the Columbia River Plateau fall along the northern boundary of this presumed subducted ridge. The edge of the subducted plate would be the most likely location for the release of massive quantities of magma that had built up underneath. Much more research still needs to be done on these suggested explanations.

Whopper Sand in the Gulf of Mexico

There's a huge deposit of sand in the deep Gulf of Mexico, and no one seems to know how it got there—except maybe Flood geologists.

Early in my career as a geologist for an oil company, we were told not to prospect in water deeper than 2,000 feet. Most offshore oil is found in sand layers sandwiched between thick layers of mud and clay, and our management believed no sand could get that far offshore, and drilling costs were too high.

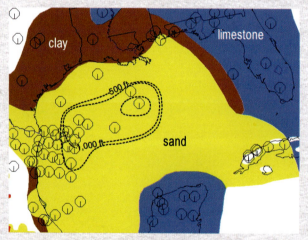

The Whopper Sand in the Gulf of Mexico

However, in 2001, the BAHA-2 well was drilled through almost 7,800 feet of water and into the Wilcox Sand at the base of the Tejas Megasequence. The drillers found over 1,000 feet of nearly continuous sand. This discovery shocked geologists, who termed it the Whopper Sand,[16] and paved the way for numerous nearby discoveries of billions of barrels of oil.

Since the drilling of the BAHA-2 oil well in 2001, over 15 billion barrels of oil have been discovered in the Paleocene-Eocene Wilcox-equivalent Whopper Sand.[8] The BAHA-2 well penetrated 335 meters (1,100 feet) of sand in the Lower Wilcox in over 2,135 meters (7,000 feet) of water within the Perdido Fold Belt of Alaminos Canyon. In Keathley Canyon, the Sardinia-1 well penetrated over 366 meters (1,200 feet) of sand, and in Walker Ridge, the Jack-2, Chinook, and Cascade-2 wells reached similarly thick Lower Wilcox sands approaching 580 meters (1,900 feet) thick.[17] Average porosity in the Whopper Sand is 18%, and permeabilities range from 10 to 30 millidarcys.[17]

The Whopper Sand extends over 40,000 square miles in water depths between 7,600 and 10,000 feet, and is over 225 miles from the nearest onshore discoveries of Wilcox-equivalent sands.[16] It is commonly more than 1,000-feet thick and can be up to 1,900-feet thick. Some layers even contain a high amount of metamorphic, volcanic, and sedimentary rock fragments, making this less like a winnowed-clean beach sand and more like a braided river sand.[16,18] And it is not just the extent and thickness of the sand that makes this section unique, it is also the lack of interbedded clay and mud layers. The Whopper Sand is nearly 70% pure.[19]

Several hypotheses, bordering on the bizarre, have attempted to explain this enigma. One idea argues that sea level fell close to 6,000 feet in the central Gulf of Mexico, leading to the deposition of the Whopper Sand in the resulting great depression.[20] Others use analogies of modern rivers and submarine canyons to explain the sand's appearance.[19] However, sea level dropping thousands of feet is not a reasonable cause. It is perhaps just as unlikely to claim that pure sand could travel 225 miles over a nearly flat basin floor. Modern deepwater deposits contain high amounts of clay that are necessary to maintain sand in suspension while traveling down a slope.[16]

So, where did the Whopper Sand come from? The answer appears to be related to the receding stage of the great Flood (Genesis 8:3). The Whopper Sand is near the base of the last worldwide sedimentary sequence formed during the Flood (Tejas).[21] Drainage across the United States changed dramatically as these layers were being deposited, with most of the water flowing toward the Gulf of Mexico.[6] It is logical that the floodwaters that inundated whole continents would have flowed off in catastrophic volumes. High-velocity, sheet-like flow would tend to transport large volumes of sand and rock fragments first, dumping the Whopper Sand into deep water.

This type of flow would only have occurred once during the recession of the Flood's water. Today, we find mere trickles of flow to the deep water, transporting a mixture of clay and sand down submarine canyons. Because the Flood was global, there are likely other whopper sands to be found in deep water worldwide.

Mammal fossil bone bed in the Ogallala Formation, Texas

Ogallala Sand Across the Great Plains

ICR scientists have led field trips through Palo Duro Canyon near Amarillo, Texas, pointing out many geological features that show clear evidence of the global Flood. The walls of the canyon display over 700 feet of strata from the rising and receding phases of the Flood.

Palo Duro is the second-largest canyon in the U.S., behind only Grand Canyon. A bright orange siltstone, the Permian Quartermaster Formation, forms the base of Palo Duro. Secular scientists claim this layer is about 250 million years old. On top of this are the multicolored Triassic Tecovas Formation (shale) and Triassic Trujillo Formation (sandstone), both claimed to be about 210 million years old. These units were deposited just prior to the layers containing most of the dinosaurs. They contain fossils of phytosaurs (large crocodile-like reptiles) and amphibians up to seven feet in length.[22]

The rocks show no evidence of the supposed 40 million years of missing time between the Quartermaster and the overlying Tecovas. We observe flat-lying rocks upon flat-lying rocks for tens of miles in all directions. These sedimentary strata look like they were deposited layer after layer, like bricks in a wall, with no time gaps in between.

Palo Duro Canyon, Texas

The cap rock that makes the upper rim of Palo Duro Canyon is the Miocene-Pliocene Ogallala Formation. Composed of a tan cliff-forming sandstone, some siltstone, and a basal conglomerate layer,[22] this uppermost unit is thought by evolutionists to have been deposited between 4 to 10 million years ago.[23] Evolutionary geologists have recognized this major problem in their uniformitarian paradigm, stating:

> You will note a lot of time is missing between the Trujillo and the Ogallala. Either the rocks representing about 200 million years of time were eroded away, or they were never deposited; whatever the case, a great unconformity is represented by the mere line between the multi-colored upper beds of the Trujillo and the lower tan beds of the Ogallala.[24]

The lack of any visible erosion is strong evidence that there were not millions of years between the deposition of the Triassic beds and the overlying Ogallala. Instead, we see a pattern—much like we see in Grand Canyon—that is best explained by continuous activity. The Ogallala is conformable to the underlying Trujillo all around the canyon rim, with no tilting of the underlying units and no erosional channels carved into the boundary surface.

The Ogallala is one of the most extensive units east of the Rockies, covering about 174,000 square miles from Texas to South Dakota.[25] While it is only 20 to 40 feet thick in the canyon, it increases to over 700 feet across much of the Great Plains. Igneous and metamorphic cobbles in the basal conglomerate of the Ogallala are sourced from the Rocky Mountains, hundreds of miles to the west.[22]

Secular scientists claim these are deposits from rivers, but a receding mega-flood explanation better fits the broad extent of the Ogallala. How else can a blanket sand layer spread across thousands of square miles with no evidence of river channelization? And localized post-Flood catastrophism cannot explain the massive extent of this deposit either, just like isolated regional processes cannot explain the huge deposit of the Whopper Sand in the deep Gulf of Mexico.[26]

The formation of the Ogallala would have required high-energy conditions over a huge area, similar to sheet-wash off a parking lot, to distribute the cobbles and sands so evenly across vast regions of the Great Plains. Visitors to Palo Duro Canyon can witness a vivid and beautiful reminder of both the rising and the receding stages of the Genesis Flood.

Upper Cenozoic Flood/Post-Flood Boundary: Introducing the N-Q

One of the most important aspects of any Flood model is definition of the boundaries. For decades, creation scientists have debated the level at which the Flood ended in the rock record. In the past, many have based their conclusions on a cursory examination of the rocks, or concentrated only on the strata across the American West.

Most agree that the Flood/post-Flood boundary is at one of two levels: 1) at the top of the Cretaceous system, known as the K-Pg (K-T) horizon,[27,28] or 2) at or near the top of the Neogene (Upper Cenozoic) at about the Pliocene level.[29,30] I have called this the N-Q boundary for Neogene-Quaternary.

> "One of the most important aspects of any Flood model is definition of the boundaries."

Our extensive global rock data from oil wells and outcrops are helping to resolve this matter. Some of our results were published in 2017 in the *Creation Research Society Quarterly*.[31]

In this paper, we presented five major geologic observations that demonstrate the Flood/post-Flood boundary is much higher than the K-Pg level. Some of these features are so large and/or unusual in scale that local post-Flood catastrophes could not have conceivably produced them. Others demonstrate geologic conditions that could only have existed while the floodwaters were still covering large portions of the continents. Collectively, they strongly refute the claim that the Flood ended at the stratigraphic level of the K-Pg boundary.[31]

First, the Whopper Sand. Oil companies discovered the Whopper Sand in the Gulf of Mexico by drilling wells

Oil rig in the Gulf of Mexico

in water depths of over 7,000 feet and over 200 miles offshore. The only reasonable explanation for this >1,000-foot-thick sand bed that covers much of the floor of the deep Gulf of Mexico is a high-energy runoff of water—something that easily fits the Flood model. This would coincide with the change in water direction described for Day 150+ of the Flood year. Initial drainage rates, coinciding with a sudden drop in sea level at the onset of the Tejas Megasequence, would correspond to geological layers after the K-Pg boundary. The forces responsible were likely high volume and highly energetic, providing a mechanism to transport the thick Whopper Sand into deep water.

Second, the tremendous amount of Tejas sediment deposited globally. The volume of Tejas sediment is second only to the Zuni Megasequence that ended with the Cretaceous system, the presumed high-water point of the Flood. The tremendous amount of Paleogene and Neogene sediments all over the world that are part of the Tejas Megasequence cannot be easily dismissed as the product of local catastrophes. These sediments and the fossils they contain are better explained by the receding water phase of the Flood as mountain ranges and plateaus were being uplifted.

Third, the thickest and most extensive coal seams are found globally in Tejas sediments. The Powder River Basin (PRB) coals, which are all within Paleogene system rock layers, contain the largest reserves of low-sulfur subbituminous coal in the world. At least six or more coal beds in the PRB exceed 100 feet in thickness, and some individual beds have been shown to extend for over 75 miles. Some of these coal beds can exceed 200 feet thick in places, such as the Big George coal layer. These coal beds are part of the receding phase of the Flood that transported huge mats of plant and tree debris. They were derived largely from angiosperms living at higher elevations, and floodwaters rapidly buried them in huge deposits.

Fourth, the tremendous amount of rapid ocean crust/seafloor spreading that continued right across the K-Pg boundary and up to the Pliocene, with no indication of a significant change in velocity. The runaway subduction model for the global Flood, described by geophysicist Dr. John Baumgardner, caused the creation of approximately one-third to one-half of the world's ocean crust to form during the deposition of the Tejas Megasequence (Paleocene through Pliocene). In addition, the huge earthquakes generated by this movement would have been devastating for any type of human civilization after the Flood if the Flood/post-Flood boundary is located at the K-Pg.

Fifth, the identification of uninterrupted water-deposited carbonate rocks from the Cretaceous (below the K-Pg boundary) and continuing upward through Miocene strata across much of North Africa and the Middle East, areas just to the south of the landing site for the Ark in Turkey. The continuous limestone beds from the time of Cretaceous deposition (Zuni Megasequence) through the top of the Miocene (Upper Tejas) in Iraq are the closest thing to proof that the Flood was not over at this point. Huge regions of the Middle East were clearly still underwater during the Tejas. If they were post-Flood deposits, it would be impossible for humans to settle there at that time and build the Tower of Babel.

In addition, our research efforts have identified more geological features that further support that the Flood/post-Flood boundary is near the top of the Tejas Megasequence, which encompasses the Paleogene and Neogene geological systems. One of these is the Ogallala Sandstone we described above and in *Acts & Facts*.[32] Local post-Flood catastrophes cannot explain this continuous sand bed that covers much of the Great Plains. It must be part of the receding phase of the Flood as well.

Collectively, these data establish that much of the Paleogene and Neogene (known previously as the Tertiary) was the receding phase of the great Flood, placing the Flood/post-Flood boundary at the top of the Tejas Megasequence (Upper Cenozoic). This can be referred to as the N-Q boundary since it marks the boundary between the Neogene and the Quaternary. The fossils of so many large mammals and the vast majority of fossil flowering plants that show up so abundantly, and for the first time, in the Tejas Megasequence rocks are best explained as a consequence of the receding phase of the Flood. These animals and plants were swept off the highest pre-Flood hills as the waters rose, and then buried as the floodwaters began to recede. Real rock data not only confirm there was a global flood as described in the Bible, but they also help us better understand its final stages of sedimentary deposition.

The Tower of Babel exhibit in the ICR Discovery Center for Science & Earth History, Dallas, Texas

Grand Canyon Carved by Flood Runoff in Tejas

Secular science has struggled to explain the timing and origin of Grand Canyon for decades. Today, the majority of secular scientists have assumed an age of less than 6 million years for the carving of Grand Canyon.[33] But they still struggle to overcome the hurdle of why it formed where it did. In particular, how did the canyon hurdle the Kaibab uplift?

One of the issues the secular scientists run into is how to get the rivers that made the canyon connected from one side of the Kaibab uplift to the other. The Kaibab uplift has warped an arch of rock about 3,000 feet above the surrounding terrain,[33] and Grand Canyon currently cuts right through it. To explain the apparent dilemma of an incised river that seemingly had to flow uphill, some secular scientists have claimed that Grand Canyon was carved by a process called stream piracy, involving headward erosion.[33] This, in itself, is very tricky because it requires two streams flowing in opposite directions to meet at the exact same point and then somehow erode away enough for one stream to take over the other (piracy). The problem with this explanation is there is still a major drainage divide in between the two rivers, even if the headwaters did touch. It doesn't remove the uplift problem. Water would still flow in opposite directions away from the divide as it does all along the Continental Divide today.

Other secular scientists claim there were caves that may have allowed the streams on either side of the uplift to connect underground first.[33] They suggest these caves later collapsed and the canyon had a ready-made path through the uplift. The problem with this is that the layer of limestone that the caves could have formed in was also warped up about 1,500 feet from the Kaibab uplift. The water would then have still had to flow uphill to pass through

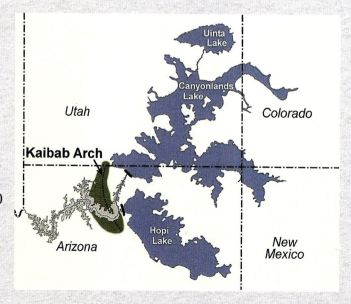

The fictional lakes that some creation geologists propose emptied in a catastrophic manner to carve Grand Canyon are based on little if any geological evidence[33]

the uplift. In addition, the carbonate layer (the Redwall Limestone) that contains caves is not thick enough to collapse and remove 1,500 feet of relief. The Redwall is about 600 feet thick, and even if the entire unit was a cave, it wouldn't remove the remaining 900 feet of relief of the Kaibab uplift. The streams on either side would still continue to flow in opposite directions.

Secular science is left without an adequate explanation for how Grand Canyon developed through a major uplift. It should have flowed around. But it didn't.

Creation scientists have also disagreed on the timing and origin of Grand Canyon, but we all agree it happened in the last 4,500 years or less. One of the most popular ideas is the breached-dam hypothesis advocated by Drs. Walt Brown[15] and Steve Austin.[34] Both have supported a post-Flood formation of Grand Canyon caused by the sudden breaching of presumed dammed lakes called Hopi Lake and Grand Lake (sometimes called Canyonlands Lake). Both believe these lakes formed a few hundred years after the Flood was over, placing their development in the Ice Age (see chapter 15). The presumed lakes are claimed to have held over 3,000 cubic miles of water,

Red Wall Limestone, Grand Canyon National Park, Arizona

roughly equivalent to three times the volume of water in Lake Michigan.[33] The explanation for the breach varies, but advocates for this hypothesis claim a catastrophic release of this dammed water carved Grand Canyon.

The biggest problem with the breached-dam hypothesis is the lack of physical evidence for these lakes. There is also not a satisfactory reason given for the cause of the breach through the Kaibab uplift. These were not ice-dammed lakes like other catastrophic breaches. In fact, Mike Oard has pointed out that the water should have taken a different direction if, in fact, it did breach, carving a canyon to the north instead.[35]

Let's examine the evidence for these massive lakes that presumably built up over a few hundred years after the Flood. Helble and Hill have pointed out that there is virtually no evidence for Grand Lake (alternatively called Canyonlands Lake), the bigger of the two lakes.[33] And the evidence for the southern and smaller lake, known as Hopi Lake, is not really strong either. So right away, the largest body of water can be eliminated from a scientific point of view. Merely drawing in a vast lake based on today's topography with no evidence to back it up is speculation, not science. In fact, Edmond W. Holroyd III, who originally plotted the outline of these lakes, no longer believes they existed.[36]

Furthermore, none of the lakes have any mapped terraces or strand lines that are common surrounding Lake Bonneville, the Ice Age lake that has since evaporated to

View of Steep Mountain and the Bonneville bench. Note the beveled lake-cut terrace covered by 150 feet of sand and gravel along the top of light-colored Oquirrh strata.

Great Salt Lake, Utah

form present-day Great Salt Lake. Many other Ice Age lakes show strand lines or wave-cut terraces also, including areas along the shores of the northern Great Lakes in my home state of Michigan.

Admittedly, the presumed Hopi Lake does contain a sedimentary unit known as the Bidahochi Formation, claimed by secular geologists to represent a lake environment. However, this formation has been dated by secular geologists as 6 to 16 million years old based on igneous intrusions that cut across the unit.[33] Although I disagree with this date and the dating methods used, these dates do imply an age older than the Ice Age, which precludes this unit as a source of water for the breached-dam hypothesis as described. Additionally, Helble and Hill pointed out that recent research has shown the Bidahochi Formation is composed of several smaller lakes that likely were not well connected.[33] So, the evidence for any major lakes that formed after the Flood to carve Grand Canyon is weak, at best.

Dr. Steve Austin has advocated for the breached-dam hypothesis because, at the time anyway, he accepted a Flood/post-Flood boundary near the top of the Cretaceous system (K-Pg).[34] His choice of a Flood boundary left him no other options. He had to somehow come up with a massive source of water since he believed the Flood runoff phase was over before Grand Canyon was carved. Picking the wrong Flood/post-Flood

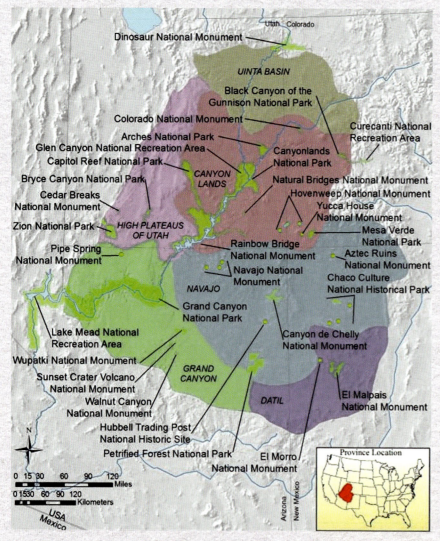

Colorado Plateau showing the location of Grand Canyon

boundary can clearly affect other interpretations even when the data tell us otherwise. This is why it is so important to pick a post-Flood boundary based on as much data as possible.

If there were no post-Flood lakes, then how can we explain Grand Canyon in a Flood model? I think the best solution is the one Mike Oard presented in his book on Grand Canyon.[35] Oard suggests that Grand Canyon formed during the latter part of the runoff phase of the Flood. According to my interpretation, this would coincide with the latter part of the Tejas Megasequence. As discussed above, after runaway subduction ceased, and even before, the thickened areas of continental crust began to rise due to isostatic adjustment. The Four Corners region and the Colorado Plateau rose about 5,000 feet late in the Tejas (Neogene).[9] Grand Canyon is on the western edge of that uplift. Oard suggested that the floodwaters receding in this area may have initially drained to the east during a sort of sheet-wash phase.[35] We saw this easterly direction of transport depositing the Whopper Sand in the Gulf of Mexico during the earliest part of the Tejas. And then later, as the Colorado Plateau rose, it reversed the direction of flow, diverting the drainage to the west and carving Grand Canyon toward the end of the Tejas. He called this latter draining the *channelization phase* of the runoff.[35]

When packed wet sand is uplifted, it will crack. Similarly, if you pack wet sand on your legs at the beach and then move your legs, the sand cracks. Water follows the easiest path. It would naturally follow the cracks and fractures in the freshly deposited sediments of the uplifted Colorado Plateau. Some of these would have undoubtedly run through the Kaibab uplift, also creating a possible path through the uplifted and stacked sediment. And since water at the surface flows downhill, the water draining off the Colorado Plateau would flow westerly toward the Pacific Ocean. Rapid uplift and surface drainage of receding floodwater provides both the path and the necessary volume of water to quickly carve out Grand Canyon. This was all accomplished before the Ice Age began. How do we know? Because there are about 150 lava flows that poured down into Grand Canyon during the Ice Age (Uinkaret volcanism), showing the canyon was already in existence before these frozen lava falls poured in. These likewise demonstrate that there was not sufficient time available for any presumed lakes to have formed.

One of ICR's volunteers, Scott Arledge, asked me why the receding floodwaters didn't form more Grand Canyons. Why is there just one Grand Canyon? The answer has to do with the right combination of several events, making Grand Canyon unique. First, Grand Canyon needed a major uplift to erode down through. This was provided by the late Flood isostatic rise of the Colorado Plateau, causing the entire Four Corners

Colorado River in Grand Canyon

region of the U.S. to move upward about 5,000 feet. Second, fractures and fissures were needed to channelize the runoff water along the western edge of the Colorado Plateau, directing the path of the water and rapidly carving a 5,000-foot-deep canyon. Third, there had to be a sufficient volume of water left in the receding phase to still carve a major canyon. Fourth, all of these events had to coincide at just the right moment. Without the massive uplift, timed perfectly with the channelization phase of the receding floodwaters, there would have been no Grand Canyon.

There were some other large canyons that formed in the U.S. during the receding phase, like Palo Duro Canyon described above, but they are not very deep by comparison to Grand Canyon. Palo Duro Canyon formed from channelized runoff water that flowed eastward away from the uplifted Rocky Mountains late in the Flood. However, the Texas Panhandle region did not experience as much uplift as the Colorado Plateau. Therefore, Palo Duro Canyon, although the second-largest canyon in the U.S., is only about 700 feet deep at its maximum.

All of these major canyons we see today are reminders of the immense power of the floodwaters. Even their recession has left the fingerprints of "grand" canyons as a witness to God's first judgment by water as recorded in the book of Genesis.

Human Perspective: Where Are the Human Fossils?

The Flood model we are developing may also help explain the lack of human fossils in the rock record. Most pre-Flood humans likely survived until close to Day 150 and were probably clinging to the highest ground even up to the last day. As the water levels crested on Day 150, humans were wiped off in high-velocity tsunami waves, spreading their bodies in all directions from a zone of concentration, transporting them great distances radially.

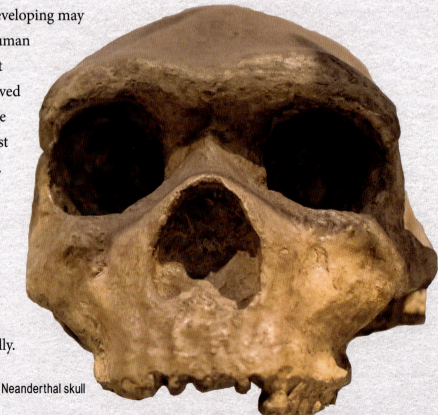

Neanderthal skull

This process would have spread their remains and lessened the likelihood of finding a concentration of human fossils. And if human remains were not buried deep enough in sediment, they would not be preserved as fossils. Subsequent post-Flood erosion, including the effects of the subsequent Ice Age, would affect the exposed Tejas strata the most and any humans buried in the uppermost few meters as a consequence.

Conclusion: The Tejas Is the Receding Phase

The geological evidence strongly supports that the Tejas Megasequence formed during the receding phase of the Flood. I have suggested between Day 150-314± of the Flood year. Floods take a long time to dissipate. They do not recede quickly. The Tejas rocks were deposited by water, just like the megasequences before.

Massive-scale Tejas deposits, like the Whopper Sand and the Ogallala Formation, can be best explained by the receding water of the Flood. Second, the age of the sea-floor shows that it was still being created at the runaway subduction velocities of the preceding Absaroka and Zuni Megasquences. There is no indication the plate motion was slowing down until possibly the Pliocene.[37]

Suggestions that the Flood was completely over at the K-Pg boundary fail to explain much of these geological data. Nor does this allow enough time for the flood-waters to drain off the continents. Picking the wrong Flood/post-Flood boundary can also lead to other misinterpretations, such as the breach-dam explanation for Grand Canyon.

And probably the best evidence, and as close to proof as we can get in geology, that the Tejas was the receding phase comes from studies of the rock columns across Africa. These columns show that the deposition of marine rocks like carbonates and salt was continuous across North Africa and the Middle East from the Cretaceous (Zuni) upward through the entire Cenozoic section, including the surface rocks of the Miocene (upper Tejas).[37] The Flood

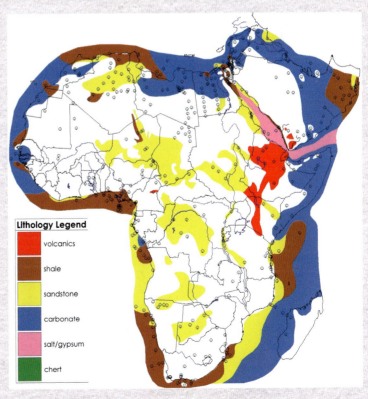

Map of the bottommost (basal) Tejas rock type across Africa

could not have been finished and still deposit these rocks. These are clearly water deposits. The area of continuous carbonate deposition includes the countries of Syria and Iraq, just to the south of the Ark landing site. How could the Tower of Babel be built if the area was still underwater?

These data clearly tell us the Flood was not over until near, or at, the end of the Tejas Megasequence. Genesis 8:13 tells us that the "waters were dried up from the earth." This was approximately Day 314 of the Flood. This was most likely when the Tejas Megasequence ended.

In addition, the advocates for a K-Pg Flood/post-Flood boundary consider all Cenozoic fossils to have formed in the window of time between the ending of the Flood and the beginning of the Ice Age.[27,28,38,39] This only allows about 100 to 200 years for the dispersal (whatever the mechanism) and incredible diversification and subsequent burial of all Cenozoic mammals, flowering plants, and other fossils on multiple continents and in nearly the exact same stratigraphic order simultaneously.[40] Therefore, the presumed local catastrophes used to explain these Cenozoic fossils seem to more closely resemble global catastrophes. Global catastrophes are better explained with a global flood event.

Picking the post-Flood boundary at the K-Pg even leads some to explain much of the Cenozoic mammal fossil record as one of hyperevolution, with four-legged animals walking off the Ark and evolving to whales in a few hundred years.[41,42] These rates of evolution are even faster than proposed by secular evolutionists!

Dr. Marcus Ross, in particular, has championed the K-Pg boundary as the end of the Flood to explain the mammalian fossil record in North America. He has determined that 22% of pre-Flood mammal baramins (kinds) in North America would have had to return again to North America after the Flood if the Flood/post-Flood boundary is chosen as the Pliocene/Pleistocene boundary.[43] He further argued that it is highly unlikely that such a high percentage of pre-Flood baramins would return to their pre-Flood locales, "display[ing] a proclivity to migrate to the graveyards of their deceased, pre-Flood baraminic kin."[43] Similarly, Ross also contended that the coincidence of kangaroo fossils found only in

Upper Cenozoic rocks in Australia, and again found only living in Australia today, as additional support for a K/Pg Flood/post-Flood boundary.[43]

However, considering there are presently five habitable continental landmasses, it is no surprise that 22% of the land animals returned to their pre-Flood locales after the Flood. They would have had about a one-in-five chance of returning to the same locations. A post-Flood "sweepstakes" model of migration easily explains these probabilities. Fossil evidence is all about probabilities and nothing more.

The same is true of the kangaroos found only in Australia today. Maybe they did migrate elsewhere but have since died off in those other areas prior to the present. We know they had to migrate across Asia from the Ark to get to Australia. It is even possible that kangaroo fossils may have been buried elsewhere and have since eroded away or have not yet been exposed. Fossils and/or the lack of fossils may not reveal the whole story.

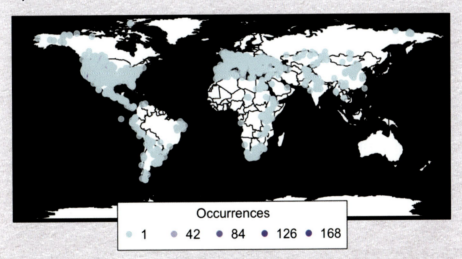
Equidae (horse) map of fossil locales in Tejas strata. Courtesy of Jeffrey Tomkins.

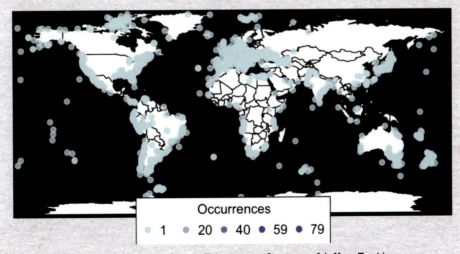
Cetacea (whale) map of fossil locales in Tejas strata. Courtesy of Jeffrey Tomkins.

An unanswered question in Ross' analysis is just how were the post-Flood mammals able to return to their respective continents?[43] How did the kangaroos get to Australia after the Flood without land bridges providing the bulk of the pathway? Those that advocate a K/Pg Flood/post-Flood boundary have not sufficiently considered the difficulties of post-Flood animal migration, particularly for the largest mammals and the hoofed animals.[28,38-40,43] Their suggestion that nearly all Cenozoic fossils were the result of post-Flood local catastrophes fails to explain how the post-Flood animals got to the various continents in the first place. They have not offered a testable, viable method for the migration of large animals to the newly separated continents after the Flood.

We will look at a solution to this migration issue in the next chapter on the Ice Age. But let's also look at the causes and the reasons for the Ice Age. God had it all planned.

References
1. Baumgardner, J. 2003. Catastrophic Plate Tectonics: The Physics Behind the Genesis Flood. In *Proceedings of the Fifth International Conference on Creationism*. R. L. Ivey Jr., ed. Pittsburgh, PA: Creation Science Fellowship, Inc., 113-126.
2. Baumgardner, J. 2005. Recent Rapid Uplift of Today's Mountains. *Acts & Facts*. 34 (3).
3. Chadwick, A. V. 1993. Megatrends in North American Paleocurrents. *Society of Economic Paleontologists and Mineralogists Abstracts*. 8: 58.
4. Humphreys, D. R. 2014. Magnetized moon rocks shed light on Precambrian mystery. *Journal of Creation*. 28 (3): 51-60. Emphasis in original.
5. Clarey, T. L. 2017. Floating forest hypothesis fails to explain later and larger coal beds. *Journal of Creation*. 31 (3): 12-14.
6. Blum, M. and M. Pecha. 2014. Mid-Cretaceous to Paleocene North American drainage reorganization from detrital zircons. *Geology*. 42 (7): 607-610.
7. Clarey, T. L. and A. C. Parkes. Use of sequence boundaries to map siliciclastic depositional patterns across North America. Search and Discovery Article # 90259. AAPG Annual Convention and Exhibition, Calgary, Alberta, Canada, June 22, 2016.
8. Higgs, R. Gulf of Mexico Paleogene "Whopper Sand" sedimentology: hypersaline drawdown versus low-salinity hyperpycnite models. Search and Discovery Article #40418. AAPG Annual Convention and Exhibition, Cape Town, South Africa, October 26-29, 2008.
9. Wicander, R. and J. S. Monroe. 2013. *Historical Geology*, 7th ed. Belmont, CA: Brooks/Cole.
10. Lowell, J. D. 1974. Plate tectonics and foreland basement deformation. *Geology*. 2 (6): 275-278.
11. Lowell, J. D. 1974. Plate tectonics and foreland basement deformation: Reply. *Geology*. 2 (12): 571.
12. Dickenson, W. R. and W. S. Snyder. 1978. Plate tectonics of the Laramide orogeny. In *Laramide Folding Associated with Basement Block Faulting in the Western United States*. V. Matthews III, ed. Boulder, CO: The Geological Society of America Memoir 151, 355-366.
13. Schmidt, C. J. 1980. Mechanical basis for deformation and plate tectonic models for the Wyoming Province. In *Manual for Field Study of Geology of the Northern Rocky Mountains*. Bloomington, IN: Department of Geology, Indiana University, 199-215.
14. Kelleher, J. and W. McCann. 1976. Buoyant zones, great earthquakes, and unstable boundaries of subduction. *Journal of Geophysical Research*. 81 (26): 4885-4896.
15. Brown, W. 2015. *In the Beginning: Compelling Evidence for Creation and the Flood*, 9th ed. Phoenix, AZ: Center for Scientific Creation.
16. Berman, A. E. and J. H. Rosenfeld. 2007. A New Depositional Model for the Deep-Water Gulf of Mexico Wilcox Equivalent Whopper Sand: Changing the Paradigm. In *The Paleogene of the Gulf of Mexico and Caribbean Basins: Processes, Events, and Petroleum Systems*. L. Kennan, J. Pindell, and N. C. Rosen, eds. Houston, TX: Proceedings of the 27th Annual Gulf Coast Section of the Society of Economic Paleontologists and Mineralogists Foundation Bob F. Perkins Research Conference, 284-297.

17. Trammel, S. 2006. Gulf of Mexico Deepwater Trends. *IHS Report*.
18. Lewis, J. et al. 2007. Exploration and Appraisal Challenges in the Gulf of Mexico Deep-Water Wilcox: Part 1—Exploration Overview, Reservoir Quality, and Seismic Imaging. In *The Paleogene of the Gulf of Mexico and Caribbean Basins: Processes, Events, and Petroleum Systems*. L. Kennan, J. Pindell, and N. C. Rosen, eds. Houston, TX: Proceedings of the 27th Annual Gulf Coast Section of the Society of Economic Paleontologists and Mineralogists Foundation Bob F. Perkins Research Conference, 398-414.
19. Sweet, M. L. and M. D. Blum. 2011. Paleocene-Eocene Wilcox Submarine Canyons and Thick Deepwater Sands of the Gulf of Mexico: Very Large Systems in a Greenhouse World, Not a Messinian-Like Crisis. *Gulf Coast Association of Geological Societies Transactions*. 61: 443-450.
20. Rosenfeld, J. and J. Pindell. 2003. Early Paleogene Isolation of the Gulf of Mexico from the World's Oceans? Implications for Hydrocarbon Exploration and Eustacy. In *The Circum-Gulf of Mexico and the Caribbean: Hydrocarbon Habitats, Basin Formation, and Plate Tectonics*. C. Batolini, R. T. Buffler, and J. J. Blickwede, eds. Tulsa, OK: American Association of Petroleum Geologists Memoir 79, 89-103.
21. Morris, J. D. 2012. *The Global Flood: Unlocking Earth's Geologic History*. Dallas, TX: Institute for Creation Research.
22. Guy, D. F., ed. 2001. *The Story of Palo Duro Canyon*. Lubbock, TX: Texas Tech University Press.
23. Spearing, D. 1991. *Roadside Geology of Texas*. Missoula, MT: Mountain Press Publishing Company.
24. Ibid, 385.
25. Ogallala Aquifer Initiative 2016 Progress Report. United States Department of Agriculture.
26. Clarey, T. 2015. The Whopper Sand. *Acts & Facts*. 44 (3): 14.
27. Austin, S. A. et al. 1994. Catastrophic Plate Tectonics: A Global Flood Model of Earth History. In *Proceedings of the Third International Conference on Creationism*. R. E. Walsh, ed. Pittsburgh, PA: Creation Science Fellowship, 609-621.
28. Whitmore, J. H. and K. P. Wise. 2008. Rapid and early post-Flood mammalian diversification evidences in the Green River Formation. In *Proceedings of the Sixth International Conference on Creationism*. A. A. Snelling, ed. Pittsburgh, PA: Creation Science Fellowship, 449-457.
29. Oard, M. J. 2013. Geology indicates the terrestrial Flood/post-Flood boundary is mostly in the Late Cenozoic. *Journal of Creation*. 27 (1): 119-127.
30. Clarey, T. L. 2016. The Ice Age as a mechanism for post-Flood dispersal. *Journal of Creation*. 30 (2): 48-53.
31. Clarey, T. L. 2017. Local Catastrophes or Receding Floodwater? Global Geologic Data that Refute a K-Pg (K-T) Flood/post-Flood Boundary. *Creation Research Society Quarterly*. 54 (2): 100-120.
32. Clarey, T. 2018. Palo Duro Canyon Rocks Showcase Genesis Flood. *Acts & Facts* 47 (7): 10.
33. Helble, T. and C. Hill. 2016. Carving of the Grand Canyon: A lot of time and a little water, a lot of water and a little time (or something else?). In *The Grand Canyon, Monument to an Ancient Earth: Can Noah's Flood Explain the Grand Canyon?* C. Hill and G. Davidson, eds. Grand Rapids, MI: Kregel Publications, 163-171.
34. Austin, S. A. 1994. How Was Grand Canyon Eroded? In *Grand Canyon: Monument to Catastrophe*. S. A. Austin, ed. Santee, CA: Institute for Creation Research, 83-110.
35. Oard, M. J. 2014. *A Grand Origin for Grand Canyon*. Chino, AZ: Creation Research Society.
36. Personal communication with Edmond W. Holroyd III, 2019.
37. Clarey, Local catastrophes or receding Floodwater? Global geologic data that refute a K-Pg (K-T) Flood/post-Flood boundary, *Creation Research Society Quarterly*.
38. Whitmore, J. H. and P. Garner. 2008. Using suites of criteria to recognize pre-Flood, Flood, and post-Flood strata in the rock record with application to Wyoming (USA). In *Proceedings of the Sixth International Conference on Creationism*. A. A. Snelling, ed. Pittsburgh, PA: Creation Science Fellowship and Dallas, TX: Institute for Creation Research, 425-448.
39. Snelling, A. A. 2009. *Earth's Catastrophic Past: Geology, Creation & the Flood*. Dallas, TX: Institute for Creation Research.
40. Snelling, A. A. and M. Matthews. 2013. When Was the Ice Age in Biblical History? *Answers* Magazine. 8 (2): 44-52.
41. Wise, K. P. 2009. Mammal kinds: how many were on the ark? In *Genesis Kinds: Creationism and the Origin of Species: Center for Origins Research Issues in Creation Number 5*. T. C. Wood and P. A. Garner, eds. Eugene, OR: Wipf & Stock, 129-161.
42. Wise, K. P. 2017. Step-down saltational intrabaraminic diversification. *Journal of Creation Theology and Science Series B: Life Sciences*. 7: 8-9.
43. Ross, M. R. 2012. Evaluating potential post-Flood boundaries with biostratigraphy—the Pliocene/Pleistocene boundary. *Journal of Creation*. 26 (2): 82-87.

15 The Post-Flood Ice Age

> **Summary:** After the Flood receded, the Ice Age began. Evidence is abundant for one Ice Age in Earth's history. Scars in the bedrock surface and rounded and out-of-place rocks show that glaciers once extended farther than they do at present. Scientists resisted the idea of an Ice Age for decades—just like plate tectonics—but eventually the data were so overwhelming they finally accepted it.

Creation scientists have developed a model for how the Flood could have caused an Ice Age. Using the acronym HEAT, the model begins with rapid plate movement during the Flood that created a hot new seafloor. This would have resulted in 1) **Hot oceans**, which would have resulted in 2) **Evaporation**. The runaway subduction of old ocean crust would have triggered much volcanic activity, causing volcanoes to spew ash called 3) **Aerosols** into the air. The aerosols would have blocked sunlight, cooling the Earth and causing surface temperatures to drop. Since volcanic activity continued for hundreds of years after the Flood, aerosols would have continuously spewed into the air, causing the earth to be cold for centuries. The final step in the model is 4) **Time**. Much snowfall and long winters would have resulted in a worldwide Ice Age in the high latitudes.

The Ice Age was critical in order to get the animals and humans from the Ark's

> "The Ice Age was critical in order to get the animals and humans from the Ark's landing site to the separated continents."

landing site to the separated continents. The massive buildup of ice lowered sea level and created land bridges that lasted for several hundred years after the Flood, enough time to populate the globe. God had a plan for everything.

The Tejas Megasequence Ends with a Bang

At the end of the Tejas, near the top of the Pliocene Epoch (at the N-Q), there is a final massive change in the fossil record. Some secular scientists call this the sixth Great Extinction.[1] Dr. Andrew Snelling has previously pointed out that a "biostratigraphic break expected to characterize the Flood/post-Flood boundary" was never identified at the Pliocene/Pleistocene level.[2] He used this to argue in favor of a K-Pg Flood/post-Flood boundary instead. However, the more recent discovery of a sixth global extinction event at the top of the Pliocene makes Snelling's argument less compelling. Pimiento et al found that 36% of Pliocene genera failed to survive into the Pleistocene (crossing the N-Q) and that extinction rates were three times higher in the Late Pliocene relative to the rest of the Cenozoic.[1] This discovery of a hitherto unrecognized global break in fossil content (known as a paleontological discontinuity) is the likely location for the Flood/post-Flood boundary. It is also the end of the Tejas Megasequence. And it is on top of this N-Q surface that we find the Ice Age sediments, at least in the northern latitudes.

Erratics at Grass Hills at Castle Hills rock formation in Southern Alps, New Zealand

After the floodwaters drained off the land, ending the Tejas, the earth's climate had already begun to cool. Within about 100 years, snow began to build up sufficiently in the high latitudes to produce ice. This initiated the Ice Age.[3] In secular science, it has been given its own interval of time called the Pleistocene Epoch. Pleistocene sediments reside on top of Flood sediments in many locations. In other places, they rest on basement rocks of the crust, such as granites. Nonetheless, Pleistocene sediments are different from the underlying sediments of the megasequences because they do not contain great numbers of marine fossils mixed with land fossils. They are also less likely to be lithified as the sedimentary rocks of the six megasequences are.

Let's review little more about the discovery of the Ice Age. Then, we'll discuss why there was an Ice Age and even its purpose. With God, there's always a purpose.

The Ice Age Begins

Ideas that glaciers in the Swiss Alps were more vast in the past date back as far as the late 18th century. Geologist/mathematician John Playfair even suggested this possibility in 1802.[4] For many years, local peasants in Switzerland had noticed rounded rocks, out-of-place rocks, and even striated rocks far from any active glaciers. They

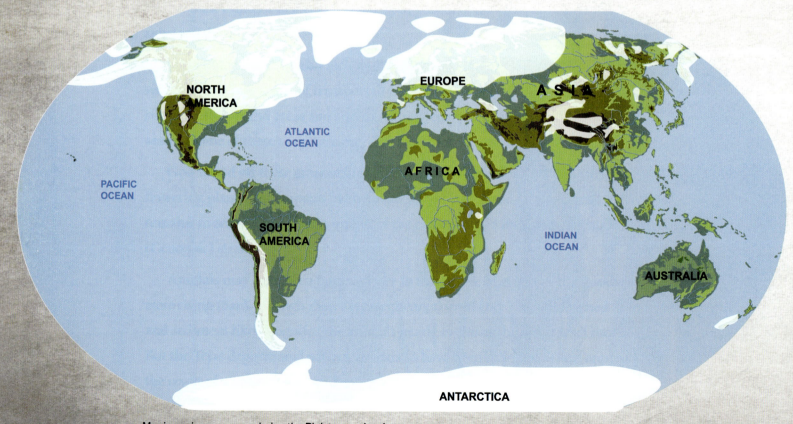

Maximum ice coverage during the Pleistocene Ice Age

White Passu Glacier with glacial moraine, Pakistan

had generally assumed that this meant the glaciers had extended farther in the past than they did at present. Then, in 1821, Swiss engineer Ignaz Venetz-Sitten read a paper to the Helvetic Society proposing that the ancient glaciers of the Alps were much more extensive than at present, becoming the first to formally propose what would later become known as the Ice Age.[4]

The first to suggest a "great ice period" on a grand scale was zoologist John Louis Agassiz when he addressed the Helvetic Society in 1837.[1] The idea soon took hold in Great Britain under the tutorship of Agassiz and in North America in 1846 when Agassiz accepted a professorship at Harvard. His concept of an Ice Age was rapidly accepted by most American and European geologists by the end of the 19th century.[4] The Ice Age created vast continental glaciers, not mere Alpine glaciers, that covered much of the northernmost landmasses of the world. Pieces of these glaciers still exist today in Greenland and in Antarctica, holding about 2.15% of the world's total water and most of its fresh water.

What convinced all these geologists of an Ice Age? The data they observed and collected. This story of the acceptance of the Ice Age is similar to the story of the acceptance of plate tectonics. It took decades for the majority of scientists to accept

Georges Island, Nova Scotia, Canada, is an example of a drumlin

both hypotheses, but the data they collected won the day. And the rock data are what convinced me of a global flood, and hopefully you too.

Some of data that convinced scientists of an Ice Age include rocks that are seemingly out-of-place known as *glacial erratics*. Other data include thick piles of unsorted material in long linear hills known as *moraines*. In addition, thick layers of drift are more evenly spread across much of Michigan's lower peninsula, my home state. Drift is composed of clay, sand, and rocks that were transported in the ice sheets as they flowed across the state. Some is unsorted by size and other layers are sorted by the meltwaters. As the ice melted, it left this glacial debris behind. The term "drift" originally was coined because early scientists thought these erratics were carried by ice floating on top of water, dropping rocks as the sea ice drifted overhead. Now, we merely use this term to describe all material transported by a glacier.

Other landforms left behind by massive continental glaciers include kames, eskers, drumlins, and kettle lakes. In many locations around the Great Lakes, the land re-

Glacial kame in Yellowstone National Park

bounded hundreds of feet after the Ice Age as the weight of the ice sheets were relieved, causing the land surface to rise. Lake terraces and beach gravels are now found hundreds of feet above the present lake levels. All of these data, and more, tell us that there were thick continental glaciers covering vast portions of North America, Europe, and Asia during the Ice Age.

What Caused the Ice Age?

Mike Oard has written extensively on the causes of the Ice Age, and I think he is entirely correct. Oard describes the first requirement being the need for cooler summers.[5] Cooler summers would allow the snow to build up from year to year, and eventually the compacted snow would transform into glacial ice. The big question is: How do you have continuous cool/cold summers for many years in a row and even for hundreds of years? The answer is volcanic activity, but only the right kind of volcanic activity.

Kettle lakes in North Dakota

Sleeping Bear Sand Dunes, Michigan

Subduction Zone Volcanoes

Volcanoes, especially those with silica-rich magmas, are essential to create the necessary conditions for an Ice Age. Not every type of volcano will do the job. They must be stratovolcanoes (composite volcanoes) that are highly explosive and generate lots of pyroclastics and release a lot of aerosols. These eruptions often emit large amounts of sulfur dioxide gas. Chemical reactions in the atmosphere form sulfuric acid droplets, which can remain in the stratosphere for two to three years. If the eruption is large enough, it can cool the earth by blocking out sunlight.

Stratovolcanoes form almost exclusively above subduction zones as the ocean lithosphere is pulled down into the earth. The internal heat in the mantle causes a distillation, or a partial melting of the crust and lithosphere, such that basaltic material is able to create a granitic (silica-enriched) magma. The first minerals to melt are those with the lowest melting points. These minerals—quartz, orthoclase feldspar, and

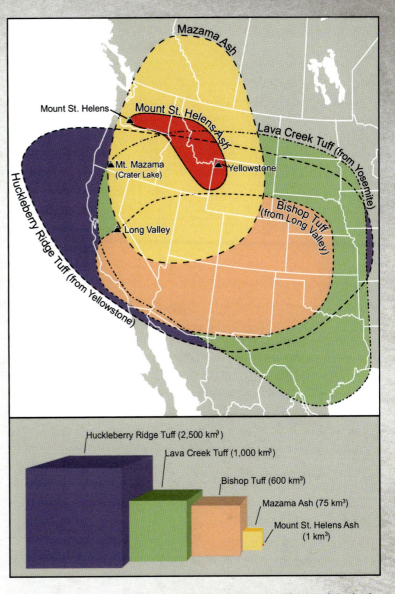

Diagrams showing the extent and volume of past volcanic eruptions in the American West. Those occurring during and soon after the great Flood were immense compared to Mount St. Helens.

biotite—are the main components in granite. Much of the remaining part of the subducted slab never seems to fully melt but descends down into the mantle intact. The subduction process also introduces a lot of water into the melt, increasing the volatility of the magma. The new granitic melt rises from buoyancy and erupts catastrophically, creating highly explosive volcanoes that send ash and aerosols into the atmosphere. The bottom line is that without subduction there is no Ice Age. God knew what He was doing.

Svínafellsjökull Glacier landscape in Skaftafell National Park, Iceland

Each eruption can cool the earth by blocking out sunlight for several years, like what the earth experienced after the 1991 eruption of Mt. Pinatubo. An Ice Age needs sustained eruptions over many decades and possibly hundreds of years. Ash and gases become trapped in the stratosphere and can have effects on the earth's climate for up to three years, cooling the northern latitudes by as much as 1°C for a single eruption.[6] For example, the 1815 eruption of Tambora, Indonesia, caused the "year without a summer" across Europe in 1816. Mike Oard refers to this temporary cooling of the earth as the "anti-greenhouse" effect.[6]

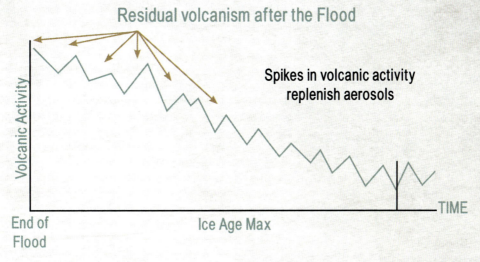

Graph showing the drop in volcanic activity with time after the Flood

There is evidence that volcanoes were erupting on a greater scale at the end of the Flood, just prior to the Ice Age. Even the secular geologists admit this.[7] Many of the largest eruptions of the Yellowstone Supervolcano occurred in the Pleistocene, with the last three dated by secularists at 2 and 1.3 million years ago, and again at 0.6 million years ago.[7] Although I respectfully disagree with the absolute dates of these eruptive events, they can be dated by relative time methods as Pleistocene age. Other volcanoes like Lassen Peak in California, most of the Cascade Mountains of Washington and Oregon, and the vast majority of the volcanoes around the Ring of Fire in the Pacific have been documented as erupting in the Pleistocene and even earlier in the Tejas.[7]

The megasequence data for the three continents in this study also show a steady increase in volcanic activity beginning in the Absaroka and continuing through the Tejas Megasequence. The Tejas in North America has 18% volcanic rocks by volume. This is

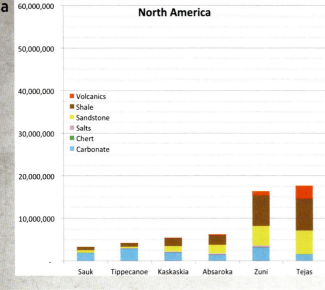
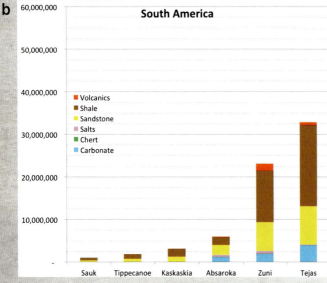
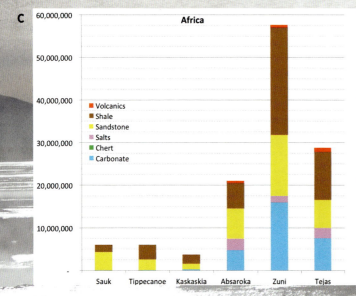

the highest percentage of volcanic rocks of any of the megasequences.

South America has its highest percentage of volcanic rocks in the Zuni Megasequence, but it still exhibits a substantial volume of volcanic rocks in the Tejas. The peak in the earlier Zuni is mostly due to large igneous provinces of basalt in the offshore regions east of South America, whereas the Tejas volcanism is from the subduction zone volcanoes associated with the Andes Mountains.

Africa has the largest amount of volcanic rocks within the Tejas Megasequence compared to all earlier megasequences. Admittedly, the Tejas is tied with the Absaroka for the highest percentage by volume, but there is much more volume of total rocks in the Tejas.

The difference between creation geologists and secular geologists is in the timing of these volcanoes, not that they weren't actively erupting in the Tejas and Pleistocene. Everyone agrees on the volcanic activity. Creation geologists just think this volcanic activity was packed into a lot shorter time frame than their secular counterparts. We feel the data support a young

Histograms of the type and volume of rocks in place across (a) North America, (b) South America, and (c) Africa for each megasequence. Note that volcanic rocks in red generally increase late in the Flood. Measurement is in cubic kilometers.

earth and a recent Flood (see chapter 19). This squeezes all those eruptions into a tighter period of time and sustains them over many centuries. Many eruptions that are spaced close together over hundreds of years provide the perfect recipe for the Ice Age. God knew what He was doing.

Hotter Oceans

An Ice Age requires a lot of snow. Sustained cooler temperature in the northern latitudes will tend to create snow instead of rain. To make snow or precipitation of any sort requires evaporation, which is primarily sourced from the oceans. How do we get enough ocean evaporation to sustain heavy snowfall for hundreds of years to cause the Ice Age? The Flood provides the answer.

Mike Oard has found that water at 30°C evaporates three times faster than at 10°C and seven times faster than at 0°C.[6] To get enough snowfall for an Ice Age requires a lot of evaporation. Evaporation requires hotter oceans. And the Flood provides the perfect conditions for hotter oceans. Recall, during the Flood, an entirely new seafloor was created as runaway subduction broke apart Pangaea and separated the continents. This new seafloor was composed of basaltic lava. Cooling of the seafloor would provide sufficient heat to warm the oceans to a much higher temperature, even as hot as 30°C, although much more research needs to be done on this topic. In addition, hot waters released as part of the fountains of the great deep likely also provided some heat to the oceans. But the creation of an entirely new seafloor during the Flood from magma/lava would most likely be the main source of hotter post-Flood oceans. This would create the evaporation needed for tremendous amounts of snowfall.

Mike Oard modeled the cooling of the oceans with time.[6] His calculations have the oceans returning to today's temperatures about 700 years or so after the Flood. Somewhere along the timeline, the snowfall would diminish and the thickening of the ice

Aguas Termales de Polques, Bolivia

sheets would cease. He estimated that it would take about 500 years to reach the peak ice thickness in the Ice Age.[6]

Snelling and Matthews have estimated a short, 250-year duration for the Ice Age.[3] To me, this seems less likely, based on the calculations by Oard[6] and the evidence for repeated ice advances and retreats observed across the northern continents.[7] The thicknesses of continental glacial depositional landforms (moraines, eskers, kames) also seem to suggest a much longer Ice Age that may have reached a glacial maximum 500 years post-Flood, as Oard has modeled.

An Ice Age that reached its peak in 500 years is also possibly supported by the book of Job. Dr. Henry Morris has pointed out that there are more references to snow and ice in the book of Job than in any other book of the Bible.[8] He realized that the continental glaciers did not extend to the land of the Patriarchs, but they had a strong effect on the climate of the Middle East, causing cooler temperatures and more rain and snow than witnessed there today, as reported in Job. Without the Ice-Age-induced climate effects in the Middle East, the area would probably have been hotter and drier than what the book of Job seems to indicate.[8] I doubt if the Egyptians built an empire

Nile River

in the desert like we find there today. At the time of the Patriarchs, this area would have been affected by the Ice Age too, creating a wetter and more humid climate. It wasn't until centuries after the Ice Age that the Egyptians began to rely increasingly on the Nile as a source of water as the desertification process took hold and the earth's climate found a new equilibrium.

Where does this lead us in terms of timing? Dr. Henry Morris points out that the book of Job is likely the oldest book of the Bible, except for possibly the first 11 chapters of Genesis.[8] Based on the lack of references in Job to Jewish Law, the judges, the prophets, the nation of Israel, or Abraham, Dr. Morris placed the age of the book before Moses and possibly even before Abraham. He suggested we place the age of the book in the time of the Patriarchs, about 2000 BC.[8]

The interpretation is supported by the climate calculation estimates of Oard[6] and by the apparent cooler and wetter conditions that were still affecting the Middle Eastern climate during the time of the Patriarchs. Exactly how long the ice sheets endured after their formation is unknown. They may have lasted an additional 500 years for a total duration of 1,000 years.

Recently, William Worracker has cautioned that stable isotope ratios of oxygen in seashells plotted against geologic time indicate a maximum ocean temperature since the Mesozoic Era of only 12° to 13°C, not the end-of-the-Flood 25° to 30°C ocean temperatures estimated by Oard.[9] However, the relationship of oxygen-18/16 ratios to temperature is only valid in conditions of equilibrium.[10] The entire Flood event would have been a time of disequilibrium, which invalidates the stable isotope temperature relationship developed by Urey.[10] It is likely that the ocean water didn't again reach equilibrium for several hundred years after the Flood. And it is most likely that the oxygen isotope ratio of the seawater would have been constantly changing during the Flood event with the continual addition of new water released by hydrothermal activity and by release from volcanic activity. Therefore, the oxygen-18/16 ratios of seashells buried in Flood sediments are not perceived to limit the ocean temperatures to 12° to 13°C.

Milankovitch Theory Refuted

Most secular geologists claim the Ice Age was caused by subtle shifts in Earth's orbit, tilt, and precession over time.[7] In the 1920s, Serbian astronomer Milutin Milankovitch suggested that slight variations in the earth's rotational tilt (up to 1.5 degrees), orbital distance from the sun, and precession of the equinoxes (wobble in the tilt) could dramatically change the amount of solar radiation reaching the high latitudes of the earth.[7] This is known as the Milankovitch theory. It only became readily accepted in the 1970s because secular scientists realized they have no other way to explain the Ice Age. When you have nothing else, the Milankovitch theory wins by default.

According to this theory, the orbital eccentricity has a roughly 100,000-year cycle, the changes in the axial tilt a 42,000-year cycle, and the precession of the axis (wobble) about a 23,000-year cycle.[7] The highly touted 1976 paper referred to as "the Pacemaker paper" used two deep-sea cores to calibrate the deep-sea sediments and the oxygen isotope ratios of the seashells to ice cores and to the ice cycles of the Pleistocene Ice Age.[11] Oxygen-stable isotopes are claimed by secular scientists to be affected by expanding and contracting ice sheets globally. These signals are preserved in the shells of the sediments and in the ice of the deep ice cores in Greenland and Antarctica.

They dated the sediments in the cores using a magnetic field reversal signature that they believed at the time occurred 700,000 years ago. They found the expected 100,000-year, 42,000-year, and 23,000-year cycles in their results, thereby "proving" the Milankovitch theory.[11]

However, in the year 2000, the age of the magnetic reversal that was used to date the sediments in the cores was changed from 700,000 to 780,000 years.[12-14] My colleague at ICR, Dr. Jake Hebert, has redone the math on the data from the two cores using the new secular age for the magnetic reversal. And wouldn't you know it, the fit to the Milankovitch cycles was considerably off from expectations. Dr. Hebert determined that the match was, in fact, generally poor.[12-14] In reality, the Pacemaker paper that so many secular scientists claim supports the Milankovitch theory really doesn't. This leaves secular science with no adequate explanation for the Ice Age or for their claimed earlier Ice Ages.

Another problem with the Milankovitch theory is the claimed frequency of the Ice Ages across secular time. If you accept the evolutionary timescale of 4.55 billion years of Earth history and the Milankovitch theory that operates on a much smaller timescale of hundreds of thousands of years to a few million, then why don't we find evidence of more frequent Ice Ages throughout geologic time? There are only five

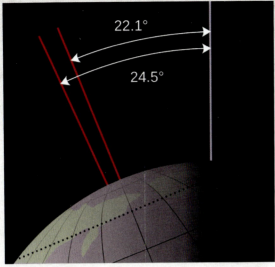
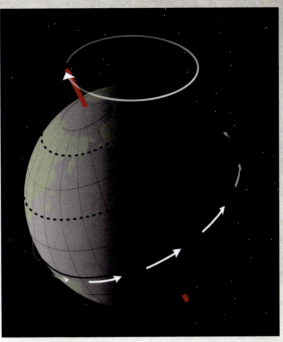

Diagrams illustrating axial tilt variation (left) and precessional wobble (right)

claimed Ice Ages by secular scientists, two in the Precambrian (about 2.2 billion years ago and 750 million years ago according to their secular time frame), one in the Ordovician (about 450 million years ago), one in the Permian (about 290 million years ago), and the recent—and likely only—one in the Pleistocene (last 2.6 million years). If Milankovitch cycles coincide on the order of a few million years, then we should expect to see evidence of Ice Ages scattered throughout the rock record, with no exceptions. But we do not. We see many hundreds of millions of years of secular time between the claimed Ice Ages. The math does not add up.

I say "claimed" because none of the prior (pre-Pleistocene) Ice Ages have any strong evidence supporting their existence. The best evidence for these so-called Ice Ages are rock striations (scratches on rock surfaces) and unsorted sedimentary strata with all sizes of clasts mixed together in one layer. However, striations can form from rapidly moving debris flows, like we observed after the 1980 eruption of Mount St. Helens, and sedimentary layers composed of all sizes of clasts can result from these aforementioned debris flows. Additionally, secular scientists have not reported finding any massive loess deposits associated with these claimed earlier Ice Ages. Loess is found all over the world today and is one of the most common sediments at the surface. In fact, about 10% of the world's land surface is still covered by loess deposits.[15] Loess is windblown, silt-size material that seems to be a leftover product of continental glaciation, likely generated by the grinding of ice against rock. In many places of the world it is over 200 feet thick.⁶ But all of the loess we observe today is from the most recent Ice Age of the Pleistocene. Where are the ancient loess deposits if there were all of these earlier glaciations in the geologic past? There should be extensive layers for each of the claimed Ice Ages. But there are few loess deposits, if any, found in the geologic record. Without loess deposits, it is harder to make the case for ancient glaciations.

None of the evidence for any of the earlier Ice Ages is compelling. The Milankovitch theory fails when tested by the math, and any ancient loess is missing from the sedimentary record. Alternatively, all the evidence for earlier Ice Ages can be interpreted as a result of rapidly deposited debris flows that occurred during the Flood. These claimed ancient Ice Ages are based only on the uniformitarian worldview that Ice Ages must have always occurred throughout all of time. Secular scientists deliberately forget that the Flood caused the one and only Ice Age.

Only One Ice Age

This is why there was only one Ice Age at the end of the Flood; there was only one global Flood. There was only one time period when subduction zone volcanoes were peaking all over the world, simultaneously, as a consequence of the Flood. These unique conditions brought on by the Flood were the perfect recipe for only one Ice Age. The hotter oceans and the peak in volcanic eruptions all had to be timed to last a few hundred years at the right time at the end of the Flood. God knew what He was doing.

System	Ice Age
Quaternary	Quaternary
Tertiary	
Cretaceous	
Jurassic	
Triassic	
Permian	
Pennsylvanian	Late Paleozoic
Mississippian	
Devonian	
Silurian	Andean-Saharan
Ordovician	
Cambrian	
Ediacaran	
Cryogenian	Cryogenian
Tobian	
Stenian	
Ectasian	
Calymmian	
Stratherian	
Orosirian	
Rhyacian	Huronian
Siderian	

Secular timescale illustrating the claimed five great Ice Ages across geologic time

What Was the Purpose of the Ice Age?

> So God blessed Noah and his sons, and said to them: "Be fruitful and multiply, and fill the earth." (Genesis 9:1)

> Now the whole earth had one language and one speech. And it came to pass, as they journeyed from the east, that they found a plain in the land of Shinar, and they dwelt there. Then they said to one another, "Come, let us make bricks and bake them thoroughly." They had brick for stone, and they had asphalt for mortar. And they said, "Come, let us build ourselves a city, and a tower whose top is in the heavens; let us make a name for ourselves, lest we be scattered abroad over the face of the whole earth." (Genesis 11:1-4)

Creation scientists have been debating the Flood/post-Flood boundary in the rock record for many years. One issue that seems to have been overlooked in this ongoing scientific debate is post-Flood animal migration. The Bible clearly tells us that humans stayed near the Tower of Babel, disobeying God's command to fill the earth for several generations after the Flood (Genesis 11:1-9). Meanwhile, the animals on the Ark had already fulfilled God's command to "abound on the earth, and be fruitful and multiply" (Genesis 8:17). But how did the animals, and the large mammals in particular, get to the individual continents after the floodwaters receded?

Creation researcher Dominic Statham has made a great case for the rafting of many of the groups of smaller animals as a possible means of dispersal after the Flood.[16] He noted that many animals could have rafted on the numerous log mats that were likely left over after the Flood. He determined that this better explained the animal distributions we see today compared to the secular story involving slow plate motions. But even Statham admitted that mammals, in particular large mammals, would be the least raftable of terrestrial animals.[16] So, how did the large mammals, like the wooly mammoths, get to North America after the Flood?

In accordance with catastrophic plate tectonic theory, the post-Flood configuration of the continents was likely vastly different from the pre-Flood arrangement. During the Flood, the breakup of this supercontinent separated the individual landmasses to the locations we see today. As Dr. Marcus Ross has pointed out, even if the pre-Flood continental configuration was identical to today's, meaning there was no plate move-

Mt. Bromo, Indonesia, eruption of 2016

ment at all, we are still left with today's modern post-Flood configuration.[17] However, neither of these scenarios resolves how large animals were able to get from the Ark landing site in Asia/Middle East to North America, South America, and Australia, now separated by vast distances of ocean water.

There seems to be a clear explanation from the Ice Age that may answer the mystery of post-Flood large animal migration.

Land Bridges

The answer seems to be land bridges.[18] Dry land migration routes could have facilitated the movement of large animals from the Ark to remote continents. The Ice Age after the Flood provided just such an opportunity. Water stored in massive ice sheets would have temporarily lowered sea levels by 200 to 280 feet below today's level.[6,19] The resulting land bridges would have made pathways for animals to simply walk to the major continents.

The timing of the Ice Age was no accident. Mike Oard calculated that the glacial maximum and the simultaneous maximum drop in sea level could have been achieved within 500 years after the Flood from high ocean temperatures and a late-Flood and post-Flood period of intense volcanic activity. This timing coincides nicely with the division of the earth that occurred during the days of Peleg.[6]

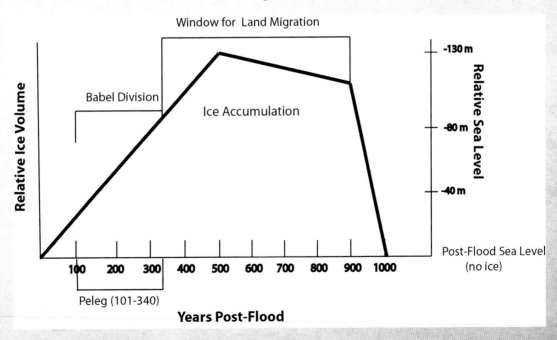

Timeline showing ice volume and sea level relationships vs. years post-Flood, including an estimated window for land animal/human migration. The ice accumulation curve is the same as the sea level curve due to their interdependency. As ice volume increases, sea level drops in an inverse relationship. Courtesy of Mary Smith.

The Bering land bridge connected Asia and North America temporarily during the Ice Age

Oard also calculated that the ice causing the Ice Age could rapidly melt away. He estimated that it probably took less than 200 years to completely melt back the continental ice sheets. Exactly how long the ice sheets endured after their formation is unknown. Once the ice melted, the sea level would have immediately risen, flooding the land bridges and closing this opportunity for intercontinental migration.

Man's Disobedience

After the Flood, mankind remained in the Middle East at "a plain in the land of Shinar" building the Tower of Babel (Genesis 11:2). This was in direct disobedience to God's post-Flood command to "be fruitful and multiply" (Genesis 9:1). God had to divide and scatter the people by confounding their languages so that they would not miss the temporary land bridge opportunity for migration. God's timing was perfect, since it may have taken several generations for humans to migrate to the Bering Sea land bridge from central Asia.

Ice Age Footprints Found Along Coasts

The discovery of Ice Age human footprints along many coastal areas demonstrates that the ocean waters were indeed warm during the migration from continent to continent during the Ice Age. Surprisingly, the footprints commonly exhibit bare feet. No shoes or sandals were evident in the 29 footprints that were documented in the recent discovery along the coast of British Columbia.[20] And what's even more startling, many Ice Age human footprint sites have been found along the world's coasts in places like Wales, England, South Korea, Australia, South Africa, Argentina, Nicaragua, and Mexico.[21]

Migration pathways from the Ark's landing site across the Ice Age land bridges

Oard's hypothesis seems to hold up that the hotter ocean water after the Flood kept the coastal temperatures warm, even during the Ice Age.[6] The thickest ice built up inland, but the coastal temperatures were moderate until the ocean water cooled sufficiently to flip the coastal climate from warm to cold. This heated ocean water kept the coasts warm so that humans and large mammals could walk across the land bridges and coastal areas free of cold and ice, even though the Ice Age was in full force many miles inland.

Conclusion: The Ice Age Was an Essential Ending to the Flood

God had a plan all along. The Flood would end, and an Ice Age would begin. The Ice Age caused a lowering of sea level at just the right time to create land bridges that were necessary to provide migration pathways for large animals and humans. Maximum ice volume was achieved at the same time that migration pathways were needed to travel from the Ark's landing site to distant continents that are now separated by water. The timing was perfect. After the oceans had cooled sufficiently and the volcanic activity began to wane, the ice sheets quickly melted, and the land bridges disappeared beneath the rising ocean waters, effectively ending the migration. God likely used this narrow window to scatter humans and animals across the globe, repopulating the earth after the Flood's destruction.

Catastrophic plate tectonics provided the conditions for an Ice Age. The rapid formation of hot new seafloor as plates rapidly subducted down into the earth's mantle would have greatly warmed the world's oceans. Likewise, the explosive subduction volcanoes that accompanied this rapid subduction would have provided the necessary aerosols needed for summer cooling. Without catastrophic plate tectonics, the formation of an entirely new seafloor from extruded lava, and the unique, explosive-style chemistry of subduction zone volcanoes, there could have been no Ice Age after the Flood. Many centuries worth of silica-rich volcanism were needed to produce the aerosols to cool the earth, and heating of the ocean water was necessary to cause higher evaporation rates and the snowfall necessary to make the massive continental ice sheets. Other suggested models to explain the Flood, like meteorite impacts, hy-

droplate theory, or flip-flops of the crust from one form to the other, fail because they have no viable mechanism to cause the Ice Age.

References

1. Pimiento, C. et al. 2017. The Pliocene marine megafauna extinction and its impact on functional diversity. *Nature Ecology & Evolution.* 1: 1100-1106.
2. Snelling, A. A. 2014. Paleontological Issues: Charting a scheme for correlating the rock layers with the Biblical record. In *Grappling with the Chronology of the Genesis Flood.* S. W. Boyd and A. A. Snelling, eds. Green Forest, AR: Master Books, 178.
3. Snelling, A. A. and M. Matthews. 2013. When Was the Ice Age in Biblical History? *Answers.* 8 (2): 46-52.
4. Flint, R. F. 1957. *Glacial and Pleistocene Geology.* New York: John Wiley & Sons.
5. Oard, M. J. 1990. *An Ice Age Caused by the Genesis Flood.* El Cajon, CA: Institute for Creation Research.
6. Oard, M. J. 2004. *Frozen in Time: The Woolly Mammoth, The Ice Age, and the Bible.* Green Forest, AR: Master Books.
7. Wicander, R. and J. S. Monroe. 2013. *Historical Geology,* 7th ed. Belmont, CA: Brooks/Cole.
8. Morris, H. M. 2000. *The Remarkable Record of Job: The Ancient Wisdom, Scientific Accuracy, & Life-Changing Message of an Amazing Book.* Green Forest, AR: Master Books.
9. Worracker, W. J. 2018. Heat problems associated with Genesis Flood models–Part 1: Introduction and thermal boundary conditions. *Answers Research Journal.* 11: 171-191.
10. Urey, H. C. et al. 1951. Measurement of paleotemperatures and temperatures of the Upper Cretaceous of England, Denmark, and the Southeastern United States. *Bulletin of the Geological Society of America.* 62 (4): 399-416.
11. Hays, J. D., J. Imbrie, and N. J. Shackleton. 1976. Variations in the Earth's Orbit: Pacemaker of the Ice Ages. *Science.* 194 (4270): 1121-1132.
12. Hebert, J. 2016. Revisiting an Iconic Argument for Milankovitch Climate Forcing: Should the 'Pacemaker of the Ice Ages' Paper Be Retracted? Part I. *Answers Research Journal.* 9: 25–56.
13. Hebert, J. 2016. Revisiting an Iconic Argument for Milankovitch Climate Forcing: Should the 'Pacemaker of the Ice Ages' Paper Be Retracted? Part 2. *Answers Research Journal.* 9: 131-147.
14. Hebert, J. 2016. Revisiting an Iconic Argument for Milankovitch Climate Forcing: Should the 'Pacemaker of the Ice Ages' Paper Be Retracted? Part 3. *Answers Research Journal.* 9: 229-255.
15. Vasiljevic, D. A. et al. 2011. Loess Towards (Geo) Tourism–Proposed Application on Loess in Vojvodina Region (North Serbia). *Acta geographica Slovenica.* 51 (2): 390-406.
16. Statham, D. 2015. Phytogeography and zoogeography-rafting vs continental drift. *Journal of Creation.* 29 (1): 80-87.
17. Ross, M. R. 2012. Evaluating potential post-Flood boundaries with biostratigraphy—the Pliocene/Pleistocene boundary. *Journal of Creation.* 26 (2): 82-87.
18. Clarey, T. L. 2016. The Ice Age as a mechanism for post-Flood dispersal. *Journal of Creation.* 30 (2): 54-59.
19. Holt, R. D. 1996. Evidence for a late Cainozoic Flood/post-Flood boundary. *Journal of Creation.* 10 (1): 128-167.
20. McLaren, D. et al. 2018. Terminal Pleistocene epoch human footprints from the Pacific coast of Canada. *PLOS ONE.* 13 (3): e0193522.
21. Curry, A. Where Our Human Ancestors Made an Impression. *Hakai Magazine.* Posted on hakaimagazine.com July 18, 2018, accessed September 4, 2018.

16 Megasequences Validate the Global Geologic Column

> **Summary:** The global geologic column contains a discernible order of fossils. Marine creatures are near the bottom (Cambrian), followed by shore-dwelling creatures, followed by terrestrial, then finally Ice Age creatures and humans. Evolutionists interpret this order as representing the evolution of life from simple to complex. However, creationists interpret this as representing the order of burial during the global Flood and the post-Flood Ice Age.

As described in chapter 3, megasequences are defined by globally recognized erosional boundaries rather than fossils. However, they do confirm the overall order of the fossil record. Many of the supposed mass extinctions correlate closely with the highest water levels of each megasequence cycle, meaning that "mass extinctions" simply reflect catastrophic Flood activity. Other confirmation of Flood activity can be found in overthrusts. These features are difficult to explain using mainstream uniformitarian thinking, but Flood geology resolves the dilemma quite easily. The global nature of the geologic column and its catastrophic features provide strong evidence of a catastrophic global flood.

Fossil bone bed in Dinosaur National Monument, Utah

The geologic column has been criticized by many creationists over the past 60 years.[1] In fact, an entire book was published by the Creation Research Society in an attempt to tackle this issue.[2] People generally question the geologic column because of its obvious ties to evolutionary theory.[3-6] Unfortunately, some of these critics still use arguments that have been invalidated in recent years, such as so-called "out of place" fossils due to overthrusting, which I will discuss more below.

The use of megasequences to study Flood sedimentation has also been criticized by some creation scientists.[7] These creationists claim, "The heart of the issue of using Sloss-based megasequences is their dependence on the geological timescale."[8] However, creation paleontologist Dr. Marcus Ross has championed the robustness of the global geologic column, arguing that it is based on physical correlations and the coincidence of paleontological data.[9] He emphasized, "The ability to correlate rocks on the basis of fossils contained is not dependent on evolutionary reasoning. Rather it is based on sound recognition of similar *patterns of fossils* found in disparate locations."[10] He explained that the type of rocks and distinctive chemical signals in some of the rocks also allow

> "The ability to correlate rocks on the basis of fossils contained is not dependent on evolutionary reasoning. Rather it is based on sound recognition of similar *patterns of fossils* found in disparate locations."

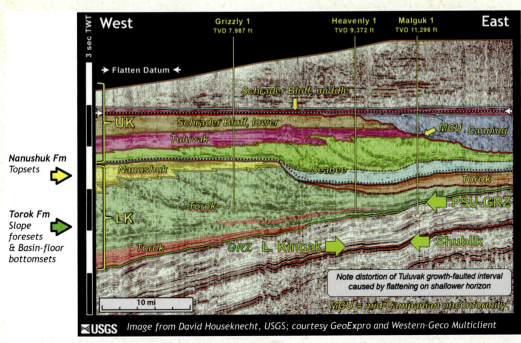

Example of reflection seismic data showing correlation of small-scale sequences

consistent correlations. It is not just the fossils that are compared from place to place.[9] For these reasons, many creation geologists do support the basic principles of the geologic column, recognizing that fossils do not reflect evolutionary patterns or time periods but are indicative of the order of burial during a one-year, global flood.[11,12] Nonetheless, the general pattern of the fossils within the geologic column remains a mainstay of secular geologic education and practice.

Using Megasequences to Study the Geologic Column

Megasequences were used in our study to examine the geologic column because they reflect major shifts in depositional patterns as sea level transgressed and subsequently regressed off the continents during the Flood.[13] Many of these depositional cycles left behind erosional surfaces at the top and base of the megasequences, changing the rock type abruptly (called *xenoconformities*).[14] These major shifts in

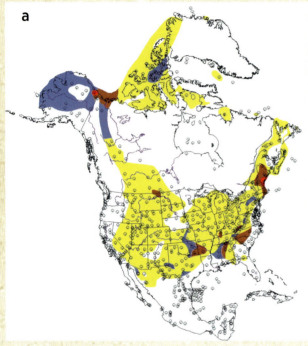
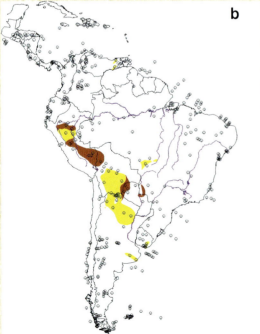
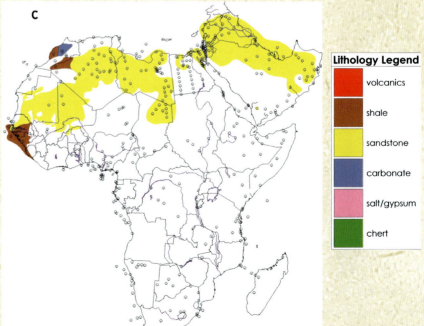

Figure 16.1. Basal rock type maps for the Sauk Megasequence, the Tapeats Sandstone equivalent, across (a) North America, (b) South America, and (c) Africa. Sandstone is shown in yellow.

depositional architecture are recognizable and traceable across continents and offshore alike using distinctive characteristics observed on seismic reflection data, such as abrupt truncations and strong reflecting horizons. Because this method concentrates on the changes in the physical attributes of the rocks, it is less dependent on the fossil record for correlations.[15]

Method 1: Construction of Basal Lithology Maps

Of particular interest were the basal rock types in each megasequence that were deposited as the ocean water transgressed across the continents. Basal rock types in each megasequence were most likely the best preserved of any interval because erosion was generally from the top down between each cycle. That is not to say that all basal rocks in each megasequence were preserved, because the regressive phase did seem to remove all of the preceding megasequence rock in some locations. We constructed basal rock type maps for each megasequence and stratigraphic cross-sections that allowed continent-scale correlations of the basal stratigraphy for each megasequence (Figure 16.1).[13]

Method 2: Construction of Maps of Unique Sediments of Semi-Regional Extent

We also compiled maps of distinctive rock types like bedded chert, salt, and gypsum-rich layers, keeping track of each by megasequence (Figure 16.2).[13] These unique

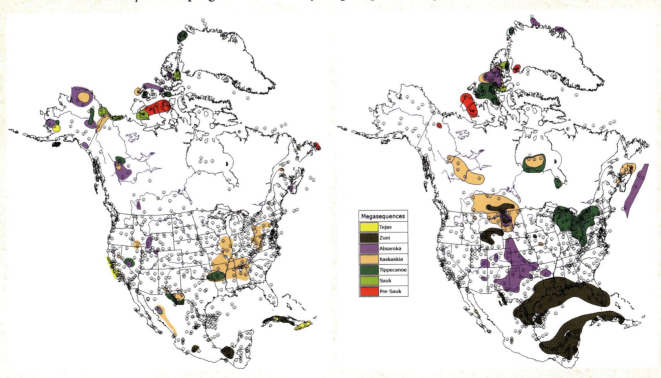

Figure 16.2. Chert-rich rocks by megasequence (left) and salt and gypsum-rich rocks by megasequence (right) for North America

lithologic units allowed us to test our megasequence boundary picks on a regional scale. For example, we assumed megasequence correlations were validated if the salt-rich or chert-rich layers remained in the same relative location within the megasequences from column to column and did not cut across the layers within each stratigraphic section. We also examined lithologically distinct rock units like the Morrison Formation and Pierre Shale that cover extensive regions of the western United States. These semi-regional formations also allowed validation of megasequence correlations.[13]

Discussion of the Results

Basal Lithology Maps

Our multi-continent study demonstrated that megasequences are related to major changes in the global sedimentary pattern and record major shifts in the global fossil record.[13] In fact, many of the claimed largest mass extinction horizons correlate closely with the highest water levels of each megasequence cycle.[16] However, Flood geologists dispute that these represent true extinction events and instead interpret them as abrupt changes in the

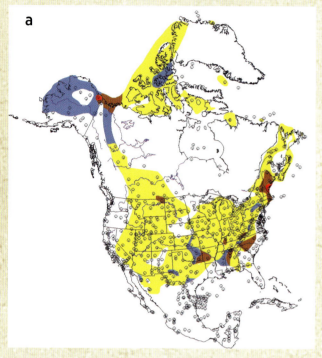

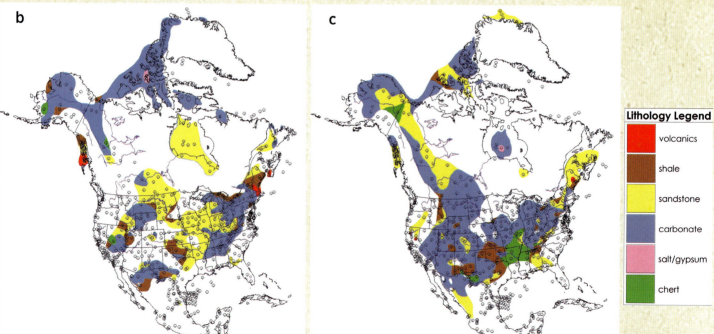

Figure 16.3. Basal rock type maps for the (a) Sauk, (b) Tippecanoe, and (c) Kaskaskia Megasequences across North America

types of fossils deposited during the Flood year. In this regard, it is no surprise a connection is observed between megasequences and the fossil record since both reflect sudden shifts in depositional pattern, including water volume and energy.

The fossil pattern observed across three continents is best explained by the systematic flooding of progressively higher and higher elevations of the pre-Flood continents as described in Genesis 7.[17] As water levels increased and coverage became more extensive, the observable pattern of fossils changed accordingly. We observe the same progressive pattern across each of the three continents in this study. In fact, one could build an independent geologic column on each of the three continents. Comparison of these would result in essentially the same global column across each continent.

The lowermost extensive Flood sediments (Sauk Megasequence) contain the same fossil taxa on each continent. And each subsequent megasequence on top of the Sauk contains the same fossil taxa and in the same order on all three continents. This is the very basis for the principle of faunal succession, the recognition of a global pattern of fossils that abruptly changes with deposition of subsequent sedimentary layers. Macroevolution is not observed since the fossils merely appear and disappear in the order of burial in the rock record.

The extent and lithology of the basal Sauk, Tippecanoe, and Kaskaskia Megasequences across North America, South America, and Africa are shown in Figures

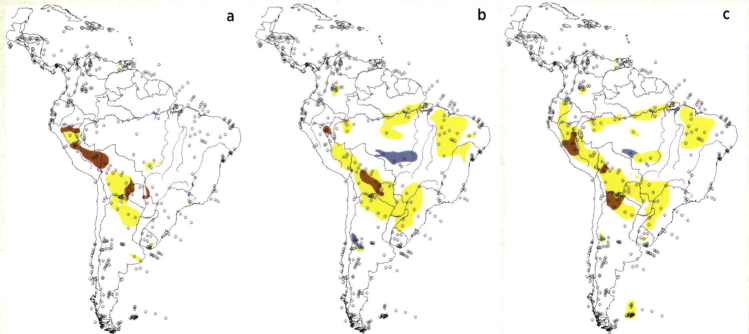

Figure 16.4. Basal rock type maps for the (a) Sauk, (b) Tippecanoe, and (c) Kaskaskia Megasequences across South America

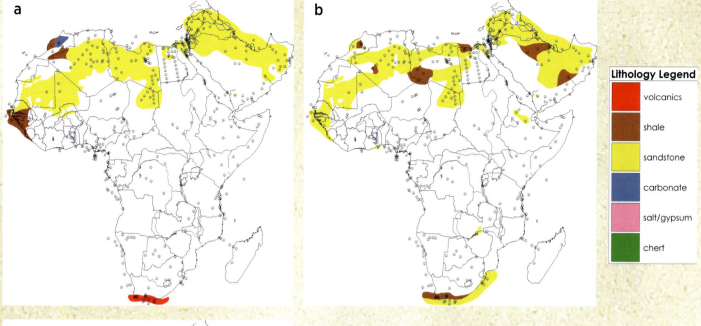

Figure 16.5. Basal rock type maps for the (a) Sauk, (b) Tippecanoe, and (c) Kaskaskia Megasequences across Africa

16.3, 16.4, and 16.5. Note that the majority of the basal rock types in each of the megasequences are sandstone layers. These basal sandstone layers are easily correlated across vast areas of the continents, helping to confirm the identification of the megasequence boundaries.

However, in contrast to the other two continents, North America has much more extensive carbonate rock in the basal Tippecanoe and Kaskaskia layers. The reason for this is not fully clear. We do observe a carbonate layer in the uppermost Sauk across much of North America (Muav Limestone and equivalent). It may be that the floodwaters did not fully drain off the North American continent at the end of the Sauk Megasequence, depositing carbonate rocks continually across the megasequence boundary and into the lower Tippecanoe. A similar process may have then repeated in the Kaskaskia where an even more extensive carbonate layer was deposited at the beginning of the third megasequence. This may imply that the floodwaters drained off to an even lesser degree at the end of the Tippecanoe, resulting in continual carbonate deposition from the upper Tippecanoe up through the lower Kaskaskia section.

Chapter 16 ★ 385

South America (Figure 16.4) and Africa (Figure 16.5) preserve much less extensive deposits of the Sauk, Tippecanoe, and Kaskaskia Megasequences compared to North America. These two continents apparently experienced much less flooding at this point.[18] Indeed, each of the first three megasequences across South America and Africa stack one on top of the other fairly uniformly. This is especially noticeable across North Africa where nearly identical locations are blanketed again and again by the Sauk, Tippecanoe, and Kaskaskia (Figure 16.5). The similar extent of each of these first

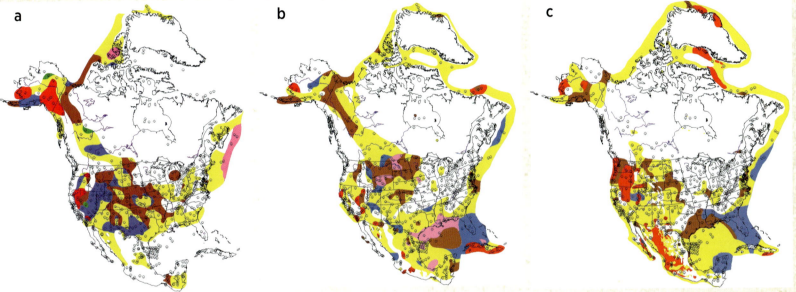

Figure 16.6. Basal rock type maps for the (a) Absaroka, (b) Zuni, and (c) Tejas Megasequences across North America

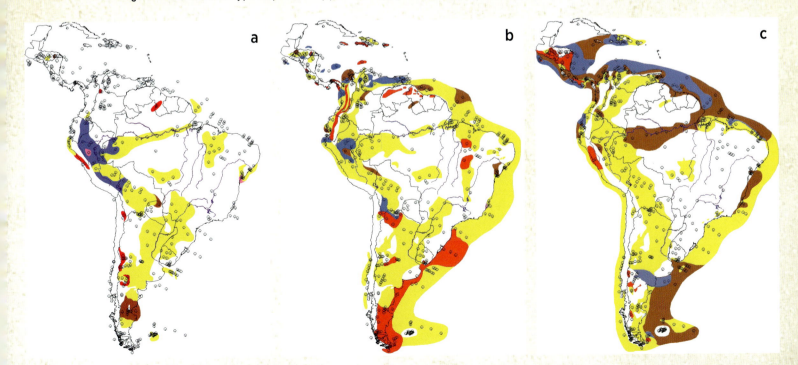

Figure 16.7. Basal rock type maps for the (a) Absaroka, (b) Zuni, and (c) Tejas Megasequences across South America

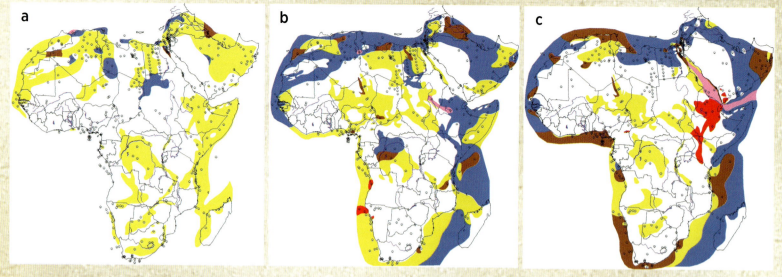

Figure 16.8. Basal rock type maps for the (a) Absaroka, (b) Zuni, and (c) Tejas Megasequences across Africa

three megasequences also argues against erosion as the major factor explaining their present distribution. Erosive processes would tend to leave randomly distributed remnants and not the near exact same locations that are observed.[18]

Figures 16.6, 16.7, and 16.8 show the Absaroka, Zuni, and Tejas basal rock types and their present extent across North America, South America, and Africa, respectively. Again, there are extensive basal sandstones that can be correlated at the base of the Absaroka across the central African and South American continents. These blanket sandstones also allow easy correlation of the latter three megasequence boundaries across vast areas of the continents.

Figure 16.8 shows a nearly continuous carbonate layer around the margins of North Africa and the Middle East at the base of the Zuni and the Tejas, indicating that the floodwaters likely never fully receded from these locations from the end of one megasequence to the start of the next megasequence. This is similar to what we observed across North America in the Tippecanoe and Kaskaskia Megasequences as discussed above.

And here again, North America seems to be a bit of an exception because it contains a mixed sandstone and shale lithology at the base of the Absaroka, Zuni, and Tejas Megasequences and not just sandstone. The reason for this difference is not immediately clear but is possibly related to tectonic activity and/or subduction along the West Coast.

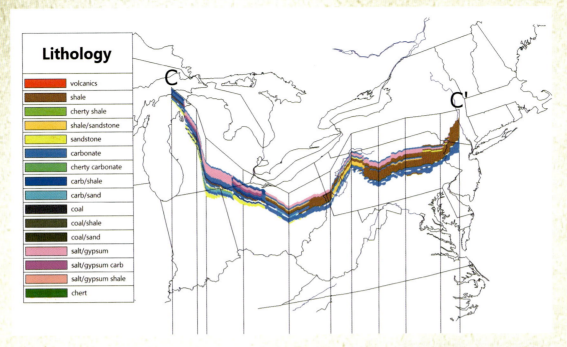

Lithologic cross-section C–C' from Michigan to New York for the Sauk and Tippecanoe Megasequences. Note the pink-colored salt/gypsum-rich layers (Tippecanoe) are continuous from column to column and remain in the same relative position between the megasequence boundaries. Also note the basal Sauk sand is continuous across Michigan, Ohio, and New York but is too thin to see past Ohio.

Unique Sedimentary Units

Our mapping identified extensive chert beds across parts of the United States at the base of the Kaskaskia Megasequence. In addition, we found massive salt beds within the Tippecanoe Megasequence across large segments of North America. These unique lithologic units allowed correlation from column to column, and verified and confirmed the megasequence boundaries and the correlations of their respective basal sandstones.[13]

Stratigraphic section C–C' shows the salt and gypsum rocks (Salina) within the Tippecanoe Megasequence.

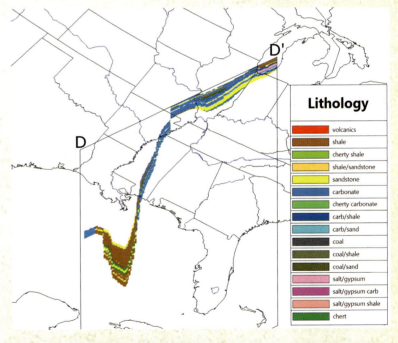

Lithologic cross-section D–D' showing the Sauk, Tippecanoe, and Kaskaskia Megasequences from East Texas to Lake Michigan. Green-colored chert-rich layers at the base of the Kaskaskia and top of the Tippecanoe Megasequences are continuous from column to column and remain in the same relative position within the megasequence boundaries.

Figure 16.9. Map of the Morrison Formation

Figure 16.10. Map of the Pierre Shale across the West

Note how the salt layer correlates to the same level within the Tippecanoe from column to column from Michigan to New York. These units independently confirm and validate the correlation of the megasequence boundaries.

Stratigraphic section D–D' shows chert-rich layers within the Kaskaskia at the base of the megasequence from Arkansas to Illinois. There are additional chert-rich layers at different stratigraphic levels elsewhere also. However, the consistent stratigraphic level of these chert-rich layers at the base of the Kaskaskia and at the top of the Tippecanoe Megasequence validates the correlation of the basal Kaskaskia boundary, independent of any fossil content.

We also correlated several recognizable and regionally extensive rock formations like the Morrison Formation (Figure 16.9) and the Pierre Shale (Figure 16.10) that extend across numerous western states.[13] In addition, we found that the Ordovician Utica Shale (Tippecanoe) and several Devonian shales (Kaskaskia) extend for hundreds of kilometers along the western flank of the Appalachians (Marcellus Shale and Chattanooga Shale).

Correlation of the Morrison Formation and the Pierre Shale (including individual bentonite-rich beds)[19] across the American West confirmed and validated the Zuni Megasequence boundaries since they also are found in the same relative positions within the megasequence interval. The Morrison Formation is always near the base of the Zuni Megasequence, and the Pierre Shale is always near the top. Each of these units can be recognized in the field and well bores by their unique characteristics and well log signals. In addition, many of the Cretaceous system (Zuni) shales found across the American West have distinctive highly radioactive well log signals that also allow correlation across vast regions. These radioactive beds fall in the same relative positions within the Zuni Megasequence, not cutting up or down or across the megasequence interval. All of these aforementioned correlations are independent of fossil content. These rocks are as empirical and factual as any data set. In fact, the results showed a remarkable match.

Overthrusts Do Not Invalidate the Geologic Column

As discussed briefly in chapter 5, overthrusts have been a source of contention among creationists for decades. Many have claimed these features are not real faults at all but merely represent locations where the fossils are out of order. However, we will see that these overthrusts are, in fact, real, and only the conditions of the Flood can explain their origin.

One of the biggest complaints levied against overthrusts is the mechanical difficulty of moving large, coherent sheets of rock great distances down fairly flat slopes. This has never been fully explained in the secular geologic literature.[20] Lithified sedimentary rock will not fold and behave plastically at surface conditions, yet we see the clear geometric results in the tight-folded rocks of overthrust belts around the globe.[12] Creationists in the past have been right to criticize secular explanations for overthrusts.[1,21-25] Their existence demands a better explanation—such as in a Flood context.

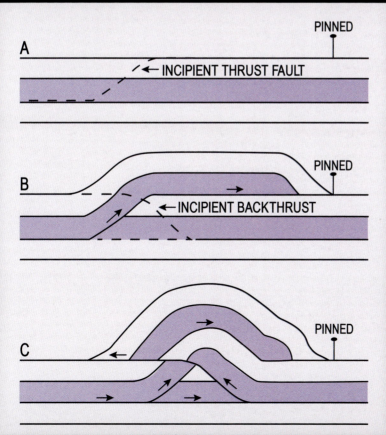

Diagram showing how complex faulting can occur in overthrusts

In 1961, when John Whitcomb and Henry Morris wrote their classic book *The Genesis Flood*, no one understood overthrust faults.[1] The rules of overthrusting were not established until the 1970s by the oil industry geologists after hundreds of wells were drilled and thousands of miles of seismic data were collected.[26,27] Overthrusts are observed to generally get younger in the direction of transport, often folding and further deforming the earlier-emplaced thrust sheets in the process. The apparent uphill movement of many overthrusts can be explained as a consequence of later folding by subsequent thrusts or by ramping uphill as the thrusting ceased. Overthrusts, generally, have a basal detachment from which all younger thrusts originate.

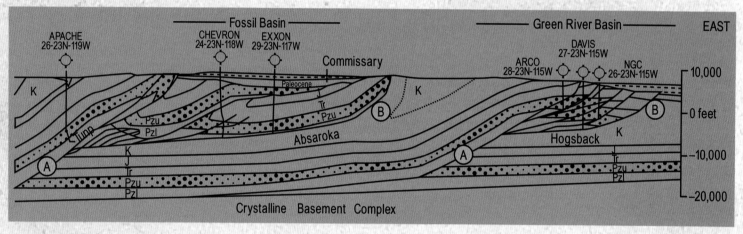

Cross-section across the frontal edge of an overthrust belt, based on oil well penetrations and seismic data

Today, creationists need to accept the results of empirical drill hole penetrations and seismic reflection data, collected since the 1970s, proving the existence of most overthrust faults.[26-31] Authors who are critical of the geologic column should no longer use the denial of overthrusts as part of their argument.[3,4] Instead, creationists should embrace these features as an opportunity to explain their unique features within a Flood context. In fact, only the Flood can explain these structures.

Prior to the development of the plate tectonics theory, most overthrusts were thought of as gravity slides.[32-35] More recently, and after the advent of plate tectonics, most secular geologists have resorted to explaining overthrusts as compressional features that are pushed. Davis et al[36] and Chappel[37] have pointed out that thrust belts are commonly wedge-shaped and move only when the wedge reaches a critical taper angle. Davis and Reynolds explain that "the critical shear stress required for sliding to occur is equal to the product of the coefficient of sliding friction and the effective stress."[38]

The Matterhorn in the Swiss Alps, formed by overthrusting late in the Flood

Heart Mountain near Cody, Wyoming, is part of an overthrust system that placed Ordovician rocks on top of Eocene strata

However, published experiments were performed with unconsolidated sediments where the basal detachment was "pulled" out from beneath the sediments.[36] Actual "pushing" of rocks from the rear, as commonly believed, results in crushing of the rocks at the point of compression with no detachments and no thrust development.[39] Gravity remains the only viable force to move overthrusts.[12] More recently, some uniformitarian geologists are again employing gravity tectonics to explain overthrusts. Alvarez, in his discussion of the development of the Alps, believes gravity spreading of uplifted areas drives collapse: "Gravity carries the rising mountains away, thrust sheet by thrust sheet."[40]

> "Gravity carries the rising mountains away, thrust sheet by thrust sheet."

This leaves the secular explanation in a conundrum. Secular scientists cannot create and maintain over-pressured layers for thousands and/or millions of years as they like to think. They cannot push the rocks. Instead, overthrusts must be gravity slides and move rapidly as well. Only a Flood model will resolve this mechanical dilemma.

Flood Solution to Overthrusts

Many authors have shown that high fluid pressures—developing during rapid sedimentation and rapid compaction, causing dewatering reactions—have the ability to create over-pressured zones and "float" large thrust sheets downslope.[32,41,42] The formation of supercritical carbon dioxide may be an additional method to move carbonate-rich sediments rapidly for overthrusts like the Heart Mountain Fault system[43] and the Lewis overthrust in Montana.

In the past, I have demonstrated that the vast majority of overthrusts are real features that have been drilled and imaged seismically for decades by oil companies. I also maintain that the necessary requirements for overthrusting can only be explained by the conditions of the global Flood.[44] Overthrusts move by slippage along surfaces of high fluid pressure that are sufficient to "float" the rocks downslope. Rapid sedimentation and rapid compaction during the Flood provide the perfect conditions to create these over-pressured zones. Water pressure will tend to accumulate along nearly impermeable surfaces such as clay-rich shales and salt and gypsum layers. Then these surfaces can become over-pressured (above static pressure) as more sediment

Peyto Lake in the Canadian Rockies in Banff National Park showing overthrusted strata

is added above, squeezing more water out of the high permeable layers at depth like a sponge and increasing the water pressure along the low permeable layers.

Rapid slippage can then be induced by uplift, loading by more sediment, a thrust sheet, or even an earthquake, allowing the overlying sediments to detach and move along the over-pressured surface. This is similar to the way landslides move. Movement is always downhill due to gravity. It only takes a slight slope if the hydrostatic pressure is maintained over a large area. Additional thrusts commonly develop as an older thrust ends, stacking the thrusts sheets at the toe of the allochthonous (moved out of place) terrane like a landslide. The refolding of earlier thrust sheets by the later thrusts can make it look like some thrusts went uphill. This is not usually the case. These steeper thrusts and rock layers are merely folded by the later deformation.

However, large areas of over-pressure cannot last for vast periods of time. The secular story of overthrusts moving slowly over millions of years, an inch at a time, is a fairy tale. And sadly, this is still taught in most college structural geology classes. Over-pressured zones would quickly dissipate over eons of time, and there would be no overthrust movement. Clastic dikes are found commonly above many overthrust detachment horizons. These dikes formed as high-pressure fluids and sediment were released upward along fractures into the overlying thrust plate from the over-pressured zone below, releasing pressure during movement. Similar fractures would

Photos of a vertical clastic dike originating from the South Fork Fault (an overthrust) near Cody, Wyoming

allow the over-pressured zone to dissipate with time within any thrust sheet. Everything leaks given enough time. Water would find a way to escape, dropping the pressure, even within the lowest permeable layers.

The folded thrust sheets and the tight folds that commonly develop in the toe of the thrusts can only form if the sedimentary layers were still unlithified, allowing them to deform in a plastic-like manner. Rocks at surface pressures cannot fold without shattering. But the surface is exactly where these folded thrusts develop. Only at great depth in the earth, and under near-metamorphic conditions, can truly lithified rocks bend and behave plastically. The only other way to fold lithified rocks is to do it very slowly so the individual grains can slide and rotate, allowing folds to develop over vast amounts of time. However, as mentioned above, these over-pressured zones cannot be maintained for the great periods of time needed to fold the rocks. Therefore, these overthrusted layers must have been folded rapidly while still unlithified.

Overthrusts move fast and fold sediments fast. The necessary recipe to make overthrust faults requires unique events. The rapid sedimentation rates and compaction to

> "Therefore, these overthrusted layers must have been folded rapidly while still unlithified."

Tightly folded strata associated with overthrusts in Montana (left) and Wyoming (right)

create over-pressured zones, the short time frame to maintain the high fluid pressures, and the uplift of slopes for sliding were all conditions uniquely provided by the activity of the Flood.

Conclusion: The Geologic Column Is Global Because the Flood Was Global

Megasequences reflect major advances and shifts in global depositional patterns and exhibit distinctive lithologic features that allow intercontinental correlations using seismic and well data. Results show extensively consistent lithologic units (i.e., blanket sandstones) covered portions of every continent at the same time and are correlative across vast regions and continent to continent. These include layers like the Tapeats Sandstone and equivalent across North America. The presence of stacked basal megasequence units across multiple continents and the correlation of other unique rock types (i.e., salt and chert layers) within the megasequences confirm the validity of the geologic column on a global scale. The fossils contained within the megasequences are merely the passive results of these major sedimentological events as the floodwaters rose and progressively inundated new environments. Fossils do not show any hint of evolution, only sudden appearance, stasis, then disappearance. This is exactly what you would expect in a rising global flood that buried the same ecological zones across the world at the same approximate levels.

> "Only a global flood can bury the same types of fossils in the same approximate order all over the world at the same time as sea level rose higher and higher."

Instead of being critical of the geologic column, creationists should embrace it as robust evidence of a global flood. Only the Flood can provide a reason for the blanket sandstones found on multiple continents at the same time, megasequence after megasequence. Only the Flood can explain the massive extent of unique rock units like salt and chert across vast regions of the continents. Only a global flood can bury the same types of fossils in the same approximate order all over the world at the same time as sea level rose higher and higher. The true global nature of the geologic column is one of the strongest evidences we have of a worldwide flood. It confirms exactly what the Bible tells us. Why haven't we pointed this out long ago?

A bed of rapidly buried crinoids

Author conducting field work on the South Fork Fault near Cody, Wyoming

Similarly, overthrust faults are also features that can only be explained within a biblical worldview and by the conditions of the Flood. And just like their poor attempts to justify the Ice Age with Milankovitch cycles, secular scientists have only provided inadequate explanations for the origins of overthrusts.

In the next chapter, I will use megasequence data across the three continents to explain how we are unlocking the secrets of the pre-Flood world.

References

1. Whitcomb, J. C. and H. M. Morris. 1961. *The Genesis Flood*. Philadelphia, PA: The Presbyterian and Reformed Publishing Company.
2. Reed, J. K. and M. J. Oard, eds. 2006. *The Geologic Column: Perspectives within Diluvial Geology*. Chino Valley, AZ: Creation Research Society Books, 1-6.
3. Matthews, J. D. 2011. The stratigraphic column—a dead end. *Journal of Creation*. 25 (1): 98-103.
4. Matthews, J. D. 2016. The overthrusting paradox: a challenge to uniformitarian geology and evolution. *Journal of Creation*. 30 (2): 83-91.
5. Oard, M. J. 2010. Is the geological column a global sequence? *Journal of Creation*. 24 (1): 56-64.
6. Woodmorappe, J. 1999. The geologic column: does it exist? *Journal of Creation*. 13 (2): 77-82.
7. Froede Jr., C. R., A. J. Akridge, and J. K. Reed. 2015. Can 'megasequences' help define biblical geologic history? *Journal of Creation*. 29 (2): 16-25.
8. Ibid, 21.
9. Ross, M. R. 2014. Improving our understanding of creation and its history. *Journal of Creation*. 28 (2): 62-63.
10. Ross, M. 2013. The Flood/post-Flood boundary. Letter to the Editor. *Journal of Creation*. 27 (2): 43-44. Emphasis in original.
11. Austin, S. A. et al. 1994. Catastrophic Plate Tectonics: A Global Flood Model of Earth History. In *Proceedings of the Third International Conference on Creationism*. R. E. Walsh, ed. Pittsburgh, PA: Creation Science Fellowship, 609-621.
12. Snelling, A. A. 2009. *Earth's Catastrophic Past: Geology, Creation & the Flood*. Dallas, TX: Institute for Creation Research.
13. Clarey, T. L. and D. J. Werner. 2018. Global stratigraphy and the fossil record validate a Flood origin for the geologic column. In *Proceedings of the Eighth International Conference on Creationism*. J. H. Whitmore, ed. Pittsburgh, PA: Creation Science Fellowship, 327-350.

14. Carroll, A. R. 2017. Xenoconformities and the stratigraphic record of paleoenvironmental change. *Geology.* 45 (7): 639-642.
15. Sloss, L. L. 1963. Sequences in the cratonic interior of North America. *Geological Society of America Bulletin.* 74 (2): 93-114.
16. Snelling, A. A. 2017. Five mass extinctions or one cataclysmic event? *Answers.* Posted on answersingenesis.org February 12, 2017, accessed August 2, 2017.
17. Clarey, T. L. and D. J. Werner. 2018. Use of sedimentary megasequences to re-create pre-Flood geography. In *Proceedings of the Eighth International Conference on Creationism.* J. H. Whitmore, ed. Pittsburgh, PA: Creation Science Fellowship, 351-372.
18. Clarey, T. L. and D. J. Werner. 2017. The sedimentary record demonstrates minimal flooding of the continents during Sauk deposition. *Answers Research Journal.* 10: 271-283.
19. Bertog, J., W. Huff, and J. E. Martin. 2007. Geochemical and mineralogical recognition of the bentonites in the lower Pierre Shale Group and their use in regional stratigraphic correlation. In *The Geology and Paleontology of the Late Cretaceous Marine Deposits of the Dakotas.* J. E. Martin and D. C. Parris, eds. Geological Society of America Special Paper 427. Boulder, CO: Geological Society of America, 23-50.
20. Briegel, U. 2001. Rock mechanics and the paradox of overthrusting tectonics. In *Paradoxes in Geology.* U. Briegel and W. Xiao, eds. Amsterdam, Netherlands: Elsevier, B.V., 231-244.
21. Lammerts, W. E. 1966. Overthrust faults of Glacier National Park. *Creation Research Society Quarterly.* 3 (1): 61-62.
22. Lammerts, W. E. 1972. The Glarus overthrust. *Creation Research Society Quarterly.* 8 (4): 251-255.
23. Burdick, C. L. 1969. The Empire Mountains–a thrust fault? *Creation Research Society Quarterly.* 6 (1): 49-54.
24. Burdick, C. L. 1974. Additional notes concerning the Lewis thrust-fault. *Creation Research Society Quarterly.* 11 (1): 56-60.
25. Burdick, C. L. 1977. Heart Mountain revisited. *Creation Research Society Quarterly.* 13 (4): 207-210.
26. Royse Jr., F., M. A. Warner, and D. L. Reese. 1975. Thrust belt structural geometry and related stratigraphic problems, Wyoming-Idaho-Northern Utah. *Rocky Mountain Association of Geologists 1975 Symposium,* 41-54.
27. Boyer, S. E. and D. Elliott. 1982. Thrust systems. *American Association of Petroleum Geologists Bulletin.* 66: 1196-1230.
28. Jones, P. B. 1982. Oil and gas beneath east-dipping underthrust faults in the Alberta Foothills. In *Geologic Studies of the Cordilleran Thrust Belt,* vol. 1. R. B. Powers, ed. Denver, CO: Rocky Mountain Association of Geologists, 61-74.
29. Lamerson, P. R., The Fossil Basin and its relationship to the Absaroka thrust system, Wyoming and Utah, *Geologic Studies of the Cordilleran Thrust Belt,* 279-340.
30. Price, R. A. 1988. The mechanical paradox of large overthrusts. *Geological Society of America Bulletin.* 100: 1898-1908.
31. Coogan, J. C. 1992. Structural evolution of piggyback basins in the Wyoming-Idaho-Utah thrust belt. In *Regional Geology of Eastern Idaho and Western Wyoming.* P. K. Link, M. A. Kuntz, and L. B. Platt, eds. Geological Society of America Memoir 179, 55-81.
32. Hubbert, M. K. and W. W. Rubey. 1959. Role of fluid pressure in mechanics of overthrust faulting. I. Mechanics of fluid-filled porous solids and its application to overthrust faulting. *Geological Society of America Bulletin.* 70: 115-166.
33. Eardley, A. J. 1963. Relation of uplifts to thrusts in Rocky Mountains. In *Backbone of the Americas.* American Association of Petroleum Geologists Memoir 2, 209-219.
34. Roberts, R. J. 1968. Tectonic framework of the Great Basin. *University of Missouri Research Journal.* 1: 101-119.
35. Mudge, M. R. 1970. Origin of the Disturbed belt in northwestern Montana. *Geological Society of America Bulletin.* 81: 377-392.
36. Davis, D. M., J. Suppe, and F. A. Dahlen. 1983. Mechanics of fold-and-thrust belts and accretionary wedges. *Journal of Geophysical Research.* 88: 1153-1172.
37. Chapple, W. M. 1978. Mechanics of thin-skinned fold-and-thrust belts. *Geological Society of America Bulletin.* 89: 1189-1198.
38. Davis, G. H. and S. J. Reynolds. 1996. *Structural Geology of Rocks and Regions,* 2nd ed. New York: John Wiley and Sons, Inc.
39. Personal communication with John Baumgardner, 2009.
40. Alvarez, W. 2009. *The Mountains of Saint Francis: Discovering the Geologic Events That Shaped Our Earth.* New York: Norton.
41. Guth, P. L., K. V. Hodges, and J. H. Willemin. 1982. Limitations on the role of pore pressure in gravity gliding. *Geological Society of America Bulletin.* 93 (7): 606-612.
42. Clarey, T. L. 2012. South Fork fault as a gravity slide: its break-away, timing, and emplacement, northwestern Wyoming, U.S.A. *Rocky Mountain Geology.* 47 (1): 55-79.
43. Beutner, E. C. and G. P. Gerbi. 2005. Catastrophic emplacement of the Heart Mountain block slide, Wyoming and Montana, USA. *Geological Society of America Bulletin.* 117 (5/6): 724-735.
44. Clarey, T. L. 2013. South Fork and Heart Mountain Faults: examples of catastrophic, gravity-driven "overthrusts," northwest Wyoming, USA. In *Proceedings of the Seventh International Conference on Creationism.* M. Horstemeyer, ed. Pittsburgh, PA: Creation Science Fellowship.

17 Unlocking the Mystery of the Pre-Flood World

> **Summary:** Creationists don't have much concrete information about the pre-Flood world. Geological data from around the globe indicate that certain areas of the pre-Flood world were higher than others, leading to a theoretical reconstruction of God's original created landmasses. Many dinosaur bones in the U.S. are found in a kind of narrow zone running north to south, with many dinosaur footprints in the same rocks. The geology indicates this area was a lowland in the pre-Flood world. It's possible that as the floodwaters rose, the dinosaurs retreated to higher and higher elevations and were eventually overcome in mass burial sites. Additionally, the global occurrence of pre-Flood fossil stromatolites confirms the presence of springs before the Flood. Seas, shallow areas, swamps, and other features can be inferred from geologic data. The process of reconstructing the pre-Flood world is slow but fascinating.

Many secularists, theologians, and creation scientists have had direct or indirect interest in pre-Flood geography, particularly when applied to the search for the Garden of Eden.[1-5] The creation model is weakest in its knowledge of the pre-Flood world because the Bible only gives a few details of "the world that then existed" (2 Peter 3:6). Although there has been much speculation about pre-Flood geography in creationist literature, very little has been based on empirical data. Most creationists readily admit that we know very little about the actual pre-Flood world.[1,2] Some creationists have relied heavily on secular interpretations in their studies on continental configurations and pre-Flood geography.

> By which the world that then existed perished, being flooded with water.
> (1 Peter 3:6)

Our geological research is the first to tackle this issue using actual rock data from around the globe. And we are getting some fascinating results.

As described in chapter 3, most of the Phanerozoic sedimentary record can be divided into megasequences of deposition (Figure 17.1). Recall that megasequences are discrete packages of sedimentary rock bounded on the top and bottom by erosional surfaces, often with coarse sandstone layers at the base.[6,7] These sedimentary packages were stacked one on top of another as each was sequentially deposited during the year-long Flood.

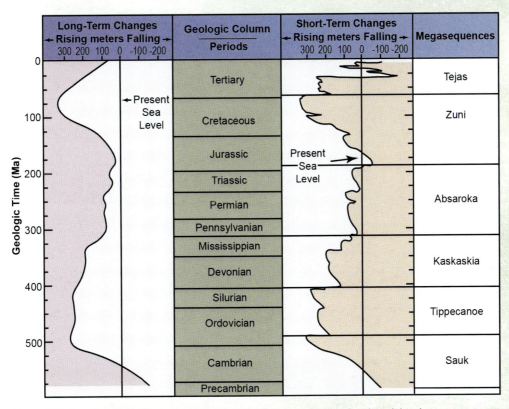

Figure 17.1. Secular chart showing presumed geologic time, global sea level, and the six megasequences (after Vail and Mitchum[7])

Our data set consists of over 1,500 stratigraphic columns from North America, South America, Africa, and the Middle East. Each rock column was compiled from published outcrop data, oil well boreholes, cores, cross-sections, and/or seismic data tied to boreholes. Rock type and megasequence data were put into a database, allowing thickness maps to be generated for all six megasequence intervals. These data were used to create a three-dimensional stratigraphic model across each of the three continents studied so far. When examined megasequence by megasequence, these

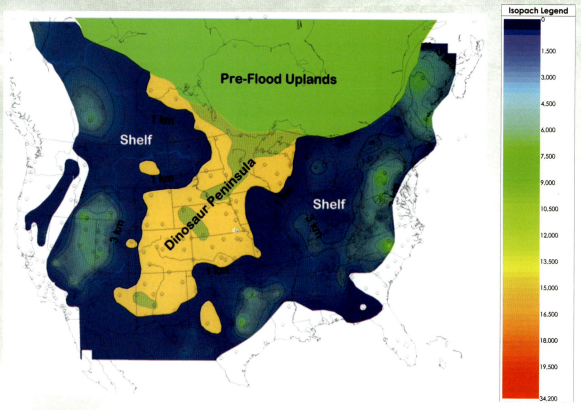

Figure 17.2. Pre-Flood map showing the proposed Dinosaur Peninsula, a lowland landmass extending from Minnesota to New Mexico likely inhabited by pre-Flood wetland plants and animals including dinosaurs. The contours show the cumulative thickness of the Sauk, Tippecanoe, and Kaskaskia Megasequences. Measurement is in meters.

models allow us to visualize the pre-Flood geographic relief.

Earlier, we identified a pre-Flood landmass across the central U.S. that we labeled Dinosaur Peninsula.[8] We found that the deposition of the earliest Flood sediments (the Sauk, Tippecanoe, and Kaskaskia Megasequences) was thickest in the eastern and far western U.S., including Grand Canyon. In contrast, the early Flood deposits across much of the center of the country are commonly less than a few hundred yards deep, and in many places there was no deposition at all (Figure 17.2). We concluded:

> It seems the dinosaurs were able to survive through the early Flood in the West simply because they were able to congregate and scramble to the elevated remnants of land—places where the related sedimentary deposits aren't as deep—as the floodwaters advanced...In this way, dinosaurs were able to escape burial in the early Flood.[8]

The Pre-Flood World

Using the thicknesses of the various megasequence intervals, we made reasonable inferences about the relative topography of the pre-Flood world. We assumed the pre-Flood lowlands would be filled in first by the earliest deposits and the uplands later as the Flood levels increased, as described in Genesis 7. By comparing the megasequence data to the fossil record, we gleaned three broad environments—pre-Flood shallow seas, lowlands, and uplands, shown in Figure 17.3.

Shallow Seas

Results indicate shallow seas existed across much of the eastern and southwestern U.S. (including Grand Canyon) and across North Africa and the Middle East where the earliest three megasequences were deposited (Figure 17.1). The areas show extensive deposition of early Flood

Figure 17.3. Pre-Flood geography map for North America, South America, and Africa combined into a Pangaea-like configuration. It is likely the landmasses continued to the east near Greenland (Europe) and Africa (India). Note that the western edge of North America does not include the majority of the West Coast states since these terranes were added later during plate motion as part of the Flood. Also, much of Central America is not shown since it was formed from activity during the Flood.

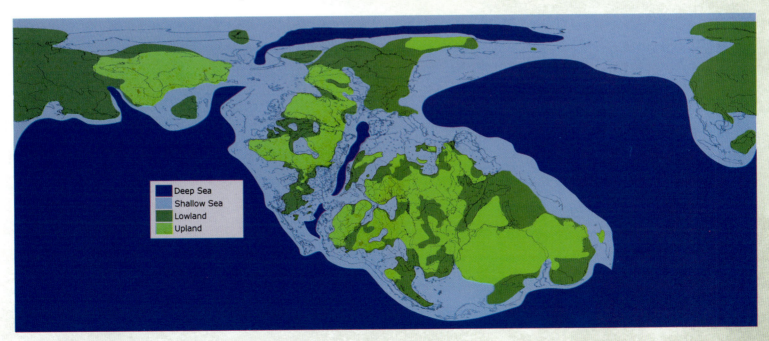

Figure 17.4. Map of the entire pre-Flood world showing the environments as determined in Figure 17.3

sediments and contain almost exclusively shallow marine fauna. There are virtually no trees or land animals in these megasequences. Apparently, only limited amounts of land were inundated at this stage in the Flood.[9]

Lowland Areas

During the deposition of the Absaroka Megasequence, the sediments began extending onto the land proper as water levels rose, starting with the lowland and wetland areas. This included the Karoo Supergroup across much of southern Africa and the Coconino Sandstone that extends across the southwestern U.S.

The first prolific deposits of coal (Pennsylvanian lycopod forests) and land animals mixed with marine flora and fauna also appear in the Absaroka. This indicates the Flood was impacting significant amounts of pre-Flood land, such as the broad lowlands in East Africa and the central U.S., including Dinosaur Peninsula.

Upland Areas

All of the megasequences thinned toward the crystalline shield areas on each of the three continents (see the diagram on page 324). The sedimentary units do not merely show evidence of erosion and truncation, they become thinner in the direction of the shields—implying they were deposited on the flanks of extensive uplands.

Many of these interpreted upland areas are completely devoid of sedimentary rock because post-Flood erosion stripped the thin Zuni sediment that might have been deposited there. According to Genesis 7:20, the highest hills were only flooded by a modest amount of water, likely leaving little room for thick sedimentary deposits to accumulate. However, there are a few Zuni remnants in Hudson Bay, Michigan, and Illinois that indicate the highest water level was achieved at this point in the Flood.

The lack of sediments preserved across the pre-Flood uplands may also help explain the lack of human fossils in the rock record. As described in chapter 14, most pre-Flood humans likely survived until close to Day 150 and probably congregated in the areas of highest ground. As the water levels peaked, the humans still there would have drowned and been washed away in all directions. Their remains would not have been clustered together or covered deeply enough in sediment for fossil preservation. Any remains that were buried most likely would have been destroyed in post-Flood erosional activity.

Pangaea, Not Rodinia

Debate exists over the pre-Flood continental configuration, with some creation geologists advocating for an initial created supercontinent called Rodinia.[10] However, as I described in chapter 7, we chose a slightly modified Pangaea because it has the most empirical geological evidence supporting it, including the best fit of the continents.[11] We placed a narrow sea (300 to 500 kilometers) between North America and Africa/Europe, allowing for limited plate subduction, an early Flood closure of the pre-Atlantic, and the formation of the Appalachian/Caledonian Mountains. The width of this pre-Atlantic is based on subducted plate remnants that diminish beneath the Appalachians below 300 kilometers, supporting this narrow-sea interpretation.[12]

OmniGlobe projection system showing the pre-Flood map and our data points (columns)

Another reason we favored Pangaea over Rodinia is that our current ocean floor was evidently created when the original creation week seafloor was consumed by subduction during the Flood event. It was the density contrast of the heavy, cold original ocean crust (the lithosphere) that allowed the runaway subduction process to begin and continue. The density difference served essentially as the "fuel." Geophysicist John Baumgardner described it as "gravitational energy driving the motion" of the plates.[13] The "runaway" process continued until the original oceanic lithosphere was consumed. There was no geophysical means or reason to stop the rapid plate motion until the density contrast was fully alleviated. At that moment, the newer, more buoyant lithosphere ceased subducting, bringing plate motion to a virtual standstill. As a consequence, today we only witness small, residual plate motions of centimeters per year.

A pre-Flood world that resembled Rodinia would require the consumption of nearly all the pre-Flood ocean crust twice. The first time would be while the continents from Rodinia moved into the configuration of Pangaea, and then a second time when Pangaea split into the present global configuration. Geophysically, the first breakup of Rodinia and reconfiguration into Pangaea would be possible, but it would also consume all of the dense pre-Flood ocean crust. A second move would be rendered impossible since any significant amount of new ocean crust created while splitting up Rodinia would not have enough density contrast to fuel a second episode of subduction. As mentioned above, it is the consumption of the cold, more dense pre-Flood ocean crust that caused runaway subduction in the first place.[13] Therefore, if there had been a Rodinia, we would still be in a Pangaea continental configuration today.

Garden of Eden

Many Christians have wondered if the current nation of Israel was the location of the pre-Flood Garden of Eden. Our maps demonstrate this is likely not the case. Figure 17.3 shows that an area of shallow seas probably existed across much of the Middle East, including Israel. Based on our research, the likely locations for the Garden are in present-day Canada, Brazil, West Africa, Asia, or even Greenland, depending on which direction was eastward of the place referenced in Genesis 2:8. The exact location of the Garden of Eden remains unknown, but these results allow us to determine the major topographic highs and lows of the pre-Flood world. And it appears that the highest areas—the so-called shield areas—were the most likely locations of human habitation.

ICR Discovery Center Globe

ICR is displaying our megasequence and pre-Flood maps on a 48-inch OmniGlobe projection system as part of an interactive exhibit in the ICR Discovery Center for Science & Earth History in Dallas, Texas. Visitors can see the pre-Flood world map and view the Flood's progression and the breakup of the landmasses over the year of the Flood. This has become a valuable visual tool demonstrating the geological effects of the Flood as God's judgment reshaped the world.

View of the Middle East from space

Fossils Confirm Pre-Flood Global Greenhouse

Sometimes, scientists find data that help our interpretation of the pre-Flood world's climate. A team of secular scientists led by Erik Gulbranson recently discovered a forest of fossil trees in Antarctica. Gulbranson, a paleoecologist at the University of Wisconsin-Milwaukee, claims this is the "oldest polar forest on record from the southern polar region."[14] The trees were found in Antarctica's Transantarctic Mountains and include a mix of evergreens, deciduous trees, and gingkoes.

This discovery should be no surprise to those who take Genesis as literal history. The Bible clearly describes a global flood that affected all landmasses. Why should Antarctica be an exception?

Dinosaur and mosasaur (massive swimming reptile) fossils occur on every continent, including Antarctica.[15] These discoveries demonstrate that a much-warmer global climate existed in the pre-Flood world. Warm-weather plants like *Glossopteris* have been found in Permian rocks on Antarctica for decades and are often used to make plate tectonic reconstructions.

Secular scientists even admit Antarctica was not always cold in the past. *LiveScience*'s Stephanie Pappas wrote:

> Antarctica wasn't always a land of ice. Millions of years ago, when the continent was still part of a huge Southern Hemisphere landmass called Gondwana, trees flourished near the South Pole.[14]

Some creation authors have tried to explain the presence of temperate climate fossils like trees, dinosaurs, and coal beds in Antarctica by making claims that Earth somehow "rolled" on its axis during the Flood, resulting in the North Pole shifting from central Asia to its present position, but this explanation appears to cause more problems than it solves.[16]

First, temperate and tropical-type flora and fauna fossils, like dinosaurs, are found buried in Flood rocks even in central China—which would have been the North Pole if the earth rolled on its axis—and everywhere else on Earth. Thus, the entire globe was warmer in the pre-Flood world. Second, the extreme forces required to explain such a radical shift in the earth's axis have never been adequately resolved; it remains a speculation.

Gulbranson and his team think the fossil trees they found are about 280 million years old, which places them in Permian system strata. The scientists discovered that the trees were so rapidly buried in volcanic ash that they contained fossilized plant cells that are virtually mummified, "preserved down to the cellular level."[14] They reported:

> The plants are so well-preserved in rock that some of the amino acid building blocks that made up the trees' proteins can still be extracted, said Gulbranson, who specializes in geochemistry techniques.[14]

How could proteins and original amino acids survive for millions of years? The secular science community has no viable answers to explain remarkable finds like these.

Rocks line the coast of Antarctica

So, the fossil trees themselves tell a different and a far more recent story—one that fits the biblical account of a global flood just thousands of years ago. These trees were buried rapidly during the global Flood described in Genesis. Temperate and tropical plants and animals were caught up and quickly buried in the ash, mud, and sand that engulfed them in this cataclysmic event. These fossils remind us that God's Word is true.

Stromatolites: Evidence of Pre-Flood Hydrology

The Bible indicates that the pre-Flood hydrology was somewhat different from today's world. It contained many springs that watered the earth, creating a mist to water the ground and also to serve as a source for surface waters.

> For the LORD God had not caused it to rain on the earth, and there was no man to till the ground; but a mist went up from the earth and watered the whole face of the ground….Now a river went out of Eden to water the garden, and from there it parted and became four riverheads. (Genesis 2:5-6, 10)

Many of the pre-Sauk sedimentary rocks of the world contain evidence of life in the form of fossil stromatolites. Stromatolites are some of the more puzzling fossils found throughout Earth's rock record. Although fairly common in the oldest-known sedimentary rocks, living stromatolites only occur today in rare isolated locations of the world. Only special conditions seem to allow these organisms to flourish, often involving unusual water chemistry and surface springs. Uniformitarian scientists have struggled to explain their abundance in the ancient rocks and their paucity today.

Polished stromatolite

The *Glossary of Geology* defines a stromatolite as "an organosedimentary structure produced

Stromatolite

by sediment trapping, binding, and/or precipitation as a result of the growth and metabolic activity of micro-organisms, principally cyanophytes (blue-green algae [cyanobacteria])."[17] The result is a finely laminated biomat that forms a mounded structure. This structure is not composed of the bacteria themselves but instead is a sediment-trapping mat formed by "biologically…mediated mineral precipitation."[18]

Stromatolites were first identified in the early 1900s in upper Precambrian (pre-Sauk) rocks in Ontario, Canada, by Charles Walcott, former director of the United States Geological Survey. He thought the mounded structures were some type of ancient reef derived from algae. It wasn't until the 1950s that paleontologists determined that stromatolites were in fact the products of biological activity.[19] This was confirmed by the discovery of living stromatolites that same decade in Australia. However, a few recent authors have again suggested that some fossil stromatolites could have had a nonbiological origin.[20,21]

Evolutionary scientists claim stromatolites were some of the earliest life on Earth, dating them back as far as 3.7 billion years.[22] The oldest undisputed stromatolites, from the Warrawoona Group in Australia, are dated by secular scientists as 3.3 to 3.5 billion years old.[19] Fossil stromatolites are found all over the world in Archean and Proterozoic carbonate rocks (usually dolomite) and to a lesser extent in Cambrian and later strata. Evolutionary scientists have tried to explain the rapid decline in stromatolites in post-Cambrian rocks by attributing it to the sudden appearance of grazing organisms that presumably eat cyanobacteria.[18]

Stromatolites Create an Evolutionary Conundrum

Because secular scientists believe stromatolites evolved about 3.5 billion years ago, it creates a significant problem for them in the timing of the origin of life on Earth.[23] How could cyanobacteria have evolved so quickly? Life would have had to originate and develop the ability to photosynthesize and colonize in less than a billion years, assuming the earth is 4.55 billion years old.

But these scientists also believe that between 4.1 and 3.8 billion years ago Earth underwent a massive bombardment by meteorites, termed the Late Heavy Bombardment.[19] This bombardment episode is supposed to have been a time of severe meteorite impacts striking Earth and the moon. These impacts would have obliterated much of Earth's crust and any forms of life that existed before 3.8 billion years ago.

Secular scientists have, in effect, painted themselves into a corner. How can they explain the formation of the atmosphere, the oceans, the mysterious process of abiogenesis, and the ability to photosynthesize in a window of just a few 100 million years? Photosynthesis alone is an exceedingly complex process. For the evolutionist, this is a ridiculously short amount of time for this cascade of events to have occurred.[23,24]

Stromatolites Are Living Fossils

Although secular science claims they go back billions of years, stromatolites show little, if any, evidence of evolution and no indication of great age. Modern stromatolites are considered an example of a living fossil, like the coelacanth. They seem to have thrived without any evolutionary change.

Stromatolite Fossils Confirm the Presence of Springs in the Pre-Flood World

Using ICR's pre-Flood world map we are developing (Figure 17.3), we plotted the locations of many of the known Precambrian stromatolite fossil locations across the

Stromatolites at Lake Thetis, Western Australia

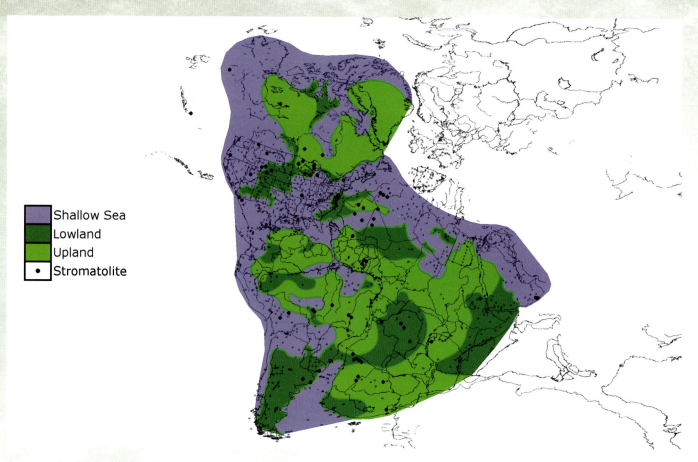

Figure 17.5. Pre-Flood geographic map of North America, South America, and Africa in an assumed Pangaea-like configuration, showing the locations of many of the Precambrian stromatolites. Stromatolite locations courtesy of J. D. Dieterle.

world. Our completed stratigraphic study only covers North America, South America, and Africa to date, but we expect the locations of stromatolites will be much the same over the remainder of the globe.

An examination of the map shows that the stromatolites seem to follow no particular environmental pattern (Figure 17.5). Their locations are found in regions that are interpreted as pre-Flood shallow seas, lowlands, and upland environments. Before the recent discovery of stromatolites on land, this interpretation would have seemed to be faulty.[25] But now that modern stromatolites have been found living in fresh water, saltwater, and also on land, it is not surprising that pre-Flood stromatolites existed in all types of environments. God tells us in Genesis 2:6 that before the Flood "a mist went up from the earth and watered the whole face of the ground."

If our data-based interpretation is correct, the presence of fossil stromatolites confirms the pre-Flood hydrology as described in Genesis.[26] That world must have had springs in great abundance to support the prevalence of stromatolites on every continent and in every type of environment. The springs watered both the uplands and the lowlands, providing mineral-rich waters in which the stromatolites thrived. God also created the perfect environments for stromatolites to grow in the pre-Flood shallow seas. These were also possibly fed by springs that provided hypersaline conditions similar to those found in Shark Bay, Australia, where stromatolites are found today.

Biblical Account Confirmed by Science

The history of stromatolites fits best with the recent creation and Flood described in the Bible. Most creation scientists believe God created stromatolites as part of the original creation, probably on Day 3 of the creation week when He made plants. Stromatolites apparently proliferated in special pre-Flood environments and grew extensively during the 1,650 years or so between creation and the Flood.

Furthermore, creation scientists have proposed that the catastrophic nature of the Flood reshaped Earth's surface sufficiently to destroy the pre-Flood environments where stromatolites formerly thrived.[27] Today, it's only in specialized environments that stromatolites are able to exercise their mat-making abilities and grow, whether on land or in the ocean.

The Flood also destroyed much of the pre-Flood stratigraphic record that contained the majority of the stromatolites. Only limited exposures of these ancient rocks are found globally in Archean and Proterozoic rocks. Of course, many likely remain covered by later Flood sediments, and others were undoubtedly destroyed by high heat and pressures associated with rapid plate motion and volcanism during the Flood event. But there are enough remnants preserved to indicate their abundance.

These recent discoveries demonstrate the accuracy and trustworthiness of God's Word. Some people claim there are errors in the Bible and that its depiction of Earth history shouldn't be trusted because it isn't a science book. Yet, again and again science demonstrates the truth of the biblical record. When the Bible discusses science, it is always shown to be correct, right down to the smallest detail.

Conclusion: The Pre-Flood World Is Coming into Focus

Stratigraphic data indicate the pre-Flood world was segregated by topography, resulting in an orderly ecological zonation, as some early creationists speculated.[28] I had earlier identified a similar topographical/ecological pattern to explain the occurrences of the dinosaurs in the American West.[8,26] It also appears that the global fossil record can be explained as a direct result of the progressive burial of higher and higher elevations during the Flood. As the floodwaters rose, new and higher areas were subsequently inundated, until all the world was covered by Day 150 of the Flood (Genesis 7:24). As described in chapter 13, the stratigraphic data seem to indicate this coincided with the end of the Zuni Megasequence. The Zuni has the most volume of rock deposited and the maximum areal coverage of any megasequence. Whereas, the Tejas Megasequence is a close second in both volume and areal extent and likely consists of Day 150+ deposits (chapter 14). Tejas fossils likely reflect the flora and fauna of the uplands areas that existed in the pre-Flood world.

Our study fills a critical need for knowledge of the pre-Flood world that is based on observable data and not mere speculation. We have developed a new pre-Flood geography map for about half of the world. This map also helps explain the

> "Our study fills a critical need for knowledge of the pre-Flood world that is based on observable data and not mere speculation."

Grand Canyon, Arizona

Rocks containing banded iron formations are also found in close proximity to rocks containing stromatolites

observable fossil record. Many previous Flood models could not explain the patterns of deposition in the rock record and the differentiation of fossils observed within the strata. The proposed ecological zonation megasequence depositional model is an important step in that direction. It may help explain why human fossils are not found with dinosaur fossils, and why dinosaurs are not found in the earliest Flood rocks (Sauk to Kaskaskia Megasequences), such as at Grand Canyon. It helps explain the major subdivisions of the fossil record in terms of their respective megasequences and their boundaries. It is data-driven because it is based on a massive set of newly compiled stratigraphic columns from across three continents.[29]

The location of the Garden of Eden will likely never be found, but these results allow the re-creation of the major topographic highs and lows of the pre-Flood world, including past continental reconstructions. Stromatolite fossils further confirm the accuracy of the Bible and its references to the pre-Flood hydrology.

References

1. Cosner, L. and R. Carter. 2016. Where was Eden? Part 1: examining pre-Flood geographical details in the biblical record. *Journal of Creation*. 30 (3): 97-103.
2. Carter, R. and L. Cosner. 2016. Where was Eden? Part 2: geological considerations-examining pre-Flood geographical details in the biblical record. *Journal of Creation*. 30 (3): 123-127.
3. Moshier, S. and C. Hill. 2016. What Is Flood Geology? In *The Grand Canyon, Monument to an Ancient Earth: Can Noah's Flood Explain the Grand Canyon?* C. Hill et al, eds. Grand Rapids, MI: Kregel Publications, 21-30.
4. Hughes, J. R. 1997. An examination of 'Eden's geography erodes flood geology.' *Creation Research Society Quarterly*. 34 (3): 154-161.
5. Munday, J. C. 1996. Eden's geography erodes flood geology. *Westminster Theological Journal*. 58: 123-154.
6. Sloss, L. L. 1963. Sequences in the Cratonic Interior of North America. *Geological Society of America Bulletin*. 74 (2): 93-114.
7. Vail, P. R. and R. M. Mitchum Jr. 1979. Global Cycles of Relative Changes of Sea Level from Seismic Stratigraphy. *American Association of Petroleum Geologists Memoir*. 29: 469-472.

8. Clarey, T. 2015. Dinosaur Fossils in Late-Flood Rocks. *Acts & Facts.* 44 (2): 16.
9. Clarey, T. L. and D. J. Werner. 2017. The Sedimentary Record Demonstrates Minimal Flooding of the Continents During Sauk Deposition. *Answers Research Journal.* 10: 271-283.
10. Snelling, A. A. 2014. Geological Issues: Charting a scheme for correlating the rock layers with the Biblical record. In *Grappling with the Chronology of the Genesis Flood.* S. W. Boyd and A. A. Snelling, eds. Green Forest, AR: Master Books, 77-109.
11. Clarey, T. L. 2016. Empirical data support seafloor spreading and catastrophic plate tectonics. *Journal of Creation.* 30 (1): 76-82.
12. Schmandt, B. and F.-C. Lin. 2014. *P* and *S* wave tomography of the mantle beneath the United States. *Geophysical Research Letters.* 41 (18): 6342-6349.
13. Baumgardner, J. R. 2016. Numerical Modeling of the Large-Scale Erosion, Sediment Transport, and Deposition Processes of the Genesis Flood. *Answers Research Journal.* 9: 1-24.
14. Pappas, S. 280-Million-Year-Old Forest Discovered in…Antarctica. *LiveScience.* Posted on livescience.com November 15, 2017, accessed November 17, 2017.
15. Clarey, T. 2015. *Dinosaurs: Marvels of God's Design.* Green Forest, AR: Master Books.
16. Brown, W. 2008. *In the Beginning: Compelling Evidence for Creation and the Flood,* 8th ed. Center for Scientific Creation. Posted on creationscience.com, updated December 2, 2017, accessed November 17, 2017.
17. Neuendorf, K. K. E., J. P. Mehl Jr., and J. A. Jackson, eds. 2005. *Glossary of Geology,* 5th ed. Alexandria, VA: American Geological Institute, 636.
18. Proemse, B. C. et al. 2017. Stromatolites on the rise in peat-bound karstic wetlands. *Scientific Reports.* 7: 15384.
19. Wicander, R. and J. S. Monroe. 2016. *Historical Geology: Evolution of Earth and Life Through Time,* 8th ed. Boston, MA: Cengage Learning.
20. Lowe, D. R. 1994. Abiological origin of described stromatolites older than 3.2 Ga. *Geology.* 22 (5): 387-390.
21. Hladil, J. 2005. The formation of stromatactis-type fenestral structures during the sedimentation of experimental slurries—a possible clue to a 120-year-old puzzle about stromatactis. *Bulletin of Geosciences.* 80 (3): 193-211.
22. Mueller, P. A. and A. P. Nutman. 2017. The Archean-Hadean Earth: Modern paradigms and ancient processes. In *The Web of Geological Sciences: Advances, Impacts, and Interactions II.* M. E. Bickford, ed. Boulder, CO: Geological Society of America Special Paper 523, 75-237.
23. Tomkins, J. P. and T. Clarey. Cellular Evolution Debunked by Evolutionists. *Creation Science Update.* Posted on ICR.org September 29, 2016, accessed March 10, 2018.
24. Awramik, S. M. and K. Grey. 2005. Stromatolites: biogenecity, biosignatures, and bioconfusion. *Proceedings of the SPIE.* 5906: 1-9.
25. Frazer, J. Stromatolites Defy Odds by A) Living B) on Land. *Scientific American.* Posted on blogs.scientificamerican.com December 21, 2017, accessed January 4, 2018.
26. Clarey, T. 2018. Assembling the Pre-Flood World. *Acts & Facts.* 47 (4): 11-13.
27. Purdom, G. and A. A. Snelling. 2013. Survey of Microbial Composition and Mechanisms of Living Stromatolites of the Bahamas and Australia: Developing Criteria to Determine the Biogenicity of Fossil Stromatolites. In *Proceedings of the Seventh International Conference on Creationism.* M. Horstemeyer, ed. Pittsburgh, PA: Creation Science Fellowship.
28. Clark, H. W. 1968. *Fossils, Flood, and Fire.* Escondido, CA: Outdoor Pictures.
29. Clarey, T. L. and D. J. Werner. 2018. Use of sedimentary megasequences to re-create pre-Flood geography. In *Proceedings of the Eighth International Conference on Creationism.* J. H. Whitmore, ed. Pittsburgh, PA: Creation Science Fellowship, 351-372.

18 Flood-Provided Energy Resources: Oil and Coal

Summary: Many people assume that oil, natural gas, and coal—also called fossil fuels—take millions of years to form, but new research shows that's not the case. A global flood could have easily formed most of the fossil fuels we find today. Oil and natural gas originate with the deposition of organic debris like marine algae and plankton. These deposits need to be buried deep enough to heat up and be transformed into oil and gas. A problem with secular formation theories is that organic debris is often consumed by scavengers—the same reason fossils are difficult to form. Bacteria consume organic material even after it is buried. Oil and gas cannot be millions of years old, otherwise the remains would have been devoured long ago. Coal was formed from plant material torn loose by the floodwaters and buried rapidly. There is no evidence that a floating forest biome ever existed. The rapid burial of organic material by a catastrophic flood easily explains the large amounts of fossil fuels around the world. God had a plan to provide energy resources for our generation even while judging the ancient earth for its wickedness.

Oil Is from the Flood

Genesis 11:3 is the first mention of oil, or more likely an oil seep, in the Bible, and it was only about 200 years after the Flood.[1] We will see below that the so-called "pitch" used to cover or seal the Ark was not a hydrocarbon, as many seem to believe. Instead, oil, natural gas, and coal are all products of the Flood. They are often referred to as fossil fuels, but most are probably younger than people think.

During the chaotic judgment of the global Flood, God was thinking ahead for future generations of humans. He was providing the energy resources for our modern world. The demand for oil really developed along with the automobile industry begin-

> Then they said to one another, "Come, let us make bricks and bake them thoroughly." They had brick for stone, and they had asphalt for mortar. (Genesis 11:3)

ning in the early 20th century. It was an almost symbiotic relationship. Cars needed gasoline, and oil needed a market.

Oil resources are in the news nearly every day, with discussions on both the pros and cons of pipelines, oil "fracking," and environmental regulations. Approximately 10% of the world's recoverable oil reserves are in shale-rich rocks that can only be accessed by hydraulic fracturing (i.e., fracking).[2] A recent study estimates there are about 345 billion barrels of recoverable shale oil globally. These same shale-rich rocks also account for up to 32% of the world's natural gas reserves.[2] The amount of gas recoverable from shale is estimated at around 7,300 trillion cubic feet in volume.

When we stop to consider the origins of the vast reserves of oil and gas, it's apparent these fuel resources are not as old as many secular scientists believe. But in order to understand the age of oil, it's important to start at its source.

Geologists have done many studies over the years testing the oil produced around the world for its chemical components. They have found that most oil and gas is derived from shale-rich source rocks—rocks abundant in organic debris trapped during deposition. The chemical signatures of both oil and gas often match, much like fingerprints. Shale is the most common sedimentary rock and can serve both as a "seal" and a source rock for oil. Liquids and gases can only pass through shale layers very slowly due to the low permeability of these clay-rich rocks, which tightly seal the oil that seeps into and becomes trapped within them. Hydraulic fracturing creates conduits that allow oil and gas to leak out of these tight formations.

Where do oil and natural gas originate? They start with the deposition of organic debris. Many oil shales commonly contain upward of 5% total organic carbon. Most organic compounds found in oils seem to match up with marine algal deposits (Type 1 oils) and marine planktonic deposits (Type 2 oils). Both types of deposits produce oil and/or natural gas as the rocks are heated by the earth's natural thermal gradient. These deposits (rocks) just have to be buried deep enough to "cook" and thus generate the oil and gas. Researchers assume that the rocks must be buried between 8,000 and 15,000 feet deep and reach temperatures of 180° to 250°F in order to generate oil from organic material. This temperature range is commonly called the *oil window*, and local variations in geothermal gradient can shift this window up and down considerably. For example, areas near volcanic activity

Laminated black shale

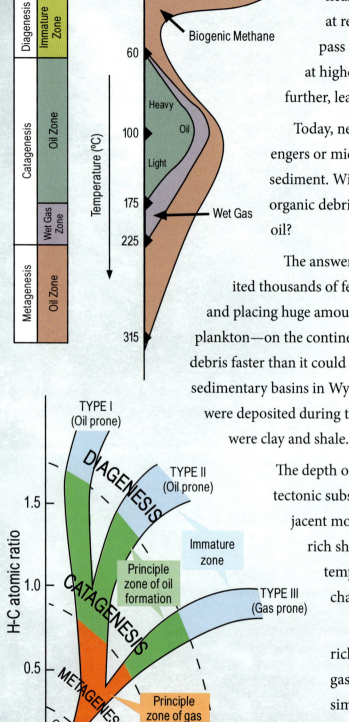

Diagrams illustrating the oil window or oil zone (top) and the three types of kerogen and their oils (bottom)

generally have higher temperature gradients, so nearby sediments may pass through the oil window at relatively shallow depths. If the organic-rich rocks pass through the oil window and continue to cook at higher temperatures, the liquid oil will break down further, leaving only natural gas deposits.[3]

Today, nearly all organic debris is consumed by scavengers or microorganisms before it becomes trapped in sediment. With this in mind, how is it possible that enough organic debris was ever trapped to produce all of the world's oil?

The answer is the great Flood, an event that rapidly deposited thousands of feet of sediments across the continents, burying and placing huge amounts of marine sediments—containing algae and plankton—on the continental crust. This process trapped the organic debris faster than it could naturally decay. In many cases, like in the deep sedimentary basins in Wyoming, up to 30,000 or 40,000 feet of sediments were deposited during the year-long Flood—and most of these deposits were clay and shale.

The depth of burial must have been enhanced by rapid tectonic subsidence, as well as simultaneous formation of adjacent mountain uplifts. These shifts placed the organic-rich shales either in the oil window or in the higher temperature range that "cooked out" the oil and changed it to natural gas, as mentioned earlier.

Unfortunately for the oil seekers, the organic-rich shales that appear to produce most oil and gas, chemically referred to as hydrocarbons (i.e., simple organic compounds), are mixed and dispersed throughout the geologic strata. This creates complexity—strata with a "club sandwich" appearance—when searching for oil deposits. Anticline

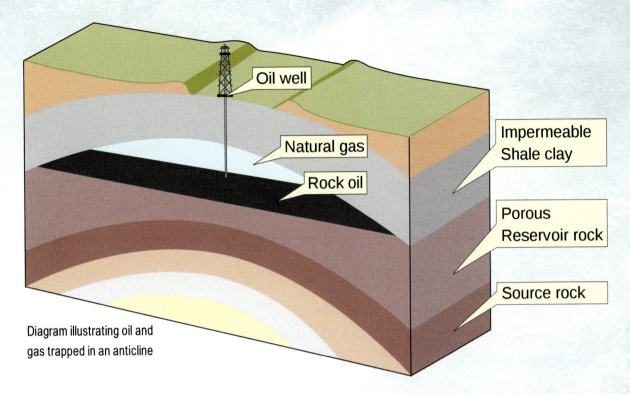

Diagram illustrating oil and gas trapped in an anticline

traps or domes need oil reservoirs (i.e., porous geologic formations where oil and gas collect and are held) to be located in the right positions above and adjacent to the organic shales, and the shales have to be in the oil window to effectively source the traps. Then, these traps need an effective seal (i.e., salt beds or low permeable shales) to hold in the oil. Unfortunately, everything leaks, and fractures allow some of the trapped oil to slip away to the surface, becoming oil seeps.

The Middle East is a prime example of an area that has greatly benefitted because it had just the right conditions to both generate and trap huge oil reserves. For instance, along with abundant, organic-rich shales that were deposited during the great Flood, this area also has sedimentary rocks with large folds that formed toward the Flood's end and trapped the oil as it was later generated. The result? Billions of barrels of oil were trapped in this region.

Geologists are constantly searching for oil traps (anticlines usually) even in the shales themselves, using increasingly more capable technology. Through wells and using seismic data, they "see" inside the earth as they search for potential new traps around the world. They also employ horizontal drilling techniques to tap into the source-rock shales, enhancing production with sophisticated, hydraulic rock fracturing. These new technologies have greatly benefitted the economies in Texas, Ohio, and

North Dakota, where shale oil is quite plentiful and accessible.

Critics of recent creation and the global Flood often try to argue that the sheer volume of oil found cannot be explained by a single ocean full of organic debris deposited in one year-long event. However, the volume of organic material in the ocean at any given time is immense.[4] By studying the organic richness of the present ocean, creation scientists have shown that all of the oil found—and yet to be found—could easily be deposited and explained by a single year-long global Flood.[5]

Nevertheless, many geologists never think through this entire process. They simply focus on searching for traps or rocks folded into domes where they know the oil is likely to be concentrated, without regard for the unusual and specific processes required to preserve the vast amounts of organic debris in the rocks in the first place.

Often, secular geologists insist that most organic oils were generated millions of years ago—even 150 million years ago—and have been preserved and trapped under great pressures ever since. This is a false assumption that is almost never thought through logically. If Earth were truly that old, the oils would have been destroyed

Trans Alaska oil pipeline

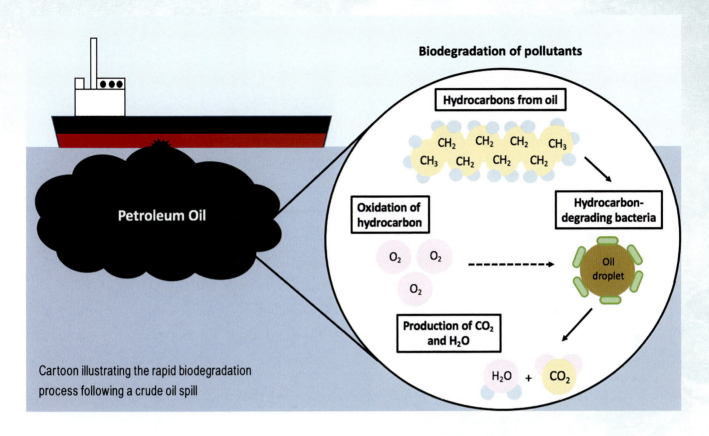

Cartoon illustrating the rapid biodegradation process following a crude oil spill

by bacterial action, and the geologic pressures would have long since dissipated. We know that oil at the surface is quickly consumed by bacterial action—literally eaten—like in the 2010 Deepwater Horizon oil spill in the Gulf of Mexico.[6] We also know that bacteria live in virtually every environment on Earth, even at great depths in the ground. So, it is reasonable to assume any oil beyond a few thousand years old would be totally degraded or consumed by bacteria by now. The oil simply can't be millions of years old—organic compounds cannot last millions of years in any natural environment. Thus, there shouldn't be any ancient oil anywhere!

In spite of false assumptions about ancient oil, secular geologists continue to drill, without asking or answering real questions, because the black gold is there waiting for them. Each year, millions of barrels continue to be found in conventional traps and in unconventional "tight" shales through fracking. This abundance and concentration of oil in our world clearly point to the hand of God and the recent Flood.

Global oil generation is another example of a process that could only have occurred because of the extraordinary burial conditions present during the recent great Flood. Most secular petroleum geologists deny the Flood, even though they are witness to this evidence every day as they search for oil. We can be thankful for God's providence in creating oil, even through a catastrophic, global judgment—oil that now provides much-needed energy for our present world.

Rapidly Forming Oil Supports Flood Time Frame

The timing of oil and gas generation and their migration into reservoirs are issues that are based mostly on assumptions and not empirical data. Unfortunately, the scientific information communicated to the public is slanted by pro-evolutionary rhetoric. The occurrence of oil is even used as an argument against a recent global flood. Evolutionary geologist David Montgomery insists all sedimentary rocks could not have formed during the year-long Flood, arguing that "a literal reading of the Bible requires that such rocks already existed at the time of the Flood because bitumen, the pitch or tar Noah used to caulk the Ark (Genesis 6:14), comes from sedimentary rock."[7]

However, as Dr. Henry M. Morris III pointed out, the Hebrew word used in this verse, *kopher*, doesn't literally translate as "pitch." He states, "The word is used 17 times in the Old Testament, and is translated 'pitch' only in Genesis 6:14. Most of

Oil rigs in the Atlantic

the time, *kopher* is translated with some term that represents money."⁸ It seems that *kopher* was some sort of expensive (hence the possible reference to money) sheathing or covering that was placed over the wood of the Ark. Dr. Morris added that "the *kopher* that sheathed or coated the Ark is not specified....The idea that *kopher* was liquid is merely assumed.... Even if the material was a liquid coating, the development of resins or other non-petroleum coating materials has long been known to man."⁸

Once the floodwaters drained off the continents, deeply buried marine algal and planktonic deposits that were disseminated in the sediments (source rocks) began to heat up, reaching the geothermal gradients we observe today. How quickly did this heating occur, and how rapidly was oil generated?

Let's first look at the biblical record. In Genesis 11:3 in the narration about the building of the Tower of Babel, God says, "They had brick for stone, and they had asphalt for mortar." The Hebrew word for asphalt is *chemar*, which is sometimes translated as bitumen, cement, or slime. So here, unlike the use of the Hebrew word *kopher*, the Bible is describing a tar or bitumen product, essentially a hydrocarbon.

Although the Bible doesn't give the specific number of years between the Flood and the Tower of Babel, we do have some time constraints. In Genesis 10:25, we read that the earth was "divided" in the days of Peleg. Assuming the word "divided" meant the division of the languages at Babel, Dr. Morris wrote, "Since he [Peleg] was born 101 (+4) years after the Flood and lived 239 years (Genesis 11:18-19), that gives a range of around 100 to around 340 years after the Flood during which the division could have taken place."⁹ This gives us a relatively narrow time window of under 400 years for oil to have generated from the Flood sediments.

Is this too short a time frame for oil to form? Not at all. It's been known for decades that crude oil porphyrin (one of the common chemicals in crude oil) can be generated in a laboratory setting in as few as 12 hours.[10,11] And in 2013, engineers at the U.S. Department of Energy's Pacific Northwest National Laboratory reported they were able to transform harvested algae into crude oil in less than one hour![12] There is no reason to think this process could not have occurred naturally in as few as 200 years after the Flood.

What about oil migrating to the earth's surface? In all likelihood, oil bubbled out of seeps at the surface near the Tower of Babel in quantities generous enough to be utilized as mortar. Moreover, Genesis 14:10 references other oil seeps during the time of Abraham in an area near the cities of Sodom and Gomorrah where "the Valley of Siddim was full of asphalt pits." Based on biblical genealogies (Genesis 11:10-28), these seeps developed in less than 500 years after the Flood.[1]

> "In 2013, engineers at the U.S. Department of Energy's Pacific Northwest National Laboratory reported they were able to transform harvested algae into crude oil in less than one hour!"

Unfortunately, little is known about oil migration from source rock to reservoir. A 2013 *AAPG Bulletin* article began by stating, "Hydrocarbon migration is by far one of the most important and yet least understood topics in petroleum geology."[13] Oil migrates as a fluid through small openings (pore spaces) in the rock layers much like water, and its flow rate is governed by the same fluid dynamics as water. Groundwater moves, on average, about 50 feet per year, but oil is a larger molecule than water and therefore struggles to pass though small openings. Although the migration of oil is relatively slow, biblical history shows oil made it to the surface within just a few centuries after deposition of the source rocks.

Oil quickly degrades from bacterial action since it is an organic compound, unable to survive for millions of years.[3] Biodegraded oils are common in reservoirs around the world, including the North Sea, the Gulf of Mexico, offshore Nigeria, and the tar sands in Alberta.[14] Other shallow reservoirs seem to be less affected by biodegradation. Although secular scientists admit these non-biodegraded oils may be the result of recent recharge, they consider this process unlikely because they insist many of these oils are millions of years old.[14]

Uniformitarian geologists attempt to explain "ancient oil" in reservoirs by invoking an unusual process known as *paleosterilization* to prevent oil from biodegrading. They hypothesize that bacterial action in oil reservoirs ceases at temperatures above 176°F, thereby preventing bacterial action in the rocks containing the oil. If reservoir rocks exceed this threshold temperature, they argue, bacterial action not only ceases but remains inactive for millions of years.[14]

However, uniformitarian scientists forget that bacteria thrive in even the most extreme conditions, such as the geothermal waters at Yellowstone National Park and hydrothermal vents in the oceans where thermophilic bacteria flourish at temperatures of 113° to 252°F. And even if the rocks were "sterilized," groundwater would quickly transport an influx of new bacteria to replenish the "dead" zone. Therefore, any non-biodegraded oil reservoirs in the world today must be recently generated and freshly recharged.[15]

The assertion that oil can survive for millions of years due to pasteurization took another hit with the discovery of vast volumes of microorganisms deep in the earth.[16] The results of a 10-year study involving 1,200 scientists from 52 countries found vast quantities of life deep in the subsurface. Scientists collected and examined numerous

Porcelain Basin Trail, Yellowstone National Park

samples from multiple wells drilled as deep as 3.1 miles (16,400 feet). They estimated that this microscopic ecosystem takes up a volume greater than double the world's oceans and weighs hundreds of times more than the combined weight of every person on Earth. They concluded that about 70% of the earth's microbiota (bacteria and archaea) lives in the subsurface at temperatures of up to 250°F.[16] *The Guardian* reported:

> "It's like finding a whole new reservoir of life on Earth," said Karen Lloyd, an associate professor at the University of Tennessee in Knoxville. "We are discovering new types of life all the time. So much of life is within the earth rather than on top of it."...
>
> The scientists have been trying to find a lower limit beyond which life cannot exist, but the deeper they dig the more life they find. There is a temperature maximum—currently 122C [251.6F]—but the researchers believe this record will be broken if they keep exploring and developing more sophisticated instruments.[16]

These findings demonstrate that life can and does thrive at extreme temperatures in the earth. Microorganisms like bacteria do not disappear deep in the earth or at high temperatures. In fact, they exist everywhere we have looked or drilled regardless of depth or temperature. There is no such thing as pasteurized oil and sterilized rocks.

The notion of old oil is a fantasy built on nothing but assumptions. And these assumptions are being shown to be false. Oil comes from marine algae buried rapidly in the global Flood that occurred just thousands of years ago.[3,15] It is consumed by bacteria. Like soft tissue in dinosaur bones, it cannot last for millions of years. And bacteria are everywhere.

> "The notion of old oil is a fantasy built on nothing but assumptions."

How long would it take to fill the numerous reservoirs that hold vast quantities of oil today? Much depends on the size of the trap that holds the oil, the amount of organic material in the source rocks, and the development of pathways (pores, fractures, and faults) to the reservoir beneath the trap. One of the few studies that tried to quantify this process was conducted in the Gippsland Basin, Australia.[17] Dr. Andrew Snelling summarized the research results, explaining, "It has been concluded that petroleum generation must still be occurring at the present time, with the products migrating relatively rapidly either into traps or even to the surface."[18] It is therefore likely that many other areas are still generating oil, and it is actively migrating to traps even today. This presents the possibility that some depleted oilfields may partially refill over the next century. Recent generation also explains the non-biodegraded oils that are found across the globe. Thus, the processes of oil generation, migration, and entrapment easily fit within the time that has elapsed since the Flood less than 4,500 years ago, even at the slow percolation rates in the subsurface.

Finally, if the shale source rocks have been in the "oil window" for millions of years, all of the oil should have either already been generated, migrated away, and/or biodegraded away millions of years ago. To put it bluntly, there shouldn't be any oil left to find when we frack. And yet, fracking in the last few decades has discovered billions of barrels of oil trapped in organic-rich shale rocks, demonstrating again that oil cannot be millions of years old.

Coal: Plant Material Buried in the Flood

For decades, it has been taught that dead plants accumulate in swamps, where they decay and eventually turn into coal over millions of years. But there are problems with this story about the unseen past. First, nowhere on Earth today does peat become coal. Second, while peats do accumulate in stagnant swamp waters, these appear nothing like coal. Swamp peats vary in elevation and are cut by numerous streams. Coals, however, usually have extremely flat surfaces above and below. Surely something other than a peat swamp was involved in the formation of coal seams.

85-foot-thick coal seam in Paleogene rocks near Gillette, Wyoming

S. E. Nevins (a.k.a. Dr. Steve Austin) explained the theories on the origin of coal:

Two theories have been proposed to explain the formation of coal. The popular theory held by many uniformitarian geologists is that the plants which compose the coal were accumulated in large freshwater swamps or peat bogs during many thousands of years. This first theory which supposes growth-in-place of vegetable material is called the *autochthonous theory*.

The second theory suggests that coal strata accumulated from plants which had been rapidly transported and deposited under flood conditions. This second theory which claims transportation of vegetable debris is called the *allochthonous theory*.[19]

Coal seams have flat tops and bottoms without visible tree roots, indicating the plant material was transported (allochthonous) and buried rapidly

These two theories diametrically oppose one another. Coal is either grown in place (autochthonous) as in a swamp, slowly adding more plant material over time, or it is from plant material that was transported rapidly by water (allochthonous). The only way to decide which of these two theories is correct is to examine the rocks themselves and see which theory better explains what we observe. In other words, what do the rocks show?

When we examine coal beds, they are pure coal with very few partings of shale or sandstone intermixed. The coal beds are often found with marine limestones either directly above or directly below. Many of these mixed coal beds and marine rocks compose the Pennsylvanian system cyclothems discussed in chapter 12 (see page 275).

Another amazing quality of nearly every coal bed is the lack of roots in the layers below. If these are from swamps, then we should expect to see roots at the base of the

Floating log mat on Spirit Lake at Mount St. Helens

coal bed penetrating into the substrate beneath it. And yet, we do not see any signs of roots at all! The coal beds are flat and planar on both top and bottom, and completely parallel to the sedimentary layers above and below. Occassionally, there are polystrate trees sticking through the seams, but even these show no true roots. If anything, they are root balls that resemble the allochthonous trees torn free and transported from the eruption at Mount St. Helens. The empirical data support the theory of allochthonous coal, or transported plant material, that became buried *en masse* quickly by another layer of sediment. These are the features and the observations you would expect to find associated with a global flood.

Dr. John Morris explained how the eruption of Mount St. Helens in 1980 may be a window into the process of allochthonous coal formation from Flood processes. He stated:

> When Mount St. Helens erupted in 1980, phenomenal processes were set in motion that instantly produced geologic results mimicking those we are taught to think required millions of years. A highly energetic blast of superheated steam was released that traveled at great speeds and devastated the surrounding forest. A ring where the trees were removed was surrounded by the "blow-down zone," with a scorched zone surrounding that. After the eruption, a charred log was found with wood on one end and material on the other that under microscopic analysis proved to be a rather high grade of coal, formed essentially instantly.
>
> The blast uprooted millions of trees and washed them into nearby Spirit Lake.…If another volcanic event were to blanket the layer with hot ash, burying the tree mat and floating plant material,

Strip-mining Cenozoic coal in the Powder River Basin, Wyoming it might rapidly heat and turn into coal. All necessary conditions would have been met.

The eruption of Mount St. Helens provided insight into processes operating during an even greater catastrophe, the great Flood of Noah's day. Observations of the eruption's aftermath have expanded our ability to understand the Flood.[20]

In addition, uniformitarian geologists have no model to explain the 100- to 200-foot-thick coal seams that are 60 miles by 60 miles in the Powder River Basin of Wyoming. What swamp today produces coal that is 100 to 200 feet thick? One of my professors at the University of Wyoming took us on a field trip to see these huge cuts of coal for ourselves near Gillette, Wyoming. As we looked at the 100-foot-thick coal bed, he reluctantly admitted that there is no swamp model to explain these thick beds. And, as well drilling has shown, these coal beds are not isolated. Most of these coal seams are found stacked on top of each other, sometimes with as many as four to eight coal beds in one location alone.

The coal-forming process doesn't require much time. In recent years, several laboratory experiments have shown that coal can form quickly, in just hours or days. Extreme conditions can accomplish it even more quickly. Heat is required, but not necessarily pressure.

Sinking the Floating Forest Hypothesis

I am all for allochthonous coal. I believe the data support the theory that plant material was torn free in the Flood, transported *en masse* by water and tsunami waves, and deposited and rapidly buried in the Flood. However, I do not think coal formed from continental-scale, pre-Flood, living floating forest biomes, as some creationists have suggested.

The concept of a pre-Flood floating forest ecosystem has been promoted in creationist literature for several decades and is often used as an explanation for the massive carboniferous coal beds found across the globe. However, this hypothesis wasn't

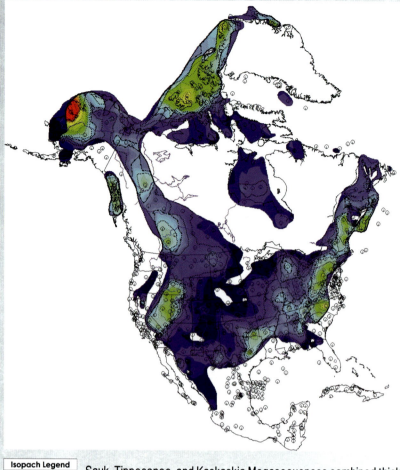

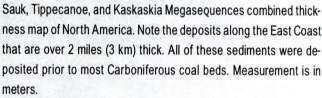

Sauk, Tippecanoe, and Kaskaskia Megasequences combined thickness map of North America. Note the deposits along the East Coast that are over 2 miles (3 km) thick. All of these sediments were deposited prior to most Carboniferous coal beds. Measurement is in meters.

adequately tested until three recent geological challenges were presented.[21] And it appears the floating forest hypothesis is failing that test.

First, floating forests are found to be incapable of maintaining a sizable freshwater lens to supply the plant life, pools, and springs as suggested.[21] Second, tsunami-like waves associated with plate movements would have likely broken up the floating forests earlier in the Flood than has been suggested, depositing coal beds throughout much of the stratigraphic column. This would contradict what is found in the rock record because little coal is found until the Upper Carboniferous (Pennsylvanian system).[21] Third, upward of 2 miles (3 kilometers) of virtually coal-free sediment (and in some places over 4 miles, or 6 kilometers) was deposited along the entire U.S. Eastern Seaboard prior to deposition of the Upper Carboniferous coal beds.[21] How did the forests remain intact atop the sea while all this deposition was occurring beneath? It seems highly unlikely that they floated as an essentially intact log mat for three megasequence cycles and were then suddenly buried completely in the fourth (Absaroka) megasequence. Finally, relatively few coal beds are found as a result of the closure of the pre-Atlantic Iapetus Ocean early in the Flood. It is not until after this pre-Flood ocean was completely consumed, pushing land against land, that we find extensive coal beds deposited on the adjacent continents.[21] If the ocean water disappeared, where did the floating forest go?

When examined against available geological data, the floating forest hypothesis is found to lack explanatory ability. Instead, a return to studies of pre-Flood paleogeography and plant zonation to explain the coal beds is suggested.

What was the floating forest? Scientists have shown that the dominant plant species of the carboniferous coal deposits were arborescent lycopods (scaly-barked trees) that could achieve heights of over 100 feet (34 meters). Advocates of the floating forest believe these now-extinct trees with their spiraling root systems somehow formed floating mats, growing more densely than do trees in modern forests. But most importantly, they believe the trees had hollow trunks and roots that provided sufficient buoyancy to enable a vast floating tree-and-plant biome to cover much of the pre-Flood oceans.[22]

In Situ or Not In Situ?

Many upright fossil trees found associated with coal seams are interpreted by secular science as being in the "growth position," commonly referred to as *in situ* trees. Secular paleontologists use this claim as evidence against the global Genesis Flood, even arguing that fossil *in situ* trees demonstrate an autochthonous (in original position) origin for coal. Creation scientists have countered with evidence supporting the allochthonous (moved from source) origin of coal, showing that many claimed *in situ* trees are better explained by active transport of trees and other vegetation

Several of the thousands of tree stumps left behind by the Mount St. Helens' eruptions. The tree trunks were transported, and many ended up in Spirit Lake.

during the global Flood after they were stripped free from the land.

Further empirical support for the allochthonous origin for upright fossil trees came soon after the 1980 eruption of Mount St. Helens. Dr. Steve Austin estimated that more than 19,000 upright and randomly spaced trees accumulated in the sediment on the bottom of Spirit Lake within just a few years. These trees became waterlogged and sank upright because of their heavier bases. He also postulated that if these trees were buried by additional sediment, in time they would give the appearance of an *in situ* forest.[23]

Criteria for In Situ Trees

In a 2016 paper published in the *Creation Research Society Quarterly*, we identified seven criteria to determine if fossil trees were transported or merely buried by Flood sediments *in situ*.[24] The identification of an *in situ* site wouldn't necessarily invalidate the allochthonous origin of coal beds; it would merely represent a location where the tops of the trees were sheared off, leaving the trunks and stumps buried in place (see page 437). Fossil trees that fulfill all, or at least most, of these criteria likely represent true *in situ* assemblages. The criteria are:

- Multiple, single-species trees spaced in the growth position in the same horizontal plane as you would find in a living forest and not merely randomly spaced
- Multiple trees in the same rock layer or along a common surface
- Trees with root systems that crosscut bedding layers
- Evidence of rapid burial by thick sediment and water
- A lack of thick sedimentary rock layers underneath the trees
- No bowing or distortion of any sedimentary layers beneath the tree stumps
- Accompanying vegetation that also crosscut the same layers as the lycopod tree stumps

Fossil Grove Site, Glasgow, Scotland

We identified one particular site in Glasgow, Scotland, that meets nearly all the criteria, including the lack of Flood sediments beneath the tree-root layer.[24] This site appears to be the remnants of a pre-Flood forest, with the fossil trees still rooted in

Figure 18.1. Fossil lycopod tree stumps buried in place (*in situ*) at Fossil Grove, Glasgow, Scotland

a pre-Flood soil layer now lithified to rock. Fossil Grove, as it is called, is located in Victoria Park (Figure 18.1). It's likely the best-preserved example of an *in situ* lycopod forest in the world[26] and possibly the first identified in a Flood context.

Fossil Grove was discovered in 1887 when a path was cut across an abandoned quarry outside Glasgow.[27] After excavation down to the common soil horizon containing the tree stumps and roots, a building was constructed to protect them and allow public viewing. The site consists of a monotypic assemblage of multiple lycopod tree stump casts with attached axial root systems.[26]

The 11 single-species stumps were found in growth position spacing as opposed

root systems penetrate downward into the soil horizon similar to modern root systems. As opposed to allochthonous deposited trees, the roots are not broken off near the trunks but instead are intact, like those of living trees. The encasing sandstone layer on top of the forest site contains ripples and oriented, broken trunk fragments indicative of a high-energy flow system directed toward the southwest. The tree stumps are also consistently distorted in a southwesterly direction, matching the paleo-flow of the floodwater currents. This indicates all of the trees were likely in place prior to burial by the encasing sandy sediments of the global Flood.

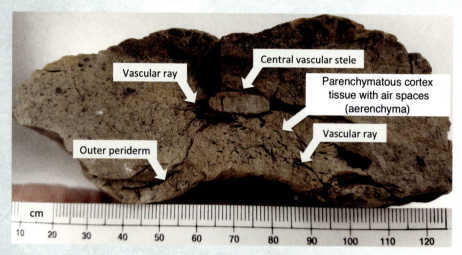

Cross-section of a lycopod tree root showing internal structures, indicating lycopods were not completely hollow

It is significant that the roots are not distorted in a southwesterly direction like the stumps. If the tree stumps, roots and all, were transported and deposited, there should be a consistent southwest distortion to both trunks and roots. The lack of directional distortion in the roots suggests that the trees were rooted in the forest soil prior to burial.[26]

Lycopod Trees Were Not Hollow

Another line of reasoning put forth in support of the floating forest hypothesis is that the arborescent lycopods were hollow in both their main aerial trunks and in their roots—a contention based primarily on speculation and not soundly supported by the scientific literature. The majority of the "hollow tree" studies do not take into account a number of key reports describing the non-hollow internal structure of lycopods. Research has demonstrated that intact, non-decayed aerial stems of arborescent lycopods clearly indicate a contiguous tissue structure across the breadth of the stem, with the same general schema found in trunks and roots.

In fact, it is now apparent that the initial stages of the global Flood would likely have caused a great deal of plant death followed by decomposition of easily destroyed tissue in the internal cortex region of lycopod trunks and roots. The aerial structures and root systems would have undergone selective tissue decay in the central cortex

Fossil tree

while retaining overall morphological shape during the hollowing process. At that time, sediments were introduced into the cavity, creating casts. In effect, it would have resulted in the hollow-looking tree fossils that are commonly observed.

Flood Model for Fossil Grove Site

The sedimentation data indicate that Fossil Grove is a preserved remnant of a pre-Flood forest that was not inundated and buried until approximately midway through the rising portion of the Flood.[24] Allochthonous layers of coal were later deposited on top of the trees as part of the Scottish Coal Measure Group. This data-driven interpretation supports the idea that as the floodwaters increased, tsunami-like waves tore the bulk of the lycopod forests free and deposited them allochthonously elsewhere as coal beds.[21] As is the case with Fossil Grove, the lycopod tree trunks were broken off, allowing substantial decay of the stumps to occur prior to burial.[26]

Sinking the Floating Forest

Fossil Grove would be the first documented *in situ* preservation of a pre-Flood soil with plants. However, it doesn't support the floating forest hypothesis since the tree roots of the 11 stumps are found embedded with intact root systems throughout a common horizon. There is strong evidence to demonstrate these stumps are in the growth position and were inundated, buried, and fossilized *in situ* by rising floodwaters.

All available geological and fossilized anatomical data support the existence of pre-Flood lycopod forests rooted in soil. These forests were likely located in wetlands and/

or coastal lowland areas along the fringes of landmasses such as the Dinosaur Peninsula.[21] Detailed analysis further demonstrates the trunks and the roots were not hollow as previously claimed. Based on these studies, we recommend abandoning the floating forest model.[21,24]

Massive Lower Cenozoic Coal Beds Are from the Receding Flood

Most of the coal beds in the Great Plains States of the U.S. are found within Cretaceous and/or Lower Cenozoic strata and contain virtually zero lycopod tree remnants.[28] The plants that make up the Cenozoic coals are commonly trees like the *Metasequoia* redwood. In contrast, the coal beds in the eastern U.S., which are composed primarily of lycopod trees, are found almost exclusively within Carboniferous rock layers. These include the Pennsylvanian system (Absaroka Megasequence) coals in Illinois, Michigan, and the Appalachian region. The Pennsylvanian coal beds in the eastern U.S. are usually 10 feet or less in thickness. In contrast, the non-lycopod-rich Cenozoic coal beds in the American West, and especially in the Powder River Basin of Wyoming, are often thicker than 50 feet and even up to 200 feet thick.[29]

Powder River Basin (PRB) coals, which are all within Lower Cenozoic (Tejas Megasequence) rock layers, contain the largest reserves of low-sulfur subbituminous coal in the world.[30] Approximately 42% of the present coal production in the U.S. comes from the Powder River Basin.[30] At least six or more coal beds in the PRB exceed 100 feet in thickness, and some individual beds have been shown to extend for over 80 miles.[30] Some of these coal beds can exceed 200 feet thick in places, such as the Big George coal layer.[31] The United States Geological Survey has estimated that the total in-place coal resources of the PRB are approximately 971 billion metric tons, with just 10 individual beds making up about 80% of that value.[30] The vast majority of the PRB coals are found in Cenozoic rocks such as the Tongue River Member of the Paleocene Fort Union Formation.[30]

Train hauling mined Powder River Lower Cenozoic coal. These Cenozoic coals are the thickest and most extensive coal beds in the world and formed in the receding phase of the global Flood.

It is quite clear that the Lower Cenozoic coal seams in the PRB were not deposited by post-Flood river systems since they are not even remotely sinuous or river-shaped. Using data from nearly 30,000 drill holes, 13 separate coal seams were found to be stacked one on top of the other through the center of the PRB.[30] Several of the seams extended over 60 miles north-south and 60 miles east-west. River systems and/or local landslides could not deposit vegetation (coal) of this extent and thickness over and over, giving the stacked coal seams found in the PRB today.

These massive Cenozoic coal beds are not exclusive to the U.S. The Cenozoic coal beds in South America are also the thickest and the most extensive.[32] Cenozoic coal beds are believed to make up about one-half of all coal in South America, and the volume is estimated to be greater than any other geologic system or combination of systems.[32]

Germany, one of the largest lignite coal producers in Europe, has approximately 65% of its reserves in Cenozoic rocks.[33] And the Rhenish Basin in Germany has coal seams of lignite that are nearly 300 feet thick within Cenozoic (Tejas Megasequence) sediments.[33]

What uniformitarian model or local catastrophe could deposit multiple layers of continuous coal seams exceeding 60 miles in extent and upward of 150 feet thick and lignite beds nearly 300 feet thick? Coal deposits this large are unexplainable within the uniformitarian worldview.[11] Only a global flood event could produce these coal seams.

How did these massive Cenozoic coal seams form? It is likely that the receding water (Tejas Megasequence) of the Flood transported massive floating vegetation mats that were torn loose from the pre-Flood land surfaces. For the Powder River Basin coals, these vegetation mats likely became trapped against rapidly rising uplifts like the Bighorn Mountains and were subsequently buried repeatedly in a succession of waves. A global Flood scenario better explains the extent and repetitive nature of these thick coal seams.[33]

Conclusion: Oil and Coal Formed by Flood Processes

The oil deposits we find today are from marine algae, buried rapidly in shales globally. Both oil and coal can form quickly under the right conditions. And oil cannot be extremely old since it is composed of organic compounds, like dinosaur soft tissue that cannot last for millions of years. Oil must be young, and some of it is likely still actively forming today.

> "Both oil and coal can form quickly under the right conditions. And oil cannot be extremely old since it is composed of organic compounds, like dinosaur soft tissue that cannot last for millions of years."

Coal beds are from allochthonous plant material that was transported by water. There were no pre-Flood floating forests. Instead, coal beds are the result of transported trees and other plant debris that were torn loose from the land by the tsunami waves of the Flood. The empirical data support this explanation. And the Flood provides the primary means to form such deposits. There are no 100- or 200-foot-thick coal beds forming today, and there are no organic-rich shales (black shales) forming today on the scale we find in the rock record. Something is definitely wrong with uniformitarianism when it comes to the origin of the so-called "fossil fuels." And the Flood better explains the cyclothems in the rock record showing marine limestones in contact with coal beds, demonstrating a rapid environmental shift from land to marine and vice versa. These rapid changes from the terrestrial to marine are better explained by the activity of the global Flood transporting marine sediments inland as the water inundated the land.

God provided these energy resources for our generation while destroying the ancient world in judgment for humanity's wickedness. He had a plan. He provided for our needs today even before we knew we would need them. He also provided for our salvation through His Son, Jesus, even though we knew Him not. God loved us so much He gave His only Son to die on a cross for our eternal salvation if we only believe in Him. And He did this even before we were born. That's providence.

In the next chapter I will review the evidence for a young earth and in particular for a recent global flood that happened about 4,500 years ago.

References

1. Morris III, H. M. 2016. *The Book of Beginnings: A Practical Guide to Understanding Genesis*. Dallas, TX: Institute for Creation Research, 357.
2. Dittrick, P. 2013. Focus: Unconventional Oil & Gas: EIA-ARI Issues Update of World Assessment of Shale Oil, Shale Gas. *Oil & Gas Journal*. 111 (7): 46-48.
3. Clarey, T. 2013. Oil, Fracking, and a Recent Global Flood. *Acts & Facts*. 42 (10): 14.
4. Woodwell, G. M. et al. 1978. The Biota and the World Carbon Budget. *Science*. 199 (4325): 141-146.

5. Woodmorappe, J. 1986. The Antediluvian Biosphere and Its Capability of Supplying the Entire Fossil Record. In *Proceedings of the First International Conference on Creationism*, vol. 2. R. E. Walsh et al., eds. Pittsburgh, PA: Creation Science Fellowship, Inc., 205-213; Technical Symposium Sessions and Additional Topics.
6. Foley, J. A. Oil From Deepwater Horizon Spill Broken Down By Hungry Ocean Bacteria, Researcher Says. *Nature World News*. Posted on natureworldnews.com April 8, 2013, accessed August 15, 2013.
7. Montgomery, D. R. 2012. *The Rocks Don't Lie: A Geologist Investigates Noah's Flood*. New York: W. W. Norton & Company, 235.
8. Morris III, *Book of Beginnings*, 255-256.
9. Ibid, 353.
10. DiNello, R. K and C. K. Chang. 1978. Isolation and modification of natural porphyrins. In *The Porphyrins, Volume 1: Structure and Synthesis, Part A*. D. Dolphin, ed. New York: Academic Press, 328.
11. Snelling, A. A. 2009. *Earth's Catastrophic Past: Geology, Creation & the Flood*, vol. 2. Dallas, TX: Institute for Creation Research, 971.
12. Rickey, T. Algae to crude oil: Million-year natural process takes minutes in the lab. Pacific Northwest National Laboratory news release. Posted on pnnl.gov December 17, 2013, accessed January 2, 2014.
13. Pang, H. et al. 2013. Analysis of secondary migration of hydrocarbons in the Ordovician carbonate reservoirs in the Tazhong uplift, Tarim Basin, China. *AAPG Bulletin*. 97 (10): 1765.
14. Wilhelms, A. et al. 2001. Biodegradation of oil in uplifted basins prevented by deep-burial sterilization. *Nature*. 411 (6841): 1034-1037.
15. Clarey, T. 2014. Rapidly Forming Oil Supports Flood Time Frame. *Acts & Facts* 43 (3): 14-15.
16. Watts, J. Scientists identify vast underground ecosystem containing billions of microorganisms. *The Guardian*. Posted on theguardian.com December 10, 2018.
17. Shibaoka, M., J. D. Saxby, and G. H. Taylor. 1978. Hydrocarbon generation in Gippsland Basin, Australia; comparison with Cooper Basin, Australia. *AAPG Bulletin*. 62 (7): 1151-1158.
18. Snelling, *Earth's Catastrophic Past*, 973.
19. Nevins, S. E. 1976. The Origin of Coal. *Acts & Facts*. 5 (11). Emphasis in original.
20. Morris, J. D. 2011. On the Origin of Coal. *Acts & Facts*. 40 (6): 18.
21. Clarey, T. L. 2015. Examining the floating forest hypothesis: a geological perspective. *Journal of Creation*. 29 (3): 50-55.
22. Wise, K. P. 2003. The Pre-Flood Floating Forest: A Study in Paleontological Pattern Recognition. In *Proceedings of the Fifth International Conference on Creationism*. R. L. Ivey, ed. Pittsburgh, PA: Creation Science Fellowship, 371-381.
23. Austin, S. A. 1986. Mt. St. Helens and Catastrophism. *Acts & Facts*. 15 (7).
24. Clarey, T. L. and J. P. Tomkins. 2016. An Investigation into an In Situ Lycopod Forest Site and Structural Anatomy Invalidates the Floating-Forest Hypothesis. *Creation Research Society Quarterly*. 53 (2): 110-122.
25. Modified from Bluck, B. J. 2002. The Midland Valley terrane. In *The Geology of Scotland*, 4th ed. N. H. Trewin, ed. London: The Geological Society, 149-166.
26. Gastaldo, R. A. 1986. An explanation for lycopod configuration, 'Fossil Grove' Victoria Park, Glasgow. *Scottish Journal of Geology*. 22 (1): 77-83.
27. Owen, A. et al. 2007. Fossil Grove to be an undercover RIGS. *Earth Heritage*. 29: 22-23.
28. Tully, J. 1996. *Geologic Age of Coals of the United States*. U.S. Geological Survey Open File Report 96-92, USGS.
29. Luppens, J. A. et al. 2009. Coal resource availability, recoverability, and economic evaluations in the United States-A Summary. In *The National Coal Resource Assessment Overview*. B. S. Pierce and K. O. Dennen, eds. U.S. Geological Survey Professional Paper 1625-F.
30. Luppens J. A. et al. 2013. *Assessment of coal geology, resources, and reserve base in the Powder River Basin, Wyoming and Montana*, Fact Sheet 2012-3143, USGS.
31. Scott, D. C. et al. 2010. *Assessment of coal resources, and reserves in the Northern Wyoming Powder River Basin*, Open-File Report 2010-1294, USGS.
32. Weaver, J. N. and J. W. Wood Jr. 1994. *Coal map of South America*. U.S. Department of the Interior, USGS, Coal Investigations Map C-145.
33. Clarey, T. L. 2017. Floating forest hypothesis fails to explain later and larger coal beds. *Journal of Creation*. 31 (3): 12-14.

19 Evidence of a Young Earth and Recent Flood

> **Summary:** Many people assume that Earth is billions of years old, but plenty of evidence challenges that belief. Rocks do not require long ages to form—they only require the right conditions. Even out in the universe, much evidence indicates that it can't be many millions or billions of years old. If spiral galaxies were ancient, they would have wound up beyond recognition. Blue stars can't be old because they burn their fuel too quickly. And no agreed-upon theory exists to explain how stars originated. Planetary magnetic fields decay too quickly to be millions of years old. Dinosaur fossils contain short-lived original biomaterials, so they must be younger than is widely assumed. There is a lack of erosion between most geologic layers, indicating that they were laid down rapidly. Subducted tectonic plates are surrounded by hot mantle but are strangely cool, suggesting they subducted recently. These evidences are only some of the reasons for believing in recent creation and a recent flood.

In this chapter, I will review the evidence for a young earth and a recent global flood that occurred about 4,500 years ago, as described in the book of Genesis. But first, I will address secular science's need for deep time and why so many scientists insist that the earth is billions of years old.

Millions or Thousands of Years: Does It Matter?

What studies have been done to prove rocks take thousands or even millions of years to form? The answer is none. It is all an assumption that has been taught to geology students and the general public for generations. Everybody seems to believe that rocks take long periods of time to form when in reality all it takes is the right conditions and rocks of any type can form very quickly. In this book we present the evidence that it was the great Flood that provided those special conditions.

What should rocks that are millions of years old look like anyway? Does anybody know? According to empirical measurements of erosion rates we observe today, rocks that are many millions of years old shouldn't even exist if exposed at the earth's surface. Recall from chapter 2 that erosion rates are about 40 feet per million years. That rate would reduce a landmass that was 1,000 feet above sea level to zero in 25 million years. Without some sort of renewed uplift, most of the world's continents should have been reduced to sea level long ago.

> "What studies have been done to prove rocks take thousands or even millions of years to form? The answer is none."

The need for deep time is based on the secular belief in slow evolution. That worldview demands it. You cannot have one without the other. Secular science hides the impossibility of macroevolution by concealing the truth behind a shrouded curtain of deep time. Has science ever documented any type of animal changing into another type of animal? And could they ever repeat that process? The answer to both questions is a resounding no!

Secular scientists make the claim that given enough time anything could happen. They say this even though they have never found any definitive fossil evidence showing a slow, gradual change from one major type of animal to another. What about fish to amphibians? Where are the documented step-by-step fossils that show these types of big changes? All fossils show sudden appearances, stasis, and sudden disappearances. But no matter the odds, or the lack of real evidence, secular scientists cling to the most remote chance that an organism could "evolve" from one type of organism into another. They blindly claim anything is possible given sufficient time. But deep time is wrong.

Everyone can grasp the concept of a hundred years. Now try and imagine 1,000 years. These are understandable time frames because we can relate to them from written human history—the Persians, the Greeks, the Romans. All these empires and the millennia when they existed are within our mental grasp.

Now try and imagine a million years, or even a billion. You cannot. No one can understand or imagine these long time frames. As finite humans, we cannot wrap our heads around the true meaning of deep time. These numbers become just words that are tossed around like they have meaning when in reality they do not. They are fictional in terms of Earth's true history. They are the result of a series of assumptions. In chapter 4, I discussed some of these assumptions and a few of the details involved in the radioactive dating game. Can these assumptions be verified? Can we go back in time to see what really took place? All we have are the resultant rocks and minerals. How they arrived at that particular point in time and space is an unresolvable mystery. This is the essence of a forensic science like geology. Most of what we observe is not repeatable.

Honesty in science is a necessity. We need to be honest about the claims of great ages for the earth and the rocks. College students are not taught about the insurmountable assumptions built into the radioisotope dating methodology. Possibly because of this oversight, the great ages derived for rocks have become entrenched in the scientific literature so that they are no longer falsifiable. Without a way to falsify our conclusions we lose one of the tenets of the scientific method. Students of deep

Milky Way galaxy

time practice what Dr. Alan Feduccia calls the "verificationist approach," which has replaced the normal activities of science.[1] Deep time advocates merely verify what is already assumed to be known.

If you don't believe me, look at the dates listed in any geology textbook. All those millions and billions of years are considered facts. Any criticism of the dates leads to cries of heresy, or worse, cries of being unscientific. We cannot criticize deep time because it is part of the uniformitarian and evolutionary religion. People get emotional about their religion. Sadly, this secular religion is considered science in our modern society.

Everyone who reads this book should think about what they believe about time and why they believe it. Is it based on empirical data or are they merely repeating what they have been taught? Like Dr. Feduccia, I hope this book will "encourage the new generation of students not to be bound by a faith-based reliance on computer-based slanted lines [i.e., isochrons] that confer an aura of precision and truth to what is in reality speculation."[2]

> "The age of the earth matters because it places the Bible on trial."

The age of the earth matters because it places the Bible on trial. If the earth is really billions of years old as secular science claims, then the biblical genealogies are wrong. If the Bible is wrong in Genesis, then who's to say it isn't wrong elsewhere? Either all of the Bible is true or God was lying to us. And God cannot lie. The Bible itself says, "The entirety of Your word is truth" (Psalm 119:160). This means right from the start in Genesis 1:1.

Evidence of a Young Solar System

Part of the resistance to the idea of a young earth is its setting in a supposedly billions-of-years-old universe. But what is the evidence the universe is that age? All observations indicate a youthful universe and solar system, especially the evidence from spiral galaxies, blue stars, comets, and planetary magnetic fields.

Spiral Galaxies

Spiral galaxies are fairly common. Our own Milky Way is a spiral galaxy. Rotation rates of spiral galaxies have been measured. Scientists have found that the inner regions rotate faster than the outer regions, causing a tightening of the galaxy arms around the center.[3] Because of the high rotation rates, a spiral galaxy can only last for a maximum of 100 million years. After a billion years, the twisted arms would become blended into concentric circles and beyond recognition.[3] These calculations give us an upper limit to the potential age of spiral galaxies, and yet secular science maintains that the universe is over 13 billion years old. And many of the spiral galaxies are claimed to be 10 billion years old.[3] The bottom line is there should be zero spiral galaxies if they are indeed billions of years old.

Blue Stars

Blue stars are also a problem for the uniformitarian scientist. Blue stars are much bigger than our sun and burn up faster, at rates that cannot last even a few million years.[4] Blue stars are very common throughout the known universe. There are blue stars in every known spiral galaxy, including the three stars that make up the constellation known as Orion's belt. To blindly claim that new blue stars form frequently is disputed by observations of the expansion of gas nebulae. Dr. Jason Lisle put it this way:

> Star formation is problematic at best.[5] Gas is very resistant to being compressed. On earth, gas always fills its container. In space, there is no container. So gas expands indefinitely. If the gas could be forced into a sphere that is very small (in comparison to a nebula) such as the sun, then the gas would be held together by its own gravity. However, in a typical nebula, the gas pressure far exceeds the miniscule force of gravity. Secular astronomers now believe that external forces, such as a shockwave from an exploding star, are necessary in most cases to trigger star

Blue stars burn up too fast to be even a million years old

Comets disintegrate in less than one million years

formation. Observations confirm that gas clouds expand; they do not appear to collapse into stars.

Even if we could compress the nebula sufficiently to the point that the force of gravity was strong enough to prevent the gas from expanding, other effects would kick in, thereby preventing the formation of a star. Clouds of gas always have a weak magnetic field, which would be concentrated if the cloud were compressed. This dramatically increases the field strength. The magnetic pressure would halt a shrinking cloud and drive it to re-expand.[6] It's a bit like trying to push the like poles of two magnets together.[4]

Blue stars continue to be a problem for secular scientists. On one hand, they cannot explain their obvious youth in an old-earth worldview without replenishing them frequently. And on the other hand, they have no viable way to form new ones. It is only their worldview that prevents them from recognizing the true solution to this dilemma: Blue stars are just not that old.

Comets

Comets are mixtures of ice and dirt that orbit the sun. They lose material by coming closer to the sun where the heat vaporizes some of the ice, blasting material into space.[3] Solar wind pushes this material away from them, creating a "tail." The problem with comets is that this process causes them to quickly disintegrate. Every time they loop around the sun they get smaller. In fact, comets cannot last even one million years.[3] So, how do old-earth scientists resolve the comet age problem?

They have come up with an untestable rescue device known as a *comet-generator* that replenishes comets in our solar system.[3] Secular scientists have hypothesized that a cloud of icy debris exists at the outer edge of our solar system that becomes occasionally disturbed somehow, spawning a new comet. Rather conveniently, this so-called "Oort Cloud" of icy debris has never been observed or detected. It is a totally fabricated region that has no scientific evidence to support its existence. It is merely assumed to be there to explain why we still have comets circling the sun.[3] But real science shows that comets are not that old. Comets fit better within the 6,000-year-old age range of the biblical worldview.

Earth's magnetic field is decaying too fast to be even 10,000 years old

Planetary Magnetic Field Decay

The earth's magnetic field protects us from a lot of harmful radiation from the sun and other stars. Most scientists assume it is generated by the earth's metallic core, but this process is not completely understood. The field is also rapidly decaying. The problem with the magnetic field is that it is too strong at its current rate of decay to be very old. Dr. John Morris put it this way:

> The strength of the magnetic field has been reliably and continually measured since 1835. From these measurements, we can see that the field's strength has declined by about seven percent since then, giving a half-life of about 1,400 years. This means that in 1,400 years it will be one-half as strong, in 2,800 years it will be one-fourth as strong, and so on. There will be a time not many thousands of years distant when the field will be too small to perform as a viable shield for earth.
>
> Calculating back into the past, the present measurements indicate that 1,400 years ago the field was twice as strong. It continues doubling each 1,400 years back, until about 10,000 years ago it would have been so strong the planet would have disintegrated—its metallic core would have separated from its mantle. The inescapable conclusion we can draw is that the earth must be fewer than 10,000 years old.[7]

In addition, planets like Jupiter, Saturn, Uranus, and Neptune all have strong magnetic fields.[3] The magnetic fields of these planets are also decaying too rapidly to be billions of years old as suggested by secular scientists.[3]

How do secular scientists make these measurements fit their old-earth worldview? They claim that the earth's magnetic field gets so weak that it reverses and somehow regenerates a stronger magnetic field in the process. The problem with this explanation is that it is unscientific. It cannot be observed, repeated, or tested. We have no evidence that the earth's magnetic field strengthens each time it reverses. And this explanation doesn't explain the similar decay rates we observe for the magnetic fields of these other planets in our solar system. Do they all reverse and strengthen simultaneously? I doubt it. The best solution to the decay of all of these magnetic fields is that they are not millions of years old. These planets, and Earth, are all just thousands of years old. This age matches perfectly with the observed magnetic decay rates of all the planets.

Evidence of a Recent Flood

Let's review some of the topics covered in earlier sections indicating that the global Flood was recent, occurring about 4,500 years ago. Collectively, this is strong evidence that tips the scale in favor of a young earth. It is not based on assumptions of deep time but is data-driven. It is the scientific data that demonstrate these sedimentary rocks cannot be millions of years old. This is not a trivial matter that can be easily dismissed or ignored by the secular community.

Original Tissues in Fossils

In chapters 4 and 5, I discussed the scores of discoveries of original proteins and other soft tissues in dinosaur fossils and other types of fossils. To me, these are the most significant discoveries in paleontology in the last 100 years. They are more significant than the first discovery of dinosaur eggs by Roy Chapman Andrews in the 1920s or the realization that dinosaurs were a separate type of reptile that walked erect by Sir Richard Owen in 1841.

Since 2005, the presence of dinosaur soft tissues has been confirmed again and again, demonstrating that these are, in fact, real original tissues from the extinct animals. Over 20 separate tests have been performed to verify these proteins by multiple scientists. And yet, no one in the old-earth community has been able to adequately explain how these original tissues could have been preserved for as long as they claim. Physical chemists tell us that these proteins, like collagen, cannot last even one million years. Degradation processes are too efficient. So, why do we keep finding real blood vessels, collagen, osteocyte cells, and red blood cell-like objects in dinosaur bones claimed to be 68 to 150 million years old? Could it be that they are not really that old? And that the Bible is right? Maybe the global Flood was only about 4,500 years ago, making dinosaur fossils only thousands of years old. But this simple solution to the preservation problem is ignored due to their biased secular worldview. They deliberately forget that there was a global flood (2 Peter 3:3-6).

The secular community has made attempts to explain how these original tissues could have been preserved for so long, but all have fallen short.[8] Conditions like the lack of oxygen and/or lack of microbes to cause decay, collagen sheltering mechanisms, and preservation with iron solutions have been suggested as ways of preserving these tissues.[8] Unfortunately for the secular scientist, all of these methods fail to adequately explain preservation for millions of years. The conditions they assume for many of these suggested methods of preservation are unrealistic over deep time. For example, the method of preserving blood vessels using an iron solution was only studied over a period lasting two years.[8] Projecting the results of a study that lasted two years into an additional seven orders of magnitude in time seems rather absurd. And the authors of the study only acknowledged making a visual estimation of blood vessel degradation at the end of those two years.[9]

Dr. Kevin Anderson summed up this failed attempt to explain deep-time preservation, stating:

> Why not use more analytical methods (such as mass spectroscopy) to analyze the vessels? Visual observation of tissue degradation can be very subjective and unrevealing. For Schweitzer and co-authors to make their claim, a more detailed analysis of the protein and tissue integrity of the vessels is essential. It is unfortunate that after years of careful and meticulous investigation, Dr. Schweitzer and her collaborators choose to take a more superficial and potentially biased approach for this study. Did it better serve their purpose?[10]

> "These discoveries are the closest thing to proof in geology (a forensic science) that fossils are not that old and instead fit exactly within the time frame of the Flood occurring about 4,500 years ago."

The bottom line is that the secular scientific community is embarrassed by these discoveries. They do not want you to know the obvious implications these discoveries have for their deep-time assumptions. Scientists need to follow data, not dogma. These discoveries are the closest thing to proof in geology (a forensic science) that fossils are not that old and instead fit exactly within the time frame of the Flood occurring about 4,500 years ago.

Carbon-14 in Fossils Also Indicates a Young Age[11]

You may wonder why many of the dinosaur soft tissues and other fossil organic molecules have not been tested for carbon-14. It is because of evolutionary dogma, not for any scientific reason. Dinosaur fossils are thought to be so old that there should be no measurable carbon-14 left even if original materials were preserved. Anything older than about 100,000 years should have no measurable carbon-14 because all of it should have decayed to undetectable levels. But what do we find when we test dinosaur fossils?

Recall from chapter 4 that absolute dates are based on the decay rate of radioactive isotopes and a whole list of unverifiable assumptions. Decay rates are converted to a half-life, which is the time it takes for one-half of the radioactive element to decay to another element. Carbon exists as three isotopes: carbon-12, the most common variety; carbon-13, the next-most common variety; and carbon-14, the rarest and only radioactive variety. All three carbon isotopes have six protons, which defines them as carbon. Carbon-12 has six neutrons, carbon-13 has seven neutrons, and carbon-14 has 8 neutrons, making the latter unstable. Carbon-12 and carbon-13 remain carbon-12 and carbon-13 forever, respectively. Carbon-14, by contrast, spontaneously decays into nitrogen-14 with time. Carbon-14 has a half-life of about 5,730 years at present, meaning that one-half of any quantity of carbon-14 will convert to nitrogen-14 in that time. By determining how much carbon-14 is left in a sample, we can estimate its age based on today's concentration of carbon-14 in the environment. If we have one-quarter of what is found in the environment today, it is assumed that two half-lives have passed, and therefore the sample is claimed to be 11,460 years old (5,730 + 5,730).

My colleague Dr. Brian Thomas at the Institute for Creation Research (ICR) sent a small piece of fossil paddlefish to an independent laboratory for carbon-14 analysis. This paddlefish was found in the Eocene Green River Formation of Wyoming, supposedly about 48 to 50 million years old. Due to its assumed old age, there should have been no detectable carbon-14 remaining in the fossil. However, the secular lab found enough carbon-14 to date the fossil at 33,530±170 years before present (BP). ICR scientists still believe this estimated age is too old for the biblical time frame, and this matter is explained below.

ICR also sent two samples from a Cretaceous hadrosaur (duck-billed) dinosaur to the same independent

A Cretaceous hadrosaur bone was found to contain measurable carbon-14, indicating it is only thousands of years old

This Eocene fossil paddlefish was found to contain measurable carbon-14, indicating it is only thousands of years old

laboratory for carbon-14 analysis. These samples were from the Hell Creek Formation in Montana, supposedly about 68 to 70 million years old, and again should have contained no detectable carbon-14. The lab found enough carbon-14 in each sample to determine dates of 28,790±100 years BP and 20,850±90 years BP.

In addition, at least 20 other dinosaur bone samples and eggshell fragments were tested for carbon-14 at independent laboratories. Some of these were from the Morrison Formation in Utah, supposedly dated at 149 to 150 million years old. These results also came back in the range of 20,000 to 50,000 years BP. The carbon-14 dates are consistently showing dates in the thousands and not in the millions.

These results are strong evidence that dinosaur fossils are not millions of years old. Critics may argue these carbon-14 dates are all due to contamination, but great care was taken to sample the bones, using bone material from the inside and not the outer edges. The commercial lab took great care not to introduce contamination, and the consistency in the carbon-14 dates from samples taken across three states and in vastly different rock layers implies there was no contamination.

As part of the carbon-14 analyses, ratios of the stable carbon-13 isotope were also reported as the standardized ratio δ13C. This variety of carbon, although more rare than carbon-12, does not spontaneously decay. The δ13C ratio can be helpful in demonstrating if contamination has occurred prior to sampling. Usually, the δ13C ratio in the mineral part of any bone, called *bioapatite*, is significantly different from the ratios of δ13C in the organic part (collagen and other organic molecules) of the bone. Preserved fossil soft tissue and fossil bone should reflect these differences in δ13C values if they are real and not contaminated in some way after burial.

The analyses showed that most of the bioapatite δ13C ratios fell in the expected ranges for carbonate minerals derived from uncontaminated bone, and the δ13C ratios for the organic component showed significantly different δ13C ratios, close to what was expected for organic material in uncontaminated samples. The measured δ13C ratio differences between the bioapatite and in the extracted organic components strongly argue against any carbon contamination.

The carbon-13 measurements provided an independent check on the validity of the carbon-14 results reported above. If everything was uniformly contaminated, all δ13C ratios should have been roughly the same. These findings indicate the measurable carbon-14 found in the dinosaur bones is real, demonstrating a young age for the dinosaurs.

> **"Creation scientists have pointed out that the pre-Flood atmosphere was likely much higher in carbon dioxide, possibly even 500 times higher than today's atmosphere."**

You may wonder why the carbon-14 ages consistently fall in the 20,000 to 30,000-year range and not in the 4,000 to 5,000-year range. Unfortunately, carbon-14 dating is based on the same assumptions of all radioactive methods, in particular the modern ratio of carbon-14/carbon-12 in the environment. We know that carbon-14/carbon-12 ratios in the atmosphere have not stayed constant even in the past few thousand years because corrections have had to be made to the carbon-14 dates to match tree ring data.

Creation scientists have pointed out that the pre-Flood atmosphere was likely much higher in carbon dioxide, possibly even 500 times higher than today's atmosphere. This estimate is based on the amount of coal, oil, and other organic debris buried in the sediments of the Flood.[12] A higher level of carbon dioxide would have greatly altered the pre-Flood carbon-14/carbon-12 ratio, assuming the cosmic rays that produce carbon-14 were not much different from today. Increasing the atmospheric carbon-12 content would have made the starting carbon-14/carbon-12 ratio much smaller in all the fossils buried in the Flood, making them seem older from the start. And when the carbon-14 in the buried fossils is measured and compared to today's carbon-14/carbon-12 ratio, it would result in inflated "ages" of up to 10 times their true age. Wrongly applying the modern ratio of carbon-14/carbon-12 gives an assigned age of 45,000 years and not the real age of 4,500 years. Unfortunately, we only have the modern ratio of carbon-14/carbon-12 to work with, so these "true" ages will always be inflated by about one order of magnitude. Regardless of which starting ratio is used, the presence of detectable carbon-14 in ancient fossils demonstrates that dinosaurs are not millions of years old!

Lack of Erosion (Time) Between Layers

When we look at the sedimentary rocks and the various megasequences, we see conformity. Even between the megasequences, we most commonly see sedimentary layers paralleling the layers below and above. There is no indication within the sedimentary strata of vast amounts of missing time. The megasequence boundaries and many sedimentary units in general can extend for tens and even hundreds of miles in all directions. The extent of the Morrison Formation across the Rocky Mountain states, for example. These units look as if they were laid down, brick by brick, in order with the oldest at the bottom and the youngest at the top.

> "Regardless of which starting ratio is used, the presence of detectable carbon-14 in ancient fossils demonstrates that dinosaurs are not millions of years old!"

Secular scientists often place many years of missing time between parallel sedimentary units such as the boundary between the Hermit Shale in Grand Canyon and the overlying Coconino Sandstone. Secular science claims there are hundreds of thousands of years missing at that particular contact.¹³ But when you examine the contact, it is nearly perfectly planar in all direction for tens of miles. Sure, there may be small, smooth undulations of a few feet in some locations, but for the most part it is planar, or nearly so, with sharp contacts from one rock type to the next. Where are the gullies and the river valleys? Where is the uneven topography that should have resulted from hundreds of thousands of years of erosion? The contact looks like brick-upon-brick with no evidence of any time delay whatsoever. There are no steep valleys or even cliffs visible along the contact as would be expected from normal erosive processes and like we observe universally on the surface of the earth today. All that is observed are flat sediments on flat sediments as far as one can see across the entire expanse of Grand Canyon and beyond.

Many other sedimentary units are also supposed to have vast amounts of time missing between their boundaries in Grand Canyon (and elsewhere too), like at the base of the Redwall Limestone where it rests on the Muav Limestone (supposedly missing 160 million years)¹³ and at the base of the Tapeats Sandstone where it rests on the crystalline basement in western Grand Canyon (supposedly missing one billion

Thick Redwall Limestone and Temple Butte Formation (red color) on Cambrian strata (tan color) in Grand Canyon. The contact is nearly planar everywhere it is observed. Where is the geomorphological evidence of over 100 million years of erosion as claimed?

Coconino Sandstone on Hermit Shale (darker reddish color) in Grand Canyon. Note the flat, planar nature of the contact. There is no physical evidence of the claimed hundreds of thousands of years of erosion between these rock units.

years).[13] The base of the Tapeats Sandstone is also part of the globally extensive Great Unconformity that was discussed in chapter 9. Secular science claims a billion years is lost at this contact, and yet the Tapeats Sandstone was deposited nearly perfectly flat across the entire expanse of Grand Canyon. And recall, the Tapeats and equivalent sandstones extend from Grand Canyon to the East Coast and across much of North America. This is no small unit. And nearly everywhere across the expanse of North America it is deposited on a nearly planar surface. How can that be if there was so much time for erosion? Instead, the evidence indicates rapid scouring and erosion, creating a planar surface, followed almost immediately by deposition of sedimentary layers as each megasequence advanced across the continents.

How can the old-earth community explain the lack of significant erosion along these surfaces globally? Where is their evidence for missing time? For all intents and purposes, the layers look like they were deposited one after the other in a short time frame as each megasequence brought in new sediments and piled them on top of one another. The only erosion would have been caused by the short time between each megasequence when the water was receding and as the new megasequence advanced across the older megasequence layers. This better explains the nearly flat and planar contacts between individual sedimentary layers and along the megasequence boundaries. Only the global Flood provides the timing, source, and conditions to create these layers on a global scale as observed.

That is not to say there are not angular unconformities between some megasequences, and sometimes even within the layers within a single megasequence. These unconformities obviously represent some missing time but do not necessitate great amounts of time, likely only days or weeks at most. How do we know? Because the rock layers above the unconformities are generally parallel to the boundary surface, and the unconformity itself is usually planar and/or nearly flat. There is no evidence of valleys, gullies, or cliffs along the unconformity surface as would be expected from normal erosional processes over long periods of time. And yes, even great amounts of time are unnecessary to form angular unconformities like the one at Siccar Point, Scotland, where James Hutton insisted vast amounts of time were needed between the rocks above and below the unconformity. As we discussed earlier, even uplifts can happen quickly, and so can erosion, especially when we are dealing with tsunami-like waves sweeping across the terrain that were likely bigger than any witnessed in recent history. Water is a powerful and efficient erosive medium.

Cold Subducted Slabs Deep in the Mantle

In chapter 6, I discussed the evidence supporting catastrophic plate tectonics and runaway subduction and rapid plate movement of meters/second during the Flood. One of the strongest evidences to support this is the seismic tomography imaging of cold, subducted oceanic lithosphere deep in the mantle. If these lithospheric slabs were really moving as slowly as secular scientists claim (at just a few centimeters/year), then they should have assimilated long

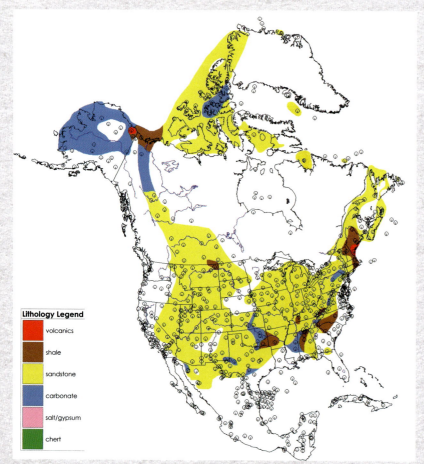

Blanket sands that are spread across vast parts of continents, like the basal Sauk sand (Tapeats equivalent), are difficult for secular geologists to explain

ago and not show such strong density contrasts (indicating a much cooler temperature) with the surrounding hot mantle. Instead, these cold lithospheric slabs indicate they were rapidly emplaced just thousands of years ago.

Dr. Jake Hebert, my colleague at ICR, summarized the findings from mantle tomography:[14]

> An imaging process called *seismic tomography* has revealed a ring of dense rock at the bottom of the mantle. Since its location corresponds approximately to the perimeter of the Pacific Ocean, it appears to represent subducted ocean crust (Figure 19.1). Located inside this ring of cold rock is a blob of less-dense rock that appears to have been squeezed upward toward the crust. If one assumes that the density of the cold ring is comparable to that of the surrounding material, which is the most straightforward assumption, this ring is 3,000 to 4,000°C colder than the inner blob. This is completely unexpected in the conventional plate tectonic model since it can take about 100 million years for a slab to descend all the way to the base of the mantle. In that time, one would expect any such temperature differences to have evened out. However, in the catastrophic plate tectonics model, such a temperature difference is to be expected if the slab rapidly subducted into the mantle just a few thousand years ago.[15]

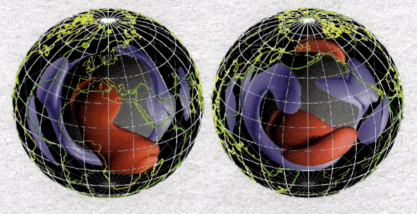

Figure 19.1. Regions of colder, more dense mantle (purple) and hotter, less dense mantle (red) in the lower mantle as shown by seismic tomography. Secular science cannot explain the cold dense material so deep in the mantle if plate movement rates are as slow as observed today.[15]

This rapid subduction of lithospheric material also may explain the apparent magnetic reversals found in the ocean crust. Dr. Hebert explained:

> Molten lava, or magma, contains minerals whose magnetic domains

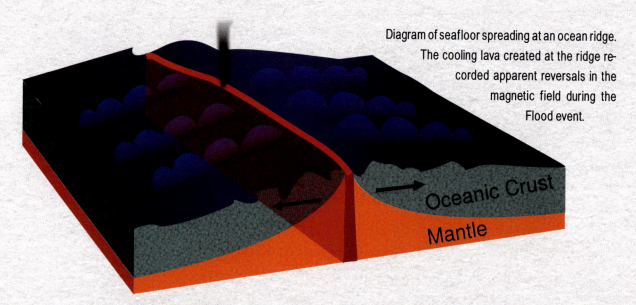

Diagram of seafloor spreading at an ocean ridge. The cooling lava created at the ridge recorded apparent reversals in the magnetic field during the Flood event.

tend to align with the direction of Earth's magnetic field. When the rock cools and hardens, this alignment is "locked" into the volcanic rock. The basaltic rocks on either side of the mid-ocean ridges depict a striped pattern consisting of alternating bands of magnetization that reverse direction as one moves away from the ridge. This striped pattern indicates that Earth's magnetic field has flipped dozens of times, with the north and south magnetic poles trading places.

If a new seafloor rapidly formed during the Genesis Flood, then the fact that these magnetic reversals are recorded in oceanic volcanic rocks (most of which were formed during the Flood) implies that the magnetic reversals must also have occurred rapidly. Uniformitarian scientists found strong evidence for rapid magnetic reversals, although such rapid reversals are very hard for them to explain.[16-18] Creation physicist D. Russell Humphreys proposed a theory that at least qualitatively explains how such rapid reversals could occur.[19] His mechanism requires strong up-and-down motions of fluids within Earth's outer liquid core due to convection. Such convection might be initiated if a cold subducting plate were to come into contact with the outer core at the core-mantle boundary, which Dr. Baumgardner argues is exactly what happened.[14,15]

Runaway subduction and rapid plate motion has been validated by these findings. Catastrophic plate tectonics did happen in the past but only during the Flood about

4,500 years ago. Once all of the original oceanic lithosphere was completely consumed and a new seafloor was created, the runaway motion ceased. Today, we merely witness the residual motion from this event.

Conclusion: The Flood Was Recent

This chapter presented strong scientific evidence that the global Flood took place just thousands of years ago, just as the biblical genealogies indicate. In fact, there is no empirical evidence to the contrary. Only biased interpretations based on unverifiable assumptions continue to argue for an old earth. By that, I mean the radioisotope dates that secular science relies on so heavily. The rocks do not show great age. The fossils do not show great age. The solar system does not show great age.

Unfortunately, you will never hear these evidences discussed in a secular classroom. They dismiss or flatly ignore them. They even often resort to calling those of us who do not accept their theories "nonscientists," regardless of whether we are collecting data, making observations, and following the scientific method.

But if the earth is really not billions of years old, then where does that leave secular scientists? It would leave them no time for their fabricated theory of evolution. Their entire uniformitarian worldview would collapse. Time and evolution are the two pillars sustaining their secular worldview. Without one pillar, the other falls. If we take away deep time, as the evidence does, they lose any chance, however remote, for any significant evolution. They would have to admit there was a Creator.

> "Time and evolution are the two pillars sustaining their secular worldview. Without one pillar, the other falls."

The secular scientific community has a tight stranglehold on science today. They have the microphone, as I like to say. They say whatever they want and call it science when it is really comparable to a dogmatic religion based on atheism. They cannot give in to any evidence to the contrary, no matter how powerful, because it would conflict with their secular worldview.

Even old-earth Christians fall victim to the falsehood of secular science. They attempt to reconcile both God and secular science but always place that "science" ahead of the Bible. Only God can change hearts and open minds. We can only keep showing the truth that real, empirical science confirms the Bible! I prefer to follow the data and what empirical observations show, not hearsay. The rocks point me to the Creator, a Creator who loved us all so much He sent His Son Jesus as our Savior so that we can be forgiven and redeemed. He stands at the door and knocks. He is our Rock and our Salvation!

References

1. Feduccia, A. 2012. *Riddle of the Feathered Dragons: Hidden Birds of China*. New Haven, CT: Yale University Press.
2. Ibid, 5.
3. Lisle, J. 2013. The age of the cosmos—what you have not been told. In *Creation Basics & Beyond*. Dallas, TX: Institute for Creation Research, 307-312.
4. Lisle, J. 2012. Blue Stars Confirm Recent Creation. *Acts & Facts*. 41 (9): 16.
5. Wiebe, D. Z. et al. 2008. Problems of Star Formation Theory and Prospects of Submillimeter Observations. *Cornell University Library*. Posted on arxiv.org July 21, 2008, accessed July 13, 2012.
6. Hartmann, L. 2008. *Accretion Processes in Star Formation,* 2nd ed. Cambridge, UK: Cambridge University Press, 57-58.
7. Morris, J. 2010. Earth's Magnetic Field. *Acts & Facts*. 39 (8): 16.
8. Anderson, K. 2016. *Echoes of the Jurassic*. Chino Valley, AZ: Creation Research Society.
9. Schweitzer, M. et al. 2013. Molecular analyses of dinosaur osteocytes support the presence of endogenous molecules. *Bone*. 52 (1): 414-423.
10. Anderson, *Echoes of the Jurassic*, 39-40.
11. Modified from Clarey, T. 2015. *Dinosaurs: Marvels of God's Design*. Green Forest, AR: Master Books, 50-51.
12. Hebert, J. 2013. Rethinking Carbon-14 Dating: What Does It Really Tell Us about the Age of the Earth? *Acts & Facts*. 42 (4): 12-14.
13. Moshier, S. and C. Hill. 2016. Missing time: Gaps in the rock record. In *The Grand Canyon, Monument to an Ancient Earth: Can Noah's Flood Explain the Grand Canyon?* C. Hill et al, eds. Grand Rapids, MI: Kregel Publications, 99-107.
14. Hebert, J. 2017. The Flood, Catastrophic Plate Tectonics, and Earth History. *Acts & Facts*. 46 (8): 11-13.
15. Baumgardner, J. R. 2003. Catastrophic Plate Tectonics: The Physics Behind the Genesis Flood. In *Proceedings of the Fifth International Conference on Creationism*. R. L. Ivey Jr., ed. Pittsburgh, PA: Creation Science Fellowship, 113-126. Figure 19.1 courtesy of Alexandra Forte.

16. Coe, R. S., M. Prévot, and P. Camps. 1995. New evidence for extraordinarily rapid change of the geomagnetic field during a reversal. *Nature*. 374 (6524): 687-692.
17. Bogue, S. W. and J. M. G. Glen. 2010. Very rapid geomagnetic field change recorded by the partial remagnetization of a lava flow. *Geophysical Research Letters*. 37 (21): L21308.
18. Sagnotti, L. et al. 2014. Extremely rapid directional change during Matuyama-Brunhes geomagnetic polarity reversal. *Geophysical Journal International*. 199 (2): 1110-1124.
19. Humphreys, D. R. 1990. Physical Mechanism for Reversals of the Earth's Geomagnetic Field During the Flood. In *Proceedings of the Second International Conference on Creationism*. R. E. Walsh and C. L. Brooks, eds. Pittsburgh, PA: Creation Science Fellowship, 129-142.

Siccar Point, Scotland, showing the angular unconformity first recognized by James Hutton. Angular unconformities do not require deep time, only the right conditions.

20 It All Makes Sense

> **Summary:** There is clear geologic evidence of a global flood just as depicted in the Bible. The megasequences show increasing thickness of deposition and surface coverage corresponding with increasing floodwaters, and then conclude with a massive drainage of the continents—just as the Bible describes. The Flood began in the Sauk, peaked in the Zuni, and receded in the Tejas. The fossil record also matches the sedimentary story of a progressive and global flood. The Flood began by burying marine creatures, then shore creatures and lowland habitats, then increased to burying large land creatures such as dinosaurs, and finally overcame the mammals and humans fleeing to higher ground. The Flood was real. It was a one-time event, and it best explains the geological data we see globally.

Review of the Rock Data

The record of the sedimentary megasequences clearly demonstrates a global flood occurred about 4,500 years ago. This book showcases the geological data compiled across three continents, amounting to about half the earth's landmass. When viewed in their entirety, the data are strongly compelling. There is clear geologic evidence of a progressive flood just as described in the Bible. Rock data do not lie.

Let's review the data this book has presented. Figure 20.1 shows all six megasequences across all three continents included in this study. This figure's isopach maps show the extent and thickness of each megasequence across each of the continents. The maps' white areas mean there are no rocks from that megasequence at that location today. Most of the time, the extent of a megasequence was lessened by erosion either immediately after the megasequence was deposited as a new megasequence advanced or at the end of the

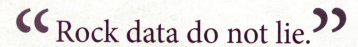

Flood after all the megasequences were deposited.

Just because none of the megasequences show complete coverage across the continents does not mean the Flood did not cover the entire globe at one point. As discussed in chapter 13, I think the maximum flooding occurred at the end of the Zuni Megasequence. This would have been Day 150 of the Flood, when all of the earth was underwater at once. However, the Bible says in Genesis 7:20 that the highest hills were only covered by 15 cubits of water (about 23 feet). This wouldn't leave a lot of room for the deposition of vast amounts of sediment as the water quickly began to recede. Twenty feet or so of sediment could easily have been eroded in the past 4,500 years, and likely much sooner, leaving blanks in the megasequence maps. Many of the areas devoid of sediment today coincide with the locations of the highest pre-Flood hills. These are areas of exposed crystalline shield rocks, like in Canada, Brazil, and sub-Saharan Africa. And as I pointed out in chapter 13, there are some erosional remnants across North America in the Zuni in particular that indicate the floodwaters were more continuous and more extensive than what is shown. I call this the "'bathtub ring.'"

> "The megasequences tell the story of a single, progressive flood event."

The megasequences tell the story of a single, progressive flood event that began slowly in the Sauk, peaked in the Zuni, and receded in the Tejas. Each of the three continents shows the same general pattern. This is what makes these data so compelling. It is not just one continent that shows this pattern but three, and three that show it simultaneously. This is the strongest evidence I have ever witnessed in my 35 years as a geologist that indicates a global flood has occurred. How can anyone look at these data, these maps, and not realize it is showing the exact same pattern and timing of global flooding? This is truly compelling evidence of worldwide activity. It should be shouted from the rooftops!

The data shown in the Figure 20.1 isopach maps are further confirmed by graphing the volume of sediment deposited during each megasequence across the three continents. Figure 20.2 shows graphs of sediment volume by megasequence for North America, South America, and Africa. Collectively, the three continents exhibit minimal volumes of sediment deposited in the earliest megasequences. Then the three

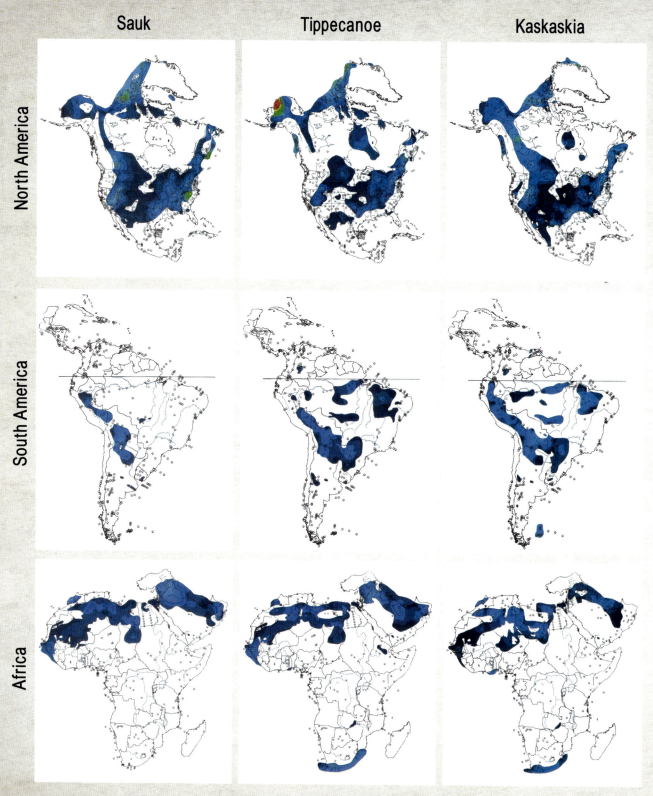

Figure 20.1. Maps showing the coverage and thickness for North America, South America, and Africa for each of the six megasequences. The colored areas represent sedimentary coverage that remains present today even after erosion between the megasequences or since the Flood. Areas with no color are devoid of sedimentary coverage. Many of these areas consist of crystalline basement rocks, like the Canadian Shield. Although the Zuni is the high point of the Flood, its rocks do not completely cover all of the continents today due to post-Flood erosion. Recall that the Bible tells us the highest hills were only covered by 15 cubits (about 23 feet) of water, so not much sediment was deposited across these areas, now consisting of the shields of each continent.

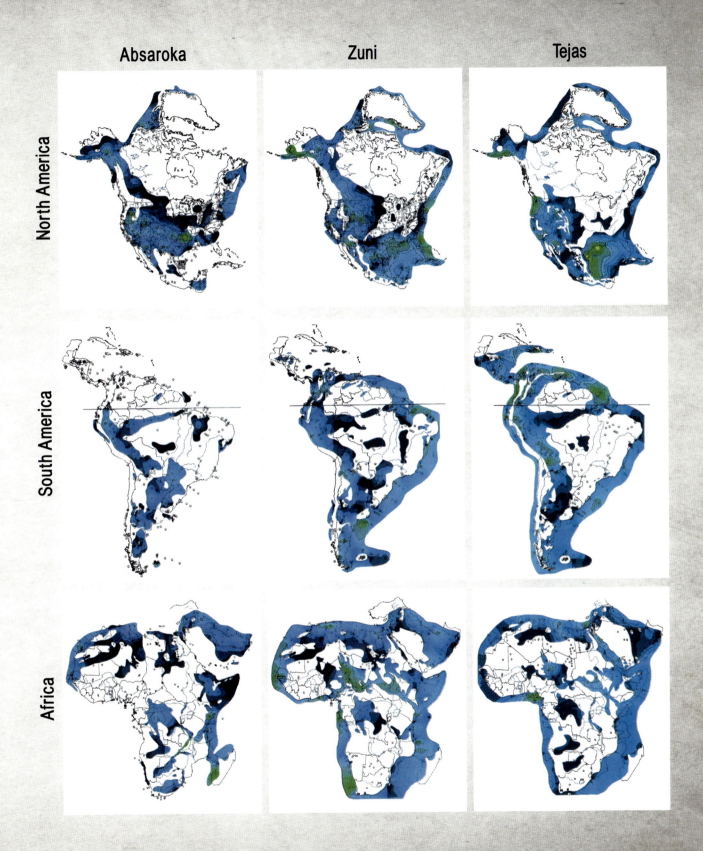

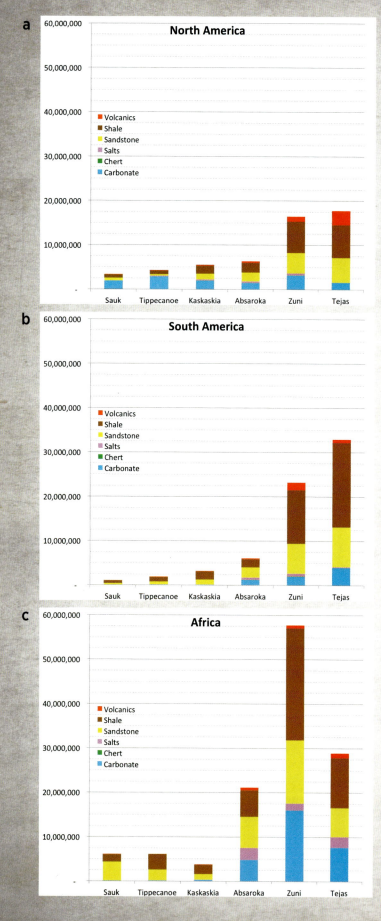

continents demonstrate greater and greater accumulations until the maximum volume is reached in the Zuni Megasequence. The subsequent Tejas, although the receding phase of the Flood, is second in volume to the Zuni globally (Table 20.1).

These sediment volumes coincide almost perfectly with the extent of coverage across the continents. This evidence for a progressive flood is probably best shown by the isopach maps of the megasequences across South America. The Sauk shows the least extensive coverage, followed by more and more coverage progressively until the Zuni, when the Flood reached its maximum level on Day 150. The Tejas, unsurprisingly, shows a nearly identical amount of coverage as the Zuni. Recall, the Tejas was the receding phase that began on Day 150 of the Flood year. The surface extent of both should be nearly identical, barring subsequent erosion.

The evidence for a simultaneous flooding event of all three continents is also shown in Table 20.1. The totals column on the far right side confirms the story. These totals are the compilation of all three continents examined in this study. These data confirm what the maps and the graphs described above show. All of these data point to a flood that began slowly, reached a maximum, and then receded. This

Figure 20.2. Histograms of sediment volume and rock type for each megasequence across (a) North America, (b) South America, and (c) Africa. Measurement is in cubic kilometers.

Surface Area (km²)	North America	South America	Africa	Total
Sauk	12,157,200	1,448,100	8,989,300	22,594,600
Tippecanoe	10,250,400	4,270,600	9,167,200	23,688,200
Kaskaskia	11,035,000	4,392,600	7,417,500	22,845,100
Absaroka	11,540,300	6,169,000	17,859,900	35,569,200
Zuni	16,012,900	14,221,900	26,626,900	56,861,700
Tejas	14,827,400	15,815,200	24,375,100	55,017,700
Volume (km³)	**North America**	**South America**	**Africa**	**Total**
Sauk	3,347,690	1,017,910	6,070,490	10,436,090
Tippecanoe	4,273,080	1,834,940	6,114,910	12,222,930
Kaskaskia	5,482,040	3,154,390	3,725,900	12,362,330
Absaroka	6,312,620	6,073,710	21,075,040	33,461,370
Zuni	16,446,210	23,198,970	57,729,600	97,374,780
Tejas	17,758,530	32,908,080	28,855,530	79,522,140
Average Thickness (km)	**North America**	**South America**	**Africa**	**Total**
Sauk	0.275	0.703	0.675	0.462
Tippecanoe	0.417	0.430	0.667	0.516
Kaskaskia	0.497	0.718	0.502	0.541
Absaroka	0.547	0.985	1.180	0.941
Zuni	1.027	1.631	2.168	1.712
Tejas	1.198	2.081	1.184	1.445

Table 20.1. Surface area, sediment volume, and average thicknesses for North America, South America, and Africa for each of the six megasequences

is why the Zuni has the maximum volume of sediment deposited globally, the maximum average thickness deposited globally, and the maximum areal extent globally. The Zuni was the high point of the Flood. There is no other reasonable way to explain these data.

Unsurprisingly, the fossil record also matches the sedimentary story of a progressive and global flood. The same types of shallow marine fossils are found globally in the earliest megasequences (Sauk through Kaskaskia), deposited before the water started to flood much of the land. Then in the Absaroka, as runaway subduction was actively creating a new global seafloor and pushing up the bottom of the ocean floor in the process, the Flood reached the land, burying the first major coal seams and many land animals in great numbers. Then the water rose higher still, flooding the lowlands and the dinosaurs, and eventually going over the top of the highest hills, stripping off everything, including the large mammals and humans clinging to the highest elevations. Finally, as the water began to recede, many of these mammals were deposited globally on top of the dinosaurs. The global Flood was real. It was a one-time event. And it best explains the geological data we see globally.

A New Global Sea Level Curve Based on Rock Data

Evolutionary geologists have compiled a global sea level curve from the Cambrian system to the present using paleo-environmental interpretations and then deep time (Figure 20.3).[1] For example, they infer global sea level was lower during the Absaroka Megasequence because they interpret many cross-bedded sandstones, like the Permian Coconino Sandstone, as being deposited on dry land. But creation geologist John Whitmore has shown that the physical and mineralogical data better support deposition in a marine setting.[2]

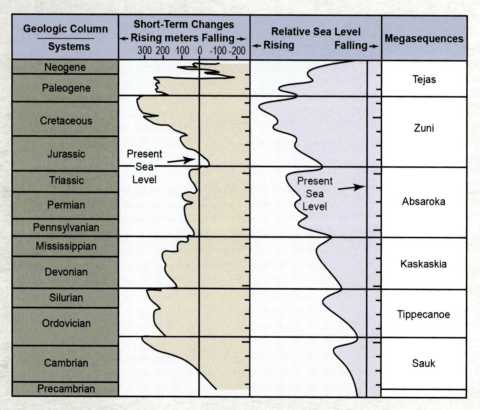

Figure 20.3. The old secular global sea level curve of Vail and Mitchum[1] (left), compared to a new, data-based sea level curve (right)

We have now compiled stratigraphic data from over 1,500 columns across North and South America, Africa, and the Middle East. If we ignore paleo-environmental speculation and uniformitarian dogma and look only at extent and volume of the rocks across these continents, we see a completely different story from that of the accepted secular sea level curve.

To create our new sea level curve, we used the aforementioned maps and megasequence data we compiled across the three continents. Collectively, these data show the least volume, extent, and thickness of sediment deposited in the earliest megasequences. Note in Table 20.1 the massive jump in sediment volume and extent in the Absaroka Megasequence across the three continents. We interpret this as about Day 40 in the Flood when the Ark began to float. It also coincides with the first extensive coal seams and the first major occurrences of land fossils in the rock record. Also note the even greater jump in volume and extent in the Zuni. This is when the dinosaurs were completely inundated and the water levels reached their peak extent across all continents.

The totals column on the far right side of Table 20.1 confirms the single Flood account. All of these data point to a flood that began slowly, reached a maximum, and then receded. There is no other reasonable way to explain these data. The secular sea level curve, based on evolutionary biases, is wrong.

Note that the new sea level curve has no numerical values because it is only a relative curve. I also started the curve a bit above the modern ocean sea level as the pre-Flood world seems to have extensive shallow seas spread across portions of many of the continents (see chapter 17). I also added a few representative fluctuations in the Absaroka, Zuni, and Tejas. These are to illustrate smaller shifts in sea level even within

Fossil *Seymouria baylorensis*

the megasequences. This is especially true in the Tejas as the water was "continually receding" or "assuaging" (Genesis 8:3).

The new sea level curve also finishes at the end of the Tejas above the current ocean level. Recall that the end of the Flood had no ice sheets initially, so sea level was over 300 feet (130+ meters) higher.[4] Of course, as ice accumulated rapidly during the ensuing Ice Age, water was removed from the ocean, causing a sudden drop in sea level. Today, the two major ice sheets on Greenland and Antarctica still keep sea level over 200 feet (70 meters) lower than at the moment the Flood ended.[4]

Cultural Evidence for a Global Flood

In addition to the geological data, there is ample cultural and even language data that independently verify a global flood occurred and that the human race spread across the globe after the Tower of Babel. There is also strong evidence that many cultures took similar building techniques with them as they scattered and eventually re-established themselves.

Restored Sumerian ziggurat temple at ancient Ur, Iraq

Flood Traditions

Dr. John Morris, president emeritus of ICR, has compiled over 200 Flood legend-type stories from people groups around the globe.[5] He found that in most of these stories, which had been passed down orally from generation to generation for thousands of years, there were similar themes. He summarized the common themes and areas of overlap within all of the stories in a single paragraph:

> Once there was a worldwide flood sent by God to judge the wickedness of man. But there was one righteous family that was forewarned of the coming flood. They built a boat on which they survived the flood along with the animals. As the flood ended, their boat landed on a high mountain, and they descended and repopulated the whole earth.[5]

Dr. Morris noted the overwhelming similarity of this account to the biblical account of Noah and the great Flood. He further elaborates:

> The only credible way to understand the widespread, similar flood legends is to recognize that all people living today—even though separated geographically, linguistically, and culturally—have descended from the few real people who survived a real global flood on a real boat that eventually landed on a real mountain. Their descendants now fill the globe, never to forget the real event.

> But, of course, this is not the view of most modern scholars. They prefer to believe that something in our commonly evolved psyche somehow forces each culture to invent the same imaginary flood legend with no basis in real history. Instead of scholarship, this is "willful ignorance" of the fact that "the world that then existed perished, being flooded with water" (2 Peter 3:5-6).[5]

Over 200 Flood legends from around the globe, from nearly every cultural group, telling roughly the same story, are compelling evidence for a universal flood that affected the entire world. This is exactly what is described in the book of Genesis. But, as Dr. Morris noted, secular scholars dismiss these data because they do not want to believe they are steeped in a factual event. That would counter their entrenched worldview.

Chinese Language

Chinese is not an alphabet language like English or most other European languages, rather it is based on both abbreviations of and combinations of picture symbols.[6] But where did many of those symbols originate? Dr. James Johnson, my colleague at ICR, explains it this way:

> During the invention of a pictographic language, creating pictographic "words" involved selecting picture symbols that were relevant and meaningful to whoever invented those written symbols. But what motifs would signify meanings that the ancient Chinese would portray about 4,500 years ago? What ideas were familiar to those who invented China's original written language?
>
> Since Chinese civilization began soon after the Tower of Babel fiasco, the first Chinese settlers still had a fresh memory of mankind's origins—from creation week to the dispersion of languages at Babel. Thus, they not only knew the history highlights in Genesis 1–11, but they would also have regarded those same events as important in human history and experience. It is unsurprising, therefore, that many of the picture-symbol characters, in the ancient Chinese language, match the thinking of a soon-after-Babel people who retained important memories of historic events reported in Genesis 1–11.[6]

"Over 200 Flood legends from around the globe, from nearly every cultural group, telling roughly the same story, are compelling evidence for a universal flood that affected the entire world."

Dr. Johnson further explained that many of the Chinese pictographs reflect events recorded in the Bible that occurred prior to the Tower of Babel and the scattering of the nations. The Chinese people took these events and molded them into their written

Create

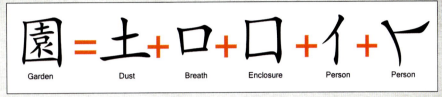

Garden

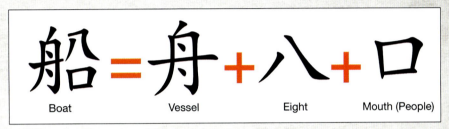

Boat

Many Chinese pictographic words are based upon events recorded in Genesis 1–11, such as the global Flood[7]

language that is still used to this day. Many Chinese pictographs reflect biblical subjects—like God's creation of the world, the Garden of Eden, and a global flood event.[6]

Dr. Johnson described how these pictographs were once used as a powerful apologetics insight with a Chinese graduate student. Until he spoke with Dr. Johnson, the student had never realized that the written language he had used all his life had its roots in the events recorded in Genesis 1–11. Dr. Johnson recounted their exchange below:

> In 1990, a graduate student from communist China—raised on atheistic evolution—asked me the following question: "Why should I believe in the Bible God, the Bible is true, and God is fair, when China was never given Bible truth about God to believe?" Simply put, this young

> **"His own language contained latent clues that the Bible's early history was once well known to the Chinese people."**

man was asking: "Why should I believe in your Bible's God?" and "Why should I believe in your God's Bible?"

Recalling that I learned somewhere that the Chinese character for "flood" somehow contained the symbol for "eight," I asked my Chinese friend to write out the Chinese word for flood, and to describe what its component symbols represented. As indicated above, his description of flood included the number eight—a fact he had no explanation for, other than he guessed that it might have once been a phonetic symbol, similar to how "4" can be shorthand for "for" or "8" for "ate."

Then I read 1 Peter 3:20 to him and pointed out how Genesis 6–10 reports that *exactly eight humans survived the global Flood*, a fact that perfectly made sense of the Chinese pictographs. Then, he added that the Chinese character for "boat" also contained the number eight, and he began to realize that *his own language contained latent clues that the Bible's early history was once well known to the Chinese people.*

After further discussion about how the biblical God is a loving shepherd who seeks to secure wandering sheep into His heavenly sheepfold (Psalm 23; Luke 15; John 10), my friend concluded that, long ago, the Chinese people had known the truth about the God of the Bible, including the early history of God's dealings with mankind as Genesis records, but that somehow this precious truth had been lost or wasted. During the wee hours of the morning, with joy in knowing that God had caringly revealed Himself to the Chinese people, my friend trusted Christ as his personal Savior, and he has enjoyed belonging to Him since (Luke 15:7; Romans 4:3; Luke 10:20).[6]

The memory of the Flood is pervasive. As we saw above, it even influenced the written language of some cultures. Ask yourself, how can all of these Flood traditions from all over the globe have nearly the same themes?

Chichen Itza, Mexico

Globally Similar Building Techniques

In addition, ancient cultures all over the globe built similar-looking step pyramids. These structures, called *ziggurats*, were commonly constructed by the ancient Assyrians, Sumerians, and Babylonians who lived in the Middle East, exactly where the Tower of Babel was built, according to the Bible. These ancient structures were usually made of bricks just like the Bible described for the Tower of Babel (Genesis 11:3-4).

Ziggurats are also found across Central America, South America, and Egypt.[8] Mounds and pyramids were also commonly constructed by the cultures spread throughout ancient China and in North America.[3] The common nature and construction style of pyramids across the globe, and step pyramids in particular, are strong evidence that these distant cultures shared a common memory and a common origin. What are the odds that so many diverse cultures would build the same types of temples or religious shrines from continent to continent without the means of a direct communication system? Only the spreading of peoples from a single common site, as at Babel, can explain the universal construction of the ziggurats and pyramids globally. They must have taken the memories of the construction plans/concepts for building a

Tower of Babel with them. In my opinion, the biblical account of the Tower of Babel is confirmed by these archaeological discoveries.

The Geological Data All Make Sense

There was one pivotal moment over the course of this project when I realized I was onto something big, and it was because of the grace of God and a little bit of understanding He gave to me. This was my so-called "ah-ha" moment. I was out for a run one day and the column data I was accumulating and mapping were going through my head, when it struck me: It all makes sense! The rock data I was compiling matched exactly with the description of the great Flood in Genesis. My research was confirming the truth of the biblical account! There really was a global flood in recent history, and we can see its effects everywhere yet today.

Geological Evidence Shows a Global Flood

All geology we observe only makes sense in the context of the Flood. In fact, the enormity of the Flood event was beyond anything any human could comprehend. We merely see a small sliver of God's genius in our study of Flood geology. Only our omniscient God could have thought through the mechanism of plate tectonics and planned the processes of the Flood down to the smallest details. Only our omnipotent God could have brought into action the judgment of the Flood, caused the breakup of the fountains of the great deep, shifted the locations of the continents, and planned for an Ice Age to get the animals to the newly separated landmasses. Only our omnipresent God could have monitored the processes of the Flood at every corner of the globe simultaneously while saving Noah and seven others aboard an ark, along with at least two of every kind of land-dwelling creature.

> "The rock data I was compiling matched exactly with the description of the great Flood in Genesis."

Catastrophic plate tectonics was the key to the 314-day global Flood. Movement of the plates generated tsunamis that impacted the continents. The creation of progressively more and more seafloor pushed the seafloor up higher and higher, causing the tsunami waves to reach higher and higher. This progression eventually caused the floodwaters to go over the top of even the highest pre-Flood hills. At each new level, the Flood buried and entombed different organisms based on ecology. It was a global phenomenon, so we find the same fossils buried at the same time on different continents.

The water was then drawn back off the continents as the seafloor cooled and sank. So again, only the process of catastrophic plate tectonics can explain the complete flooding of the continents by making hot new seafloor that pushed up the water, followed by the subsequent cooling and sinking of the seafloor, draining the water off the landmasses and back into the deep ocean basins.

Catastrophic plate tectonics also set up the conditions for the Ice Age after the Flood. The seafloor that was created hot and new from mantle magmas provided heat to the oceans to cause high evaporation rates for centuries. Then the subduction zone volcanoes, because they are formed by the partial melt of oceanic lithosphere and are concentrated in silica and extremely explosive, spewed ash and aerosols into the sky for several centuries also. These aerosols cooled the earth after the Flood by blocking

out a small portion of normal amounts of sunlight. High ocean evaporation rates resulted in condensation that fell down to the earth as snow and quickly developed into thick ice sheets in just a few hundred years. Not any old volcano would do, though; only the subduction-generated volcanoes have the right chemistry to be explosive. All of these conditions, working in perfect sync, could only have been linked together by our great God.

The sedimentary rocks clearly illustrate a global pattern. The first megasequence began with only minimal coverage of the continents, progressing steadily to more and more coverage as each subsequent megasequence was deposited upon the previous layers. And each continent showed the same high-water point near the end of the Zuni Megasequence. By following the data, we were able to map out the progression of the Flood from Day 1 until Day 314, when the earth was again dry (Genesis 8:13). There is an undeniable match to biblical history!

The order of the fossils is best explained by ecological zonation. Humans didn't evolve from sea creatures. The rocks simply became a record of what was buried by the sediments of the Flood, starting with the Cambrian Explosion. As the waters were

pushed higher due to increasing ocean floor production, more and more ecological zones were inundated, creating more and different fossils, including land animals and plants. And we noticed that the mixing of land and sea fossils is a global phenomenon. This is entirely what you would expect in a global flood.

The geological data presented in this book are different from anything previously published elsewhere. This research is as data-based as possible. When I began this project, I had no idea where the data would lead. Like a lot of scientists, I began to gather and collect data, hoping some sort of meaningful result would materialize. Little did I know that the data would lead me on a journey of unforeseen discovery, a journey that led directly to seeing the hand of God in action.

My interpretations came after the data were input and the maps were made. I tried to remain as unbiased as possible and follow the data. However, these data show the Flood was a real, one-time event that actually happened in history. We cannot use uniformitarianism to study a one-time event like the Flood.

God was the only witness outside of the Ark to the power of the Flood. And yet, we see clear evidence that the whole earth was affected, including the crust. This book provides insight into the devastation that actually occurred. It bears witness to a flood that progressively destroyed the earth, ecological zone by ecological zone, as the water rose higher and higher. And it all matches and confirms the narrative in the book of Genesis.

Evidence Further Supports a Recent Flood

The geological evidence presented in this book also shows the Flood was a recent event, occurring just thousands of years ago. Sedimentary rocks are not old. Fossils cannot be millions of years old and contain original tissues and proteins. Oil cannot be millions of years old either. The rocks and fossils tell us there was a recent global flood that completely destroyed the original earth. This was a judgment for the wickedness of humanity.

The Ark as a Type of Salvation

Although God judged the wickedness of the pre-Flood world, He provided a way of salvation for those who chose to believe Him. He had Noah build a massive ark that could have held many more people, but only eight people believed a flood was imminent. No one else had the faith to believe something that they had never seen. Noah preached and tried to convince many people to no avail. Entering the Ark through the open door was the only way of physical salvation for anyone who believed.

Conclusion: It All Points to Christ

Only Christ provides salvation. Today, Christ's shed blood and resurrection have become God's way of salvation for all mankind. He has provided redemption for the sin of Adam and Eve and for all of us. God Himself has opened the door to salvation through His shed blood on the cross. He is the open door we enter through to gain eternal life through His grace. We just need a little faith (Ephesians 2:8).

God is not an impersonal God as some old-earth adherents seem to imply. He did not start His creation with a Big Bang and let it run its course through eons of evolution while standing back passively and watching. The data do not support this. The geology presented in this book does not support this. And the Bible does not support

this. God is an active, personal God who cares for every human. He loves every human equally. We all come from Adam and Eve and from the eight who entered the Ark. We were all created in His image. He is a loving God. He has a plan for everyone (Jeremiah 29:11). He desires everyone to believe in Him, repent, and be saved.

I pray this book challenges unbelievers and reassures believers. The evidence of a global flood is clearly there to be seen in the rocks around the world. We just need to cast off the blinders of uniformitarianism and deep time and look at the real rock data with unbiased eyes. Following the data leads to a confirmation of the biblical worldview. There was a first global judgment, and God has promised to come again and judge the world a second time. Are you ready to meet God face to face?

References
1. Vail, P. R. and R. M. Mitchum Jr. 1979. Global cycles of relative changes of sea level from seismic stratigraphy. *American Association of Petroleum Geologists Memoir.* 29: 469-472.
2. Whitmore, J. H. and R. Strom. 2010. Petrographic analysis of the Coconino Sandstone, Northern and Central Arizona. *Geological Society of America Abstracts with Programs.* 41 (7): 122.
3. Clarey. T. 2017. South America Shows the Flood Progression. *Acts & Facts.* 46 (3): 9.
4. Clarey, T. L. 2016. The Ice Age as a mechanism for post-Flood dispersal. *Journal of Creation.* 30 (2): 48-53.
5. Morris, J. D. 2014. Traditions of a Global Flood. *Acts & Facts.* 43 (11): 15.
6. Johnson, J. J. S. 2015. Genesis in Chinese Pictographs. *Acts & Facts.* 44 (3): 18-20. Emphasis in original.
7. Kang, C. H. and E. R. Nelson. 1979. *The Discovery of Genesis: How the Truths of Genesis Were Found Hidden in the Chinese Language.* St. Louis, MO: Concordia Publishing House; Nelson, E. R., R. E. Broadberry, and G. T. Chock. 1997. *God's Promise to the Chinese.* Dunlap, TN: Read Books Publishers.
8. Hodge, B. 2013. *Tower of Babel: The Cultural History of Our Ancestors.* Green Forest, AR: Master Books.

IMAGE CREDITS

t: top; b: bottom; l: left; r: right

AAAS: 140

Alataristarion (via Wikipedia): 327b

Scott Robert Anselmo: 301

Arkansas Geological Society: 272t

Astroskiandhike via Wikipedia: 332

John Baumgardner: 138, 315

Bigstock: 7bm, bl, r, 8bl, br, 9bl, 10b, 11-12, 13br, 14, 19, 22, 24, 29, 31-32, 37, 41, 44, 58-59, 71, 74, 82, 88-89, 95, 104-107, 109t, 112, 115, 122-123, 129, 131, 145, 154-155, 158t, 162-163, 166, 168-170, 173, 175, 182, 188, 191, 195, 214, 230, 232, 235, 257, 264-265b, 271, 277, 281, 283, 294, 309, 328-329, 331, 333, 337, 339, 343, 345, 347-348, 350, 355-358, 360b, 362, 364-365b, 366, 375-376, 391, 393, 397, 404, 408, 409t, 419-420, 424, 426-427, 429, 435, 437, 442, 448, 452, 460, 475-477, 481, 483-484, 486

Art Chadwick: 202, 224, 242, 264l, 290, 322

Lisa Christiansen/Caltech Tectonics Observatory: 142

Timothy Clarey: 9br, 10t, 13t, bl, 26-27, 36, 42-43, 45-46, 52, 55, 67, 69, 77, 96r, 108, 135t, 136, 157, 165, 183-184, 203-204, 206, 209, 244, 314, 335, 392, 394-395, 398, 405, 409b, 421, 431, 434

J. D. Dieterle: 160-161

Energy Information Administration: 251

Alexandra Forte, used by John Baumgardner in 2003 ICC publication. Used here in accordance with federal copyright (fair use doctrine) law: 463

Glasgow Museums and the Glasgow City Council (reproduced courtesty of): 439

GoogleEarth: 34

Magnus Hagdorn: 466-467

Heritage Auctions: 274tr

Carol Hill et al: 342

ICR: 57, 91, 94, 144, 181, 248, 291, 302, 304, 341, 361, 388-390, 456-457

iStock Photo: 130, 192, 252, 372, 411, 413

International Seismological Centre: 121

Nathaniel Jeanson: 111, 243, 278

Lawrence Livermore National Laboratory: 73

MagentaGreen (via Wikimedia Commons): 423

MC Dinosaur Hunter (via Wikipedia): 109b

John Morris: 8tl, 23, 38, 62b, 97t, 99, 272b, 432, 461

NASA: 21, 64, 146, 360t, 369, 407, 450-451, 464

NOAA: 119, 124, 148, 270b, 306, 313

NPS: 96l, 207, 299b, 346, 374, 378

NPS/Jo Suderman: 359

NSF: 310

PawełS: 267

Will Perry: 336

Poozeum (via Wikipedia): 97b

Public domain: 7tl, 61, 62t, 65, 79, 116-117t, 118, 274tl, b, 297

Public domain (taken from Nelson et al): 479

C. D. Rowe, F. Meneghini, C.J. Moore, A. Yamaguchi, and A. Tsutsumi (modified from unpublished work): 135b

Virgil L. Sharpton/Lunar and Planetary Institute: 299t

Martin R. Smith (via Wikipedia): 93

Mary Smith: 373

Andrew Snelling (adapted image): 103

Dave Snowden: 276

Dave Souza (via Wikipedia): 68

Raimond Spekking: 110

James St. John: 416, 441

Timmer26 (via Wikimedia Commons): 425

Jeffrey Tomkins: 351, 440

USAP: 158b

USGS: 117b, 125-127, 132, 139, 141, 159, 330, 379, 433

Utah Geological Survey: 344

Davis J. Werner: 153, 177b, 178-180, 185-187, 189, 196-201, 205, 210-211, 217-223, 225-226, 228-229, 236-241, 245, 247, 250, 258-263, 268-269, 270t, 284-289, 292-293, 295, 317-321, 324-325, 327t, 334, 349, 364l, 380-387, 402-403, 412, 436, 462, 470-472

Wiemann et al, *Nature Communications*, Figure 1 excerpt, creativecommons.org/licenses/by/4.0: 85

Susan Windsor: 33, 50, 70, 78, 92 (with elements from Bigstock), 156, 249, 363, 415, 422

INDEX

Abraham, 87, 367, 428

Acadian Orogeny, 234, 244

Adam, 60, 64-65, 486-487

Agassiz, John Louis, 357

Alaminos Canyon, 334

allochthonous, 394, 432, 434-435, 437-438, 440-441, 444

Alpine-Himalayan Belt, 329

Alps, 126, 297, 329, 355-357, 391-392

Alvarez, Walter, 297, 392

Andes Mountains, 125, 199, 291, 298, 318, 326, 328, 332, 364

angiosperm(s), 109-110, 170, 323, 340

angular unconformity(s), 66, 68, 266, 462, 467

anticline(s), 422-423

Antrim Shale, 250

Appalachian Basin, 185, 250

Appalachian Mountains, 34, 186, 207, 235, 266

Archaeopteryx, 83, 89, 108, 110

Archean, 129-130, 161, 163-165, 410, 414

arthropod(s), 105-107

asphalt, 371, 419, 427-428

asthenosphere, 123, 125, 132, 143

Austin, Steve, 80-81, 132, 136, 153, 160, 165, 343, 345, 431, 438

autochthonous, 432, 437

Bacon, Francis, 21-22, 116

banded iron formations (BIFs), 160-162, 164, 168, 416

bathtub ring, 284, 293, 308, 469

Baumgardner, John, 33, 131, 134, 138-139, 142, 144, 146, 155, 159, 216, 245, 257, 269-270, 315-316, 329, 340, 399, 406, 464

Beartooth Mountains, 81, 85

Becquerel, Henri, 61-62

Belt Supergroup, 166-167, 180, 182

Bidahochi Formation, 345

Bighorn Basin, 329

biochemical sedimentary rocks, 36

biochemical sediments, 32

biodegradation, 425, 428

biodegraded, 428, 431

Black Hills, 330

black shale, 93, 250-252, 421, 444

blanket sand, 338, 462

blanket sandstone(s), 51, 197, 201, 222, 240, 305, 386, 396

blue stars, 446, 449-451

bracketing, 78

Brazilian Shield, 186

Brown, Walt, 343

Bryce Canyon, 249

Burgess Shale, 93

Caledonian Orogeny, 217, 234

Cambrian Explosion, 90, 104-107, 176, 194, 202, 213, 229, 484

Canadian Rockies, 112, 186, 330, 393

Canadian Shield, 157, 161, 470

Canyonlands Lake, 343-344

Cape Fold Belt, 222, 240

carbon-12, 72-73, 456, 458-459

carbon-13, 72, 456-458

carbon-14, 72-75, 77-78, 80, 455-459

carbon-14 analysis, 72, 456-457

carbon-14 dating, 72-73, 458

carbonization, 95, 97

Cardenas Basalt, 81, 85, 182

Cascade Mountains, 332, 363

Casper Sandstone, 272

casts, 95, 97, 439, 441

catastrophic plate tectonics, 24, 28, 114-115, 128, 131-132, 138-147, 149-150, 174, 217, 333, 376, 462-464, 482-483

Catskill Delta, 244-245

Cenozoic, 73, 75, 95, 102, 109-110, 249, 283, 318, 324, 326, 333-334, 339, 341, 349-352, 355, 434, 442-443

Chadwick, Art, 91, 201-202, 224, 242, 264, 290, 322

Chattanooga Shale, 38, 388

chemar, 427

chemical fossils, 82

chemical sedimentary rocks, 36-37

Chicxulub crater, 298-300

Chinese language, 478-480

Chueng, Stephen, 272

Circum-Pacific Orogenic Belt, 329

clastic dike(s), 298, 394

clastic rocks, 31, 35

clastic sedimentary rocks, 35

clastic sediments, 31-32

coal beds, 108, 273, 340, 408, 432, 434-436, 438, 441-444

Coconino Sandstone, 45-46, 258, 272, 306, 404, 459, 461, 474

coelacanth, 277, 411

Coelophysis, 69, 295

collagen, 77, 82, 85, 99, 454, 458

Columbia River Plateau, 330, 334

comet(s), 65, 449, 451-452

Congo Basin, 189

continental drift, 116-120

convergent boundary(s), 121, 123-125, 128, 142

corals, 96, 104, 108-109, 168

COSUNA, 24, 55

craton(s), 54, 57, 158, 187, 189, 206-207

cross-bed(s), 18, 31, 37, 42-46, 201, 272, 306

cross-bedded sandstone, 275, 279, 474

cyanobacteria, 163, 410

cyclothems, 275-276, 432, 444
Darwin, Charles, 61, 109
de Buffon, Georges, 116
Deepwater Horizon, 425
Dinosaur Peninsula, 185, 197, 218, 236, 247, 249-250, 293, 295, 402, 404, 442
disconformity(s), 66, 68-69
East Pacific Rise, 268, 333
ecological zone(s), 112, 299, 396, 414, 416, 484-485
Ediacaran, 106, 176, 370
Eoconfuciusornis, 84
extinction(s), 101-104, 168-169, 273, 298-301, 355, 378, 382
Falkland Islands, 278
Farallon Plate, 296
faunal succession, principle of, 70, 80, 87, 383
floating forest(s), 273, 418, 435-437, 440-442, 444
Flood traditions/legends, 477-478, 480
flute(s), 43-44
footprint(s), 97, 374-375, 400
Fort Union Formation, 442
fossil fuels, 28-29, 40, 418, 444
Fossil Grove, 438-439, 441
fossil record, 24, 28, 48, 51, 87, 90, 92-93, 101, 104-105, 107-110, 112-113, 170, 273, 350, 355, 378, 381-383, 403, 414, 416, 468, 473
fountains of the great deep, 28, 31, 37, 143, 156, 168, 172, 175-176, 184, 194, 208-209, 312, 314-315, 365, 482
Four Corners, 346-347
fracking, 251, 419, 425, 431
Franciscan Complex, 296
Front Range, 329
Garden of Eden, 152, 400, 406, 416, 479
Gippsland Basin, 430
glacial erratics, 358
Glacier National Park, 330

Glasgow, 273, 438-439
Gomorrah, 428
Gondwana, 245, 256, 262, 407
Gorman, James, 90, 98
graded bedding, 43-44
Grand Canyon, 23, 38, 42, 67, 69, 81, 165, 181-182, 202-203, 248-250, 306, 312, 336, 338, 342-349, 402-403, 415-416, 459-461
Grand Canyon Supergroup, 181-182, 203
Grand Lake, 343-344
Great Lakes, 345, 358
Great Salt Lake, 345
Great Unconformity, 67, 69, 165, 194, 202-204, 213, 229, 338, 461
Green River Formation, 277, 456
Greenland, 117, 181, 186, 278, 284, 296, 316, 324, 357, 368, 403, 406, 476
Grenville Orogeny, 185
Gulf of Mexico, 36, 189, 284, 293, 300, 304-306, 316, 325, 334-335, 338-340, 346, 425, 428
gymnosperms, 108, 169
Hawaiian Islands, 34-35
Heart Mountain Fault, 112, 393
Hebert, Jake, 368, 463
Hell Creek Formation, 95, 301, 303, 457
Hermit Shale, 459, 461
Hess, Harry, 119-120, 124, 142
human fossils, 348-349, 405, 416
Hopi Lake, 343-345
Horner, Jack, 90, 98, 248
Hudson Bay, 208, 218, 284, 291, 405
Humphreys, Russell, 323, 464
Hutton, James, 66, 266-267, 462, 467
hydrocarbon(s), 251, 418, 422, 427-428
hydroplate theory, 128, 150, 333
hydrothermal, 37, 40, 162, 176, 189, 208-209, 367, 429

Iapetus Ocean, 216-217, 234, 244-245, 253, 268, 436
ichthyosaur(s), 83, 273-274
Illinois Basin, 208, 236, 250, 275, 291
iridium, 300
intracratonic basins, 205-209
Jeanson, Nathaniel, 110-111, 243, 278
Job, 366-367
Johnson, James, 478-480
Kaibab uplift, 342-344, 347
Keathley Canyon, 334
keratin, 83-84
Keweenaw Peninsula, 183
Kodiak Island, 26-27, 135-137
kopher, 426-427
K-Pg, 298, 300, 339-341, 345, 349-350, 355
K-T, 298, 339
Lake Bonneville, 344
Lake Michigan, 344, 387
land bridge(s), 352, 355, 373-376
Laramide Orogeny, 298, 316, 328
Late Heavy Bombardment, 410
Laurasia, 116, 262
Lewis Overthrust, 112, 167, 393
Lisle, Jason, 450-451
lithosphere, 121-128, 134, 139-140, 142-147, 149, 159, 190, 194, 217, 225-226, 232, 255, 258, 269, 314-315, 361, 406, 462, 465, 483
living fossil(s), 164, 411
loess, 369-370
Louann Salt, 189, 284, 293, 304
low-velocity zone, 123
Lower Wilcox, 316, 325, 334
lowland(s), 273, 279, 293, 295, 400, 402-404, 412-413, 442, 468, 473
lycopod(s), 108, 110, 273, 279-280, 404, 437-442
Lyell, Charles, 66
Lyons Sandstone, 272

magnetic field(s), 65, 368, 446, 449, 451-453, 464

magnetic reversal(s), 139-140, 368, 463-464

Maiasaura, 91, 248, 307

Marcellus Shale, 250, 388

melanosomes, 82-84

Mesoproterozoic, 30, 161, 167

Mesozoic, 72, 95, 102, 108, 166, 168, 233, 249-250, 283, 367

Metasequoia, 110, 442

Michigan Basin, 55-56, 185, 236, 250

Microraptor, 83

Midcontinent Rift, 55, 181-186, 206-209

Midway Shale, 325

Milankovitch theory, 367-368, 370

Milky Way, 448-449

Mississippi Canyon, 305

molds, 95, 97

Montgomery, David, 426

moraine(s), 357-358, 366

Morris III, Henry, 20, 426-427

Morris, Henry, 111-112, 296, 366-367, 390

Morris, John, 25, 434, 453, 477

Morrison Formation, 56-57, 299, 382, 388, 457, 459

mosasaur(s), 83, 90, 109, 282, 291, 407

Mount St. Helens, 79-80, 125, 127-128, 150, 361, 369, 433-435, 437-438

Mt. Pinatubo, 125, 148, 363

Mt. Vesuvius, 125

Muav Limestone, 384, 460

mudcracks, 43-45

natural gas, 305, 418-419, 421-422

Navajo Sandstone, 272

Neogene, 313, 316, 322, 339-341, 346

Neoproterozoic, 30, 167, 176, 186, 206

New Albany Shale, 250

New Red Sandstone, 272

Nile River, 320, 366-367

Noah, 91, 115-116, 149, 172, 192, 212, 231, 253, 280, 312, 371, 426, 435, 477, 482, 485

nonconformity(s), 15, 66, 68-69

Norphlet Sandstone, 304-305

North Sea, 278, 428

N-Q, 339, 341, 355

Oard, Mike, 344, 346, 359, 363, 365-367, 373

ocean ridge(s), 114, 120, 122-124, 126, 128, 139-140, 143, 149, 172, 174-176, 258, 268, 271, 295, 464

Ogallala Formation/Sandstone, 305, 335-336, 338, 341, 349

oil seep(s), 418, 423, 428

oil window, 421-423, 431

OmniGlobe, 405, 407

Oort Cloud, 452

organic sedimentary rocks, 36

original tissues, 81-82, 85, 90, 95, 100, 454, 485

Orion's belt, 450

Ouachita Mountains, 245

overthrust(s), 111-112, 167-168, 266, 296-298, 378-379, 389-395, 398

Pacemaker paper, 368

Paleogene, 282, 284, 300, 313, 316, 322, 340-341, 431

paleontology, 63, 90, 454

paleosterilization, 429

Paleozoic, 37, 72, 95, 101, 104, 107-108, 110, 168, 205, 207-208, 248-249, 275, 283, 370

Paleozoic basins, 207

Palo Duro Canyon, 336-338, 348

Pangaea, 116, 118, 128-129, 157-160, 178-179, 190, 204-205, 226, 244, 247, 254, 256-257, 262, 266, 271, 283, 296, 308, 324, 365, 403, 405-406, 412

Pasagshak Point, 135, 137

Peleg, 147, 149, 373, 427

Perdido Fold Belt, 334

Permian Basin, 208, 250

petrifaction, 95-96

Phanerozoic, 72, 101-102, 283, 401

phytosaurs, 336

Pierre Shale, 56, 382, 388

pitch, 418, 426

plate tectonics, 28, 114-115, 120-121, 127-128, 131, 149, 152, 296, 315, 354, 357, 390, 464, 483

Playfair, John, 356

Pleistocene Epoch, 356

Powder River Basin, 91, 340, 434-435, 442-443

precession, 367-368

Proterozoic, 163-164, 178, 190, 410, 414

pseudotachylyte (PST), 135-137, 142

pteridophytes, 169

Purdom, Georgia, 164

pyramids, 481

Quartermaster Formation, 336

Queenston Delta, 216-218, 244

radioisotope(s), 23, 61, 63, 72-74, 76, 78, 80-81, 84-86, 164, 448, 465

radioisotope dating, 61, 74, 78, 80, 84, 86, 448

radioisotope method(s), 63, 73-74, 80-81, 84

raindrop prints, 43-44

Radioisotopes and the Age of the Earth/RATE, 72, 81, 156

recrystallization, 95-96

Red Sea, 20, 36, 183, 189, 320, 327

Redwall Limestone, 38, 243, 248, 343, 460

Reed, John, 184-185

Rift Valley, 140, 184-185, 188, 327

Ring of Fire, 142, 146, 332, 363

ripples, 39, 43-44, 201, 255, 440

Rockworks 17, 25, 55-56

Rocky Mountains, 171, 207, 250, 284, 316, 318, 320, 326, 328-330, 332, 338, 348

Rodinia, 157-160, 405-406

Ross, Marcus, 51, 350-352, 371, 379

Rowe, Christen, 137

San Andreas Fault, 124, 126-127, 330, 332

Schweitzer, Mary, 82-83, 98, 100, 455

scour marks, 43-44, 201

sea level curve, 228-229, 231, 373, 474-476

seafloor spreading, 119-120, 122, 140, 190, 266, 283, 314, 340, 464

sedimentary structures, 18, 31, 43, 45, 47, 165

seismic tomography, 134, 138, 140, 150, 316, 462-463

Sevier Orogeny, 298

shale oil, 419, 424

shallow sea(s), 29, 194, 204, 212-213, 225, 229, 249, 254, 256, 269, 295, 403, 406, 412-413, 475

sheet sand/sandstone, 210-211

Siccar Point, 68, 266-267, 462, 467

siliciclastic(s), 199, 220, 238, 258, 260, 262, 284, 286, 288, 316, 318, 320

Sinosauropteryx, 83, 94, 302

Sixtymile Formation, 165, 167, 182

Sloss, Laurence, 48, 53, 228, 243, 349

Smith, William, 71

Snelling, Andrew, 143, 164, 355, 366, 430

Snider-Pellegrini, Antonio, 116

Sodom, 428

soft tissue(s), 65, 82-85, 92, 96, 98-99, 100, 113, 430, 443-444, 454-455, 458

solar radiation, 368

solar system, 65, 114, 449, 452-453, 465

South Fork Fault, 394, 398

Spinosaurus, 277, 303

spiral galaxy(s), 446, 449-450

St. Peter Sandstone, 211, 218, 225-226, 243

Steno, Nicolas, 65-66

stream piracy, 342

stromatolite(s), 101, 161-164, 168, 189, 400, 409-414, 416

subduction initiation, 190

subduction zone(s), 123, 125, 130, 133, 135, 137, 142, 144, 146, 150, 216, 268, 316, 329, 332-333, 361, 364, 370, 376, 483

Suess, Eduard, 116

Taconic Orogeny, 216, 226

Tapeats Sandstone, 42, 51-52, 67, 165, 197, 202, 210, 212, 248, 380, 396, 460-462

Taylor, Frank Bursley, 118-119

Tecovas Formation, 336

Tensleep Sandstone, 272

Tertiary, 103, 313, 316, 322, 341, 370

Tomkins, Jeffrey, 273, 351

Tower of Babel, 341, 350, 371, 374, 427-428, 476, 478, 481-482

Transantarctic Mountains, 98, 407

Transcontinental Arch, 197, 218, 236, 250

transform boundary(s), 124, 126-127, 190

Triceratops, 69

trilobites, 104-108, 158-159, 168, 182, 245

Trujillo Formation, 336, 338

Tyrannosaurus rex (*T. rex*), 82, 95, 98-100, 109, 276-277, 301

unconformity(s), 48, 66-69, 87, 165, 194, 202-204, 213, 266-267, 338, 462, 467

upland(s), 248, 283, 324, 403-405, 412-414

Ussher, James, 60-61, 63

Valley of Siddim, 428

Venetz-Sitten, Ignaz, 357

Vishnu Schist, 202

von Humboldt, Alexander, 116

Walcott, Charles, 410

Walker Ridge, 334

Warrawoona Group, 410

weathering, 31-33, 36, 88, 203

Wegener, Alfred, 114, 116-120, 123

Wentworth scale, 35

wetland(s), 108, 163, 273, 402, 404, 441

Wheeler, Harry, 53

Whitcomb, John, 111-112, 296, 390

Whitmore, John, 46, 272, 474

Whopper Sand, 304-305, 316, 325-326, 334-335, 338-340, 346, 349

Williston Basin, 185, 236

Wilson, J. Tuzo, 120

Wise, Kurt, 162, 165

Woodford Shale, 250

Yellowstone, 316, 330, 333-334, 359, 363, 429

Yucatán Peninsula, 299-300, 305

ziggurat(s), 476, 481

ABOUT THE AUTHOR

Dr. Timothy Clarey received a Master of Science in Geology in 1984 from the University of Wyoming and a Master of Science in Hydrogeology in 1993 from Western Michigan University. His Ph.D. in Geology was received in 1996 from Western Michigan University. From 1984 to 1992, Dr. Clarey worked as an exploration geologist at Chevron USA, Inc., developing oil drilling prospects and analyzing assets and lease purchases. He was a professor at a public college for 17 years before leaving in 2013 to join the science staff at the Institute for Creation Research, having earlier conducted research with ICR in its FAST program. He has published many papers on various aspects of the Rocky Mountains and has authored two college laboratory books. He is the author of *Dinosaurs: Marvels of God's Design* and a contributor to *Guide to Dinosaurs*. He and his wife, Reneé, are co-authors of the children's book *Big Plans for Henry*.

ABOUT THE INSTITUTE FOR CREATION RESEARCH

After 50 years of ministry, the Institute for Creation Research remains a leader in scientific research within the context of biblical creation. Founded by Dr. Henry Morris in 1970, ICR exists to conduct scientific research within the realms of origins and Earth history and then to educate the public both formally and informally through professional training programs, through conferences and seminars around the country, and through books, magazines, and media presentations. ICR was established for three main purposes:

Dr. Henry Morris

Research. ICR conducts laboratory, field, theoretical, and library research on projects that seek to understand the science of origins and Earth history. ICR scientists have conducted multi-year research projects at key locations such as Grand Canyon, Mount St. Helens, Yosemite Valley, Santa Cruz River Valley in Argentina, and on vital issues like Radioisotopes and the Age of the Earth (RATE), Flood-Activated Sedimentation and Tectonics (FAST), the human genome, soft tissue in fossils, and other topics related to geology, genetics, astro/geophysics, paleoclimatology, paleobiochemistry, and much more.

Education. ICR offers formal courses of instruction and conducts seminars and workshops, as well as other means of instruction. With 40 years' experience in education, first through our California-based science education program (1981–2010) and now with programs offered through the School of Biblical Apologetics, ICR trains men and women to do real-world apologetics with a foundation of biblical authority and creation science. ICR's online programs include a one-year, non-degree training program for professionals called the Creationist Worldview and an Origins Matter Short Course series. Additionally, ICR scientists and staff speak to numerous groups each year through seminars and conferences, as well as offering live science presentations at the ICR Discovery Center for Science & Earth History.

Communication. ICR produces books, videos, periodicals, and other media for communicating the evidence and information related to its research and education. ICR's central publication is *Acts & Facts*, a free full-color monthly magazine with a readership of over 250,000, providing articles relevant to science, apologetics, education, and worldview issues. ICR also publishes the daily devotional *Days of Praise* with over 500,000 readers worldwide. Our website at ICR.org features regular and relevant creation science updates. The three radio programs produced by ICR can be heard on outlets around the world, and we make our materials available through multiple social media outlets.

Headquartered in Dallas, Texas, ICR's latest outreach is the ICR Discovery Center for Science & Earth History, with cutting-edge exhibits, planetarium shows, and live science presentations. The Institute for Creation Research continues to expand its work and influence, endeavoring to encourage Christians with the wonders of God's creation.

P. O. Box 59029
Dallas, Texas 75229
ICR.org

800.337.0375 (main) | 800.628.7640 (customer service)

RESOURCES FROM ICR

Creation or evolution? This debate is one of the most vital issues of our time. ICR's original DVD series present the evidence that confirms the biblical account of creation and provide defensible answers to questions of faith and science. Ideal for group study, these compelling and engaging presentations demonstrate that not only does the scientific evidence *not* support evolution, it strongly affirms the accuracy and authority of God's Word.

Find out more about these DVD series and other resources at **ICR.org/store**.

ICR.org

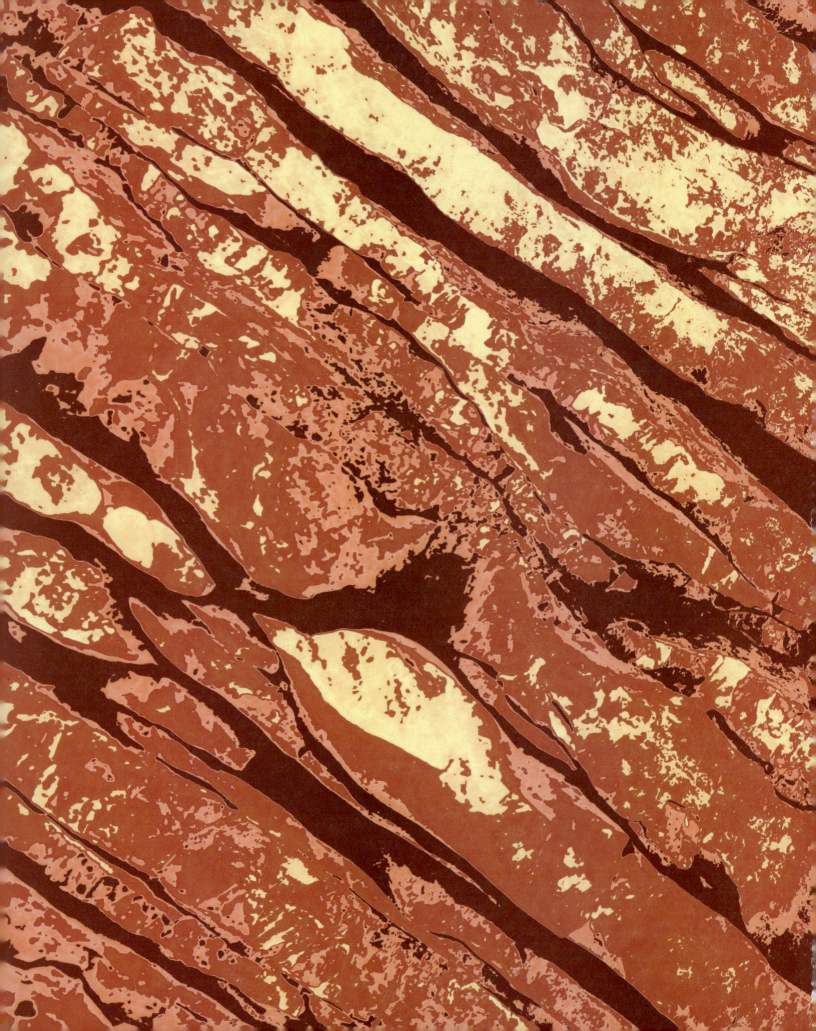